EXPLORATIONS IN LANGUAGE AND SPACE

Series editor
Emile van der Zee, University of Lincoln

Published

1 Representing Direction in Language and Space
Edited by Emile van der Zee and Jon Slack

2 Functional Features in Language and Space
Insights from Perception, Categorization, and Development
Edited by Laura A. Carlson and Emile van der Zee

3 Spatial Language and Dialogue
Edited by Kenny R. Coventry, Thora Tenbrink, and John A. Bateman

4 The Spatial Foundations of Cognition and Language
Edited by Kelly S. Mix, Linda B. Smith, and Michael Gasser

5 Interpreting Motion
Grounded Representations for Spatial Language
Inderjeet Mani and James Pustejovsky

6 Motion Encoding in Language and Space
Edited by Mila-Vulchanova and Emile van der Zee

7 The Construal of Spatial Meaning
Windows into Conceptual Space
Edited by Carita Paradis, Jean Hudson, and Ulf Magnusson

8 Representing Space in Cognition
Interrelations of Behaviour, Language, and Formal Models
Edited by Thora Tenbrink, Jan Wiener, and Christophe Claramunt

Representing Space in Cognition: Interrelations of Behaviour, Language, and Formal Models

Edited by
THORA TENBRINK, JAN WIENER,
AND CHRISTOPHE CLARAMUNT

OXFORD
UNIVERSITY PRESS

OXFORD
UNIVERSITY PRESS

Great Clarendon Street, Oxford, OX2 6DP,
United Kingdom

Oxford University Press is a department of the University of Oxford.
It furthers the University's objective of excellence in research, scholarship,
and education by publishing worldwide. Oxford is a registered trade mark of
Oxford University Press in the UK and in certain other countries

First Edition published in 2013

Impression: 1

Published in the United States of America by Oxford University Press
198 Madison Avenue, New York, NY 10016, United States of America

British Library Cataloguing in Publication Data
Data available

ISBN 978-0-19-967991-1

As printed and bound by
CPI Group (UK) Ltd, Croydon, CR0 4YY

Contents

Part 3: Human-centred assistance

Preface

This book emerged from a collaborative interdisciplinary specialized workshop in 2010, located at the HWK Hanse Institute for Advanced Studies in Delmenhorst (Germany), <http://www.wayfinding.uni-bremen.de/workshop/>. The workshop concluded the Tandem project on 'Wayfinding strategies in behavior and language', funded by the Volkswagen Foundation, and furthermore followed up on excited debates at the Conference of Spatial Information Theory (COSIT) 2009. Based on the invitation of renowned experts in the various related fields, the workshop led to energetic discussion and a broad variety of inspirations for future interdisciplinary research in the field of spatial cognition. Our aim in compiling this book was to capture this spirit and enhance progress in this direction. Following the workshop we circulated a call among experienced researchers in the fields of cognitive science, psychology, linguistics, and computer science, inviting submissions and triggering a range of excited responses. We firmly believe that interdisciplinary research in this area carries considerable promise for enhancing state-of-the-art spatial assistance technology, based on solid empirical insights and specialized formal modelling techniques.

We would like to thank the Volkswagen Foundation, the COSIT committee, and the Hanse Institute for Advanced Studies for jointly financing the workshop, and the Volkswagen Foundation for funding the foregoing Tandem project. Further research funding by the DFG for the SFB/TR 8 Spatial Cognition Research Centre is acknowledged. We are grateful to our reviewers for substantially contributing to the quality of this book, and to the OUP editorial team—most particularly Julia Steer and Kate Gilks—for making the final editing steps a smooth and pleasant experience.

Thora Tenbrink
Jan Wiener
Christophe Claramunt

The contributors

Malika Auvray is a CNRS researcher in cognitive sciences at the Laboratoire d'Informatique pour le Mécanique et les Sciences de l'Ingénieur (LIMSI-CNRS, Orsay, France). Her main research interests are in sensory substitution and broader cross-modal interactions. She investigates experimentally the learning of sensory substitution devices and the involved mechanisms of distal attribution, challenges the distinction of sensory modalities through the examples of these devices, and collaborates on the development of two visual-to-auditory conversion systems: The Vibe and NAVIG.

Marios Avraamides is an Associate Professor of Cognitive Psychology at the University of Cyprus (Nicosia, Cyprus). Previously he has worked as a research scientist at the Max-Planck-Institute for Biological Cybernetics (Tübingen, Germany) and at the University of California, Santa Barbara (USA). His research interests lie within the field of spatial cognition and include among others the organization of spatial memories from different inputs and the spatial updating of information during movement.

Michael Barclay is currently developing natural language system software and doing associated consulting work for Life Inside Ltd. He graduated in engineering science from Trinity College Cambridge and spent twenty years designing and modelling radio systems. Needing a change he went back to University and obtained a PhD in computer science from Exeter. His principal focus is on machine learning of human language and in particular spatial language.

Mehul Bhatt is a researcher in the Cognitive Systems (CoSy) group within the Faculty of Mathematics and Informatics at the University of Bremen, Germany. He also serves as project leader within the Spatial Cognition Research Center (SFB/TR 8) at the Universities of Bremen and Freiburg. Mehul studied economics (bachelors, India), Information Technology (masters, Australia), and computer science (PhD, Australia). Mehul publishes on spatial cognition, spatial representation and reasoning, commonsense and non-monotonicity, and applied ontology. He addresses applications in computational design analysis, behaviour interpretation, cognitive robotics, and dynamic GIS. Mehul has edited for the journals *Spatial Cognition and Computation*, *Human-Centred Computing and Information Sciences*, *Ambient Intelligence and Smart Environments*, and *Automation in Construction*.

Tad Brunyé is a Senior Cognitive Scientist with the US Army and a Research Assistant Professor at Tufts University (Medford, MA, USA). His primary interests include understanding the nature and integration of spatial and non-spatial representations in human memory. He has authored or co-authored over 45 publications in this area, including experimental work and integrative reviews, and has presented at over 60 international conferences. In 2011, he received the Presidential Early Career Award for Scientists and Engineers, the highest honour bestowed by the US Government on science and engineering professionals in the early stages of their independent research careers.

Eric Chown is a Professor of Computer Science at Bowdoin College (Maine, USA). He is the head of the Bowdoin RoboCup team, the Northern Bites. RoboCup is a competition involving autonomous teams of robots playing soccer. The Northern Bites were world champions in 2007, finished third in the world in 2008, and second in the world in 2009. His main research interest is cognitive modeling, especially the areas of spatial reasoning and learning.

Christophe Claramunt is a Professor in Computer Science and Chair of of the Naval Academy Research Institute in France. He was previously a Senior Lecturer in Computing at Nottingham Trent University and a Senior Researcher at the Swiss Federal Institute of Technology in Lausanne. He holds a PhD in computer science from the University de Bourgogne in France. His main research interests concern theoretical and multi-disciplinary aspects of geographical information science, including spatio-temporal and computational models, alternative models of space, semantic GIS, integration of GIS and simulation systems, and the spatial Web.

Michel Denis is an Emeritus Senior Research Scientist at the French National Centre of Scientific Research (CNRS). He is member of the Laboratoire d'Informatique pour la Mécanique et les Sciences de l'Ingénieur (LIMSI-CNRS) situated on the Orsay campus of the Université de Paris-Sud, a major computer science institute with a strong pluri-disciplinary orientation, where he established a research group dedicated to interfacing cognitive psychology, ergonomics, and human–machine communication. His research focuses on the mechanisms involved in the construction of mental images and mental representations of space. His recent work includes the use of virtual reality to study the cognitive processes responsible for the construction of spatial knowledge.

Gilles Fernandez, who graduated in psychology at the Université René-Descartes (Boulogne-Billancourt, France), has completed a PhD in cognitive psychology as a member of LIMSI-CNRS (Université de Paris-Sud, Orsay). His academic interests encompassed the relationships between linguistic and visuospatial representations; creativity and human intelligence; psychometry and child giftedness; Bayesian statistical methods. His is now a private writer and lecturer in the domain of 'why people believe weird things' (such as pseudosciences, UFO, superstitions, modern myths, urban legends).

Christian Freksa is a Professor of Cognitive Systems at the University of Bremen. He directs the Spatial Cognition Research Center SFB/TR 8 at the Universities of Bremen and Freiburg. His research concerns representation and reasoning with incomplete, imprecise, lean, coarse, approximate, fuzzy, and conflicting knowledge about physical environments. His interdisciplinary research group employs formal, computational, and empirical studies using diverse cognitive technologies. Freksa is an ECCAI Fellow and a member of the Advisory Board of the journal *Spatial Cognition and Computation*.

Mathieu Gallay is a Lecturer in Cognitive and Developmental Psychology at Paris Sorbonne University (France). He worked for five years as PhD student and junior lecturer at Paris Descartes University (France). In a project aiming at confronting neuropsychological and developmental approaches of face perception (with CSGA, Dijon, France), he was in charge of infant studies. He also occupied a postdoctoral position for two years at LIMSI (Orsay, France), as part of a National Research Agency funded project, labelled NAVIG, whose objective is to improve the autonomy of visually impaired individuals in navigation and object localization tasks.

Antony Galton is a Reader in Knowledge Representation at the University of Exeter, and has worked in the Computer Science department there since 1987. His research interests are focused on the analysis and representation of spatial and temporal knowledge, with particular reference to philosophical and ontological issues. He is the author of *The Logic of Aspect* (Oxford University Press, 1984) and *Qualitative Spatial Change* (Oxford University Press, 2000), as well as numerous papers in journals and edited collections.

Nathan Greenauer is an Assistant Professor of Applied Psychology at Pennsylvania State University, Berks (USA). Previously he was a postdoctoral research fellow in the Experimental Psychology Laboratory at the University of Cyprus (Nicosia, Cyprus). His research interests lie in the area of spatial cognition and include, among others, the study of the organization and structure of long-term spatial memory and spatial category formation. In particular, his research examines the role of egocentric and allocentric cues in the integration of spatial knowledge across experiences.

Joana Hois is a PhD student at the University of Bremen, Germany. She is a developer of the spatial module for the 'Generalized Upper Model', a linguistic ontology, and works on combining formal models of space with spatial language. Her current research activities are focused on modular ontologies of space in different domains and applications as well as on combining ontologies with different kinds of uncertainties.

Parisa Kordjamshidi is a PhD student at Katholieke Universiteit Leuven. She did her bachelor and masters degrees in computer science at the University of Isfahan and Tarbiat Modares University in Iran. Afterwards she was a university lecturer for computer science in Iran. In December 2008 she started her PhD on spatial language understanding from an artificial intelligence and machine learning point of view. Her ongoing PhD research is on structured machine learning for mapping natural language to formal spatial representations. She has formulated the first semantic evaluation task and benchmark of 'Spatial role labelling' to consider spatial semantics in SemEval2012.

Catherine Mello is an independent research consultant and translator. Prior to this, she was a postdoctoral fellow in the Experimental Psychology Laboratory at the University of Cyprus (Nicosia, Cyprus). As a consultant and instructor, she has provided workshops and training on the use of virtual reality in psychological research. Her research interests reside in multimodal spatial perception, navigation, and spatial updating. She also examines the roles and influence of technology on human spatial reasoning and performance.

Marie-Francine Moens is a full-time professor at KU Leuven. She is the author of more than 230 international publications among which are two monographs published by Springer. She leads the research group of Language Intelligence and Information Retrieval at KU Leuven. She is involved in the programme committee of major conferences in the fields of natural language processing and information retrieval. She is the current chair of the European Chapter of the Association for Computational Linguistics (EACL), a member of the executive board of the Association for Computational Linguistics (ACL) (2011–12), and a member of the coordinating committee of ACL 2010 and 2013.

Nhung Nguyen was a PhD student and research assistant in the Artificial Intelligence Group at Bielefeld University. In her PhD thesis, conducted in the Collaborative Research Center 673—Alignment in Communication in the project C1—Interaction Space, she investigated how artificial, embodied (virtual) agents' different bodily sensor modalities form a representation of the agent's body and the space immediately surrounding it. Her research interests include cooperative and social behaviour in artificial agents, as well as perception and sensation for virtual humans, and spatial perspective taking.

Martijn van Otterlo lectures on Artificial Intelligence at the Radboud University Nijmegen (Netherlands). His PhD thesis at the University of Twente (Netherlands) was on the combination of reinforcement learning and logic (partially as a Marie Curie Fellow at the University of Freiburg, Germany). Recently he undertook a postdoctoral project (K.U. Leuven, Belgium) on computational logic for language, vision, and robotics. He has published the books *The Logic of Adaptive Behavior*

(IOS Press, 2009) and (with Dr Wiering) *Reinforcement Learning: State-of-the-Art* (Springer, 2012). His current research includes reinforcement learning and knowledge representation for perception and control, and privacy implications of smart adaptive algorithms.

Carl Schultz is a postdoctoral researcher working in the Cognitive Systems Group (CoSy) at the University of Bremen as a member of the DesignSpace project. His current research focus within DesignSpace is on modelling and reasoning about high-level spatial knowledge for architectural design. His primary research area focuses on methodologies for facilitating the development of qualitative spatial reasoning software applications; on this topic he completed his PhD in July 2010 at the University of Auckland, New Zealand.

Inessa Seifert is a postdoctoral research scientist at the German Research Centre for Artificial Intelligence (DFKI GmbH), where she works in the Intelligent Interfaces department. She holds a doctoral degree in Informatics from the University of Bremen with a thesis in artificial intelligence and spatial cognition on 'Spatial Planning Assistance: a Cooperative Approach'. While at Bremen, Inessa worked within the Cognitive Systems Research Group and the Transregional Collaborative Research Centre SFB/TR 8 Spatial Cognition. She holds a masters degree from the Technical University Berlin. Inessa is interested in spatial cognition, interactive maps, schematization, and information visualization approaches.

Holly A. Taylor is a Professor of Psychology at Tufts University (Medford, MA, USA). Her main research interest is to examine the mental representation of information, sometimes referred to as mental models or situation models, and how information sources influence the representation. She is particularly interested in the domains of spatial cognition and comprehension. She is the author of numerous empirical publications, as well as integrative book chapters. She was named a Fellow of the Association for Psychological Science (APS) in 2012.

Thora Tenbrink is a Senior Lecturer in Cognitive Linguistics at Bangor University (Wales, UK). She worked for ten years as a research scientist at the Faculty of Linguistics at Bremen University (Germany), and is a principal investigator in two projects in the Collaborative Research Centre SFB/TR 8 Spatial Cognition (Bremen/Freiburg). Her main interest concerns the relationship between cognitive processes and linguistic representations. She is the author of *Space, Time, and the Use of Language* (Mouton de Gruyter, 2007), and editor (with Kenny Coventry and John Bateman) of *Spatial Language and Dialogue* (Oxford University Press, 2009).

Ipke Wachsmuth is a Professor of Artificial Intelligence at Bielefeld University, a principal investigator in the Centre of Excellence 'Cognitive Interaction Technology' (CITEC) and an adjoint member of Bielefeld University's Research Institute for

Cognition and Robotics (CoR-Lab). He has also been director of the Center for Interdisciplinary Research (ZiF) and coordinator of the Collaborative Research Center 'Alignment in Communication'. His research focuses on human–machine interaction and virtual reality, including gestural and multimodal interaction, and the modelling of communication in embodied artificial agents.

Jan Wiener is a Senior Lecturer in Psychology at the University of Bournemouth (UK). Previously he has worked as a research scientist the University of Freiburg (Freiburg, Germany), the CNRS (Paris, France), and the Max-Planck-Institute for Biological Cybernetics (Tübingen, Germany). His research focuses primarily on the cognitive processes and strategies that underly navigation and wayfinding behaviour.

List of abbreviations

AAL	Ambient Assisted Living
AI	Artificial Intelligence
CAAD	Computer-Aided Architecture Design
CMP	Cognitive Model of Planning
CRF	conditional random field
DC	disconnected (RCC-8 relation)
DD	direction down
DR	direction right
EC	externally connected (RCC-8 relation)
EQ	equal (RCC-8 relation)
EST	Event Segmentation Theory
FOR	frame of reference
GIS	Geographic Information System
GPS	Global Positioning System
HP-TSP	human performance on the Travelling Salesman Problem
HPI	Haptic Pointer Interface
ICM-9	Internal Cardinal Direction Model
NN	nearest neighbour
NTPP	non-tangential proper part (RCC-8 relation)
NTPPI	non-tangential proper-part inverse (RCC-8 relation)
PO	partial overlap (RCC-8 relation)
PP	prepositional phrase
PrP	peripersonal
QSR	qualitative spatial representation
RCC	Region-Connection Calculus
RFID	Radio Frequency Identification
RIAS	Remote Infrared Audible Signage
SAS	spatial assistance system
SpQL	spatial qualitative labelling
SpRL	spatial role labelling

SSE	Structured Spatial Environment
SVM	support vector machine
TPP	tangential proper-part (RCC-8 relation)
TPPI	tangential proper-part inverse (RCC-8 relation)
TSP	Travelling Salesman Problem

1

Representing space in cognitive science: from empirical insights via computational models to human-centred assistance

THORA TENBRINK, JAN WIENER,
AND CHRISTOPHE CLARAMUNT

Research in the areas of spatial language, spatial cognition, and spatial information theory has rapidly developed over the last few decades. This is reflected by the steadily growing number of researchers addressing spatial topics, national and international conferences and meetings, and major interdisciplinary research projects spanning a wide variety of issues from spatial behaviour, interaction, and knowledge representation to formal and computational models. Recent developments have now highlighted a widespread interest in establishing a reliable basis for combining insights and representation models from various sources and applications. The combination of perceptual and language-based representations, for example, has led to a better understanding and knowledge of how humans process space in various strands of research. Examples are findings as diverse as functional components associated with spatial prepositions (Coventry and Garrod, 2004; Carlson and van der Zee, 2005), the representation of direction (Klippel and Montello, 2007) and distance relations (Montello, 2009), a rich diversity of spatial planning processes reflected in language (Tenbrink and Wiener, 2009; Hölscher, Tenbrink, and Wiener, 2011), and the integration of functional (affordance-related, cf. Gibson, 1977) factors (Gorniak and Roy, 2007; Moratz and Tenbrink, 2008). Findings such as these highlight the potential of systematically combining the existing manifold approaches to interpreting spatial representations in language with behavioural navigation and action data, associated cognitive processes and strategies, and formal representation systems capturing cognitively relevant structures, categories, and features of space.

Gradually, a rich diversity of formal representation approaches become available, as the empirical basis in behavioural research grows tighter, questioning the general validity of earlier layout-based formalisms (Davies and Peebles, 2010). Early influential models were collected by Frank, Campari, and Formentini (1992). Later on, Freundschuh and Egenhofer (1997) formalized six types of space, distinguished by manipulability, locomotion, and size; Raubal and Egenhofer (1998) computationally compared the complexity of wayfinding tasks; Mark et al. (1999) discussed cognitive models of geographical space, and Klippel et al. (2005) presented a language for modelling conceptual route knowledge, to mention just a few examples.

One particularly intriguing result of joint efforts in spatial cognition is the development of assistance systems that can support humans in increasingly challenging spatial tasks. In this context, spatial descriptions serve as a common medium to infer and communicate spatial information about an environment. Assistance systems for blind people are one example in which such spatial descriptions have been applied. Other relevant research directions address how language provides a means for the description of the forms and structures that emerge from environment scenes (Le Yaouanc, Saux, and Claramunt, 2010) and the development of route planning services (Winter and Wu, 2008). In numerous ways, spatial language serves as a bridge between internal representation and external behaviour to be supported according to current information needs. For this reason, a deep understanding of the principles underlying spatial language use along with a clear concept of the interrelationships to other forms of spatial representation is essential for a variety of purposes.

The collected chapters in this book reflect the current progress in understanding, analysing, and modelling of spatial behaviour and associated language, with a specific emphasis on interrelationships between, and combinations of, interdisciplinary approaches and methods. A range of practical application issues emerge from these diverse efforts. Detailed knowledge about human spatial cognition in terms of representing and structuring space as well as knowledge about how this information is used and communicated during wayfinding and planning decisions supports remarkable system development in many directions, including assistance for the blind, navigation planning, and architectural design. It is our conviction, and our hope, that this promising development as reflected in the current collection will be taken up by an increasing range of experts across the world, to jointly pursue the exciting goal of making space intuitively accessible for everybody in many challenging situations.

1.1 Overview of this book

The authors of the book's chapters span various disciplines related to spatial cognition, such as psychology, linguistics, and computer science. Each of these disciplines draws on a research tradition with established methodologies of its own, which regularly leads to fundamental problems when relating findings from one area to those of another. In this sense, building bridges in itself is a major feat, which this book has aimed to accomplish from the very first planning steps. The result is a respectable breadth of perspectives adopted by experts taking different issues and approaches as their starting point. Jointly, they pursue the aim of establishing clear connections between perception, language, behaviour, modelling approaches, and practical applications.

The book is composed of three sections—*Empirical insights, Computational models,* and *Human-centred assistance*—reflecting the authors' different perspectives and approaches. In the following we introduce each section, outline the contribution of each chapter, and highlight connections between chapters both within and between sections.

1.2 Empirical insights

How people act in space is closely related to how they perceive and think about their spatial surroundings. Their actions and decisions are guided by a range of underlying principles that can be captured by controlled experimentation across various scales. The analysis of spatial language in relation to spatial behaviour is one way of highlighting the cognitive processes that mediate these principles. A number of questions arise in this regard. For example, what kinds of cognitive representations are involved when relating language to action? Can language substitute for perceptual information when behaviour is at stake? What is the best way of conveying such information? The first three chapters of the book address these questions as follows.

Holly A. Taylor and Tad T. Brunyé set the stage in their chapter 'Describing the way out of a cornfield: Understanding cognitive underpinnings of comprehending survey and route descriptions'. They summarize the main cognitive mechanisms involved in transforming verbally conveyed information into navigation actions. The chapter presents a thorough analysis of the particular characteristics of survey and route descriptions, and thus contributes to a better understanding of the cognitive properties of linguistic descriptions. The authors finally sketch the ways in which insights such as these have been formalized and build the basis for practical applications.

In their article 'Spatial representations for described and perceived locations', Marios Avraamides, Catherine Mello, and Nathan Greenauer discuss the extent to which spatial language can convey sufficient information in order to be able to act and make appropriate decisions in spatial environments. They focus on a direct comparison between verbal as opposed to perceptual information and discuss how the resulting spatial representations compare, concluding that language is well capable of substituting for the arguably richer or more direct perceptual resources, which may not always be available. These findings have important implications for the understanding of human conceptions of space, which can be modality-independent and transferable. Moreover, human conceptual flexibility allows for successful assistance and across-task support in more than one way, using the methods and modalities available and conveyable at any given moment.

The next chapter by Michel Denis and Gilles Fernandez, 'The processing of landmarks in route directions', spells out in detail how this effect of conveying sufficient navigation information verbally may be achieved in practice, thus supporting the development of computational representations. The authors provide a comprehensive overview focusing particularly on the role of landmarks in a wayfinder's understanding of the environment. Starting from empirical research in this area, they highlight the impact of the variability of landmark descriptions on navigation performance. Building on these insights, they establish the relation to computational modelling as well as the implications for the development of automated systems and robotics, pointing out the fundamental importance of establishing suitable methods for offering landmark-related information to wayfinders as needed.

From these chapters, the general insight emerges that the ways in which language is structured as a system, and how it is put to use within a spatially situated discourse context, reflect the underlying cognitive representations of spatial relationships. Moreover, the empirical section illustrates the necessity of a deep understanding of modality-transcending conceptual categories and processes, to be used as a foundation for, and validation of, modelling approaches and cognitively adequate user support systems.

1.3 Computational models

Building on empirically derived principles of how concepts are structured, how language reflects underlying spatial representations, and how humans understand their environment and act in it, formal and computational approaches implement human verbalizations and perceptions of space, rendering cognitive representations accessible for automatic processing and reasoning. The four chapters in the second section—Computational models—deal with this issue in various ways as follows.

In their chapter 'Selection of reference objects for locative expressions: The importance of knowledge and perception', Michael Barclay and Antony Galton present a spatial knowledge representation for computational purposes that enables the modelling of object location descriptions by humans. Their formal model successfully encompasses a wide range of cognitive principles such as those outlined in the first three chapters, as proved by a statistical comparison of human and machine performance. Their approach is based on a modelling construction that studies how spatial objects are used in locative expressions, the emerging spatial properties, and variables that describe these objects, at different levels of abstraction. Reference objects are used as modelling abstractions to interpret in a given environmental scene. The originality of their proposal is to retain several spaces, from table-top, to room-scale to street and vista scales. This diversity of scales provides a large number of research perspectives and potential developments to their approach. Indeed, the machine model developed and experimented by a large panel is far from generating a complete replica of human perceptions, but it is convincing enough to emphasize the role of spatial properties and the environment as well as opening several application perspectives for further research, and surely many possible connections with semantic and ontological approaches.

Building on empirical work on how humans perceive their environment, Eric Chown proposes a theoretical contribution for representing the ways in which humans perceive, visualize, and schematize their spatial environment. The work is grounded on a large body of contributions from cognition to neuropsychology and qualitative spatial reasoning. Spatial prototypes are used as building blocks and modelling primitives that constitute spatial scenes whose dynamic properties are studied using the notions of events and changes, and that generate networks of change. The model that emerges is somehow egocentric and generates several snapshots of space that capture the state of a given scene at a given time. The whole approach is implemented in a mobile robot and constitutes a promising basis for causal reasoning. Still the question of the identification and interpretation of changes is a problematic issue (how to measure change?), and forms part of the further work of the author and its long-term experiments. Overall this chapter provides a clear formal progress on the perception and visual schematization of spatial environments, and proposes a set of abstractions that open up many opportunities for further research exploration. It represents a fine instance of a theoretical modelling framework firmly based on empirical evidence, providing a comprehensive representation of the interplay of perception and cognition.

In 'Learning to interpret spatial natural language in terms of qualitative spatial relations', Parisa Kordjamshidi, Joana Hois, Martijn van Otterlo, and Marie-Francine Moens explore a possible machine-learning mapping between linguistic features and qualitative region-based abstractions derived from a given corpus. The authors introduce the notion of spatial role that qualifies the respective contributions of the spatial abstractions used in textual expressions, and that provide the basis for a mapping toward qualitative RCC8 topological relations. The computational experiment developed shows how both the extraction and mapping processes can be machine learned successfully. Their experiments confirm the influential roles of the objects described, landmarks and spatial constraints in a scene description, the relatively minor role played by the dynamism of the scene, and the role of the frame of reference, and the complications involved in mapping RCC relations to real textual descriptions. This work opens up many further possibilities, from extensions to other qualitative spatial relations to mapping towards different formal computational languages, as well as closer connections with ontological languages.

Computational spatial models most often simulate the way people abstract their neighbourhood, but also how artificial objects perceive and behave in their environment. In their chapter, Nguyen and Wachsmuth model the space perceived by a virtual human interacting with several partners. Their work is grounded on cognitive neurosciences, biology, and social interaction studies. The modelling target here is the interaction space, the one surrounding the body, that favours cooperation between other agents, and is perceived by touch sensors, vision, and proprioception. The approach developed is a learning and acting machine that should help an agent to model and structure its environment, and interact with its neighbouring space and other potential partners. The model is supported and implemented on a humanoid called Max, and is original as it proposes different levels of interaction spaces such as visual space, touch-space, and lean-forward space. These correspond to different sensor modalities, providing useful abstractions for structuring the spatial environment. The work offers many perspectives for the study of cooperative interaction between humanoids (and humans), as well as simulation and anticipation of their behaviour.

Taken together those computational approaches show the diversity of possible interpretations and structuration of space. Indeed language and perception of space are complex and diverse, and the computational representations reflect this extraordinary variability by the large extent of formal proposals suggested in this section.

1.4 Human-centred assistance

Once formal and computational models of spatial understanding are available, the ideas that emerge can be integrated into assistance systems that can do far more than simply provide information to users confronted with diverse spatial challenges. In particular, the aim of such systems is to support navigation and other kinds of spatial decisions by presenting spatial information as intuitively as possible, at the most suitable point in time, and in the easiest processing fashion available given the abilities of the users, thereby centring on the human user rather than technological features. The final three chapters of this book address these challenges as follows, ranging from targeted empirical work to primarily application oriented perspectives.

In their chapter 'The "space" in spatial assistance systems: Conception, formalization, and computation', Mehul Bhatt, Carl Schultz, and Christian Freksa address the concept of space from a formal modelling and computational viewpoint. By relating such formal conceptions to spatio-linguistic conceptions of space which occur within a range of spatial assistance systems, this chapter provides a thorough overview of the ways in which cognitive principles can—and should—be incorporated in spatial assistance systems for a wide range of application purposes such as spatial design, virtual games, and training.

Inessa Seifert and Thora Tenbrink's chapter 'Cognitive operations in tour planning' addresses human route planning in the context of an extended holiday trip, ultimately aiming at the development of a tour planning assistance system based on cognitive principles. Following a detailed review of previous work addressing conceptual strategies involved in such tasks that are similar to Travelling Salesman Problems (TSP), they present an empirical study of human tour planning. The tours created by participants were strikingly similar to typical TSP solutions. The verbal reports collected in this study provide the basis for the identification of cognitive operators decisive for this task, again highlighting systematic similarities to planning strategies described in the TSP literature. Seifert and Tenbrink's study reveals new insights into how people solve complex spatial planning tasks and provides a basic framework for a planning assistance system that implements the cognitive planning principles found in this study.

Finally, in their chapter on 'Navigation assistance for blind pedestrians: guidelines for the design of devices and implications for spatial cognition', Mathieu Gallay, Michel Denis, and Malika Auvray review the state of the art in navigation assistance systems for people with visual impairments with respect to the accumulated knowledge in the wayfinding literature. They describe the general functioning of several electronic navigation aid devices for route planning and navigation and their potential when used by blind pedestrians. Gallay, Denis, and Auvray then suggest ways to improve these navigation aids and discuss how blind people's spatial representations are influenced by their use. This chapter is particularly informative in the way it meticulously relates findings from empirical work to the practical application in assistance systems.

This book by no means covers exhaustively the complex relationships between space and language, perception and conception, behaviour, computational representations, and practical applications. Its objective is rather to present a set of recent findings, written by experts in their fields, that together provide a coherent contribution to the field. This book is also part of the series 'Explorations in Language and Space' that has recently explored other fundamental issues closely related to cognition and language, from the interpretation of motion by language constructs (Mani and Pustejovsky, 2012) to the representation of objects and places in dialogue (Coventry, Tenbrink, and Bateman, 2009a) and the close relationship between the physical properties of space and cognitive processes (Mix, Smith, and Gasser, 2009). This clearly shows the dynamism of the field, and how critical it is to search for a better understanding and integration of cognition, language, and computational representations. One major emerging trend concerns the importance of cross-disciplinary research, and its role in the development of successful spatial information systems. Overall there is an exciting time of development before us, as illustrated by the research directions presented in the contributions that follow and the many perspectives remaining for future work.

Part 1

Empirical insights

2

Describing the way out of a cornfield: understanding cognitive underpinnings of comprehending survey and route descriptions

HOLLY A. TAYLOR AND TAD T. BRUNYÉ

Spatial descriptions can convey different perspectives, notably within-environment or route and bird's-eye or survey perspectives. Spatial perspective, in turn, has implications for how people mentally represent the described space. We suggest four cognitive mechanisms implicated in this regard. First, route descriptions may have greater working memory demands than do survey descriptions. Second, and strongly related to working memory, route and survey descriptions differ in the extent to which they convey a configural framework. Survey descriptions convey configural information directly, but it must be inferred from route descriptions. Third, route descriptions promote active egocentric imagery; they describe an ego moving through an environment. Fourth, route descriptions involve temporally sequencing information (e.g. turn right *after* going through the intersection). Sequence information may vie for some of the same visuospatial resources as the spatial information (e.g. Zwaan & Radvansky, 1998). In this chapter, we discuss empirical support for these contributing factors.

2.1 Introduction

Imagine having recently moved to a new town. As is your habit, you head off one morning for a run with your partner, who happens to have more experience navigating through this new environment than you. Your run takes you through unnamed paths that wend their way through cornfields and woods. Three-quarters of the way through your run, your partner asks if he can give you directions to return so that he can continue on a longer run. You agree and receive the following directions: 'Continue along the main path until you come to where we turned left earlier, go straight through that intersection and then you'll dead-end where you'll have to turn right. After you turn you'll see the tunnel under the road and you'll know where you are.' The description has accompanying gestures, including one pointing straight down the main path as your partner says, 'Continue along the main path.' You have mentally followed this description, matching it to what little you know about the environment, and think it sounds clear. So, you head off on your own. (We will leave it to your imagination as to whether you returned home without difficulties.) What happened cognitively while you listened to this description? What led you to conclude that it made sense? In a more general sense, what cognitive processes underlie spatial description comprehension? Whether or not this description remains clear as you follow it, at this point in comprehension, you have a metacognitive sense that you understand it. People commonly encounter situations such as the one described here. That is, descriptions serve as a common medium for relating spatial information about an environment. As such, it is important to understand the cognitive processes and pitfalls involved with spatial description comprehension.

The present chapter reviews recent work examining how people develop and use mental representations built from spatial descriptions of environments. Phenomenologically, listening to or reading a spatial description and then following it is not exciting. Most of us have done this many times, perhaps while on vacation and trying to find a famous landmark. Most of us accomplish the task with little difficulty or at least can recover from any difficulties. Cognitively, however, the ability to successfully understand spatial descriptions is akin to a complex mental gymnastics routine. By this, we mean understanding spatial descriptions requires the coordination of many cognitive processes, each of which is interesting, and perhaps not completely understood in its own right, but together lead to a remarkable outcome. In studying spatial description comprehension, we strive to understand the underlying component processes (the cartwheels, flips, and round-offs) as well as how they come together (the routine). Through the chapter, we will relate work trying to explicate the component processes as well as that trying to understand how the processes work in concert. We have found that comparing different, yet common, ways of describing environments can reveal the more general cognitive underpinnings.

2.2 Ways to describe environments

Space and language have inherently different dimensionality. Thus, spatial and linguistic representations of environments also have inherently different dimensionality. The environment itself has three dimensions. Travelling through the environment adds the fourth dimension of time. Spatial representations, most frequently maps, usually have two dimensions, but can sometimes add a third through clever mechanisms such as overlays. Language has one dimension in that we can only say, read, or write one thing at a time. These differences in dimensionality mean that spatial descriptions must linearize space to match the inherent dimensional limitations of language (Habel and Tappe, 1999).

One of the first studies to examine how we describe environments is Linde and Labov's (1975) now classic study in which they asked New Yorkers, 'Could you tell me the lay-out of your apartment?' Relevant to their goal of translating thought into language, they found that 95 per cent of descriptions related an imaginary tour through the apartment. These tours had common characteristics, such as addressing the listener or reader as 'you' and using egocentric terms, such as *left*, *right*, *front*, and *back*, to locate apartment features to the addressee's current imagined position (Taylor and Tversky, 1992b). Further, the environment is described as if the addressee is within it, moving from place to place (Perrig and Kintsch, 1985; Taylor and Tversky, 1992b; Noordzij and Postma, 2005; Brunyé and Taylor, 2008a). Referred to more commonly as **route descriptions** in the research literature, the approach of taking an addressee on a linguistic mental tour provides an effective means for systematically linearizing a spatial layout into language. Further, the tour approach seems intuitive because it reflects our direct experiences navigating through the world. It reflects how we might move through an environment and what we might see along the way. Because of its intuitive appeal, some linguistic and psycholinguistic researchers have proposed that route descriptions should be the typical (e.g. Levelt, 1982, 1989) and preferred (Klein, 1982, 1983) way of structuring environment descriptions. Bhatt, Schultz, and Freksa (this volume) discuss several ways to implement route-based linearizations in automated spatial assistance systems, including *route graphs*, *spatial sequence models*, and *flow vector graphs*. In many cases, multiple route options exist through an environment based on different options to linearize. Seifert and Tenbrink (this volume) discuss linearization in route planning when multiple linearization orders exist, such as when planning a holiday tour. Importantly, the ability to consider multiple route options means information about the options exists in memory. In considering how such information is represented in memory, many models of cognitive mapping employ topological networks (e.g. Chown, 1999; this volume).

Yet, there are other common ways of experiencing environments and other approaches to linearizing space into language that appear to reflect these experiences (Tversky et al., 1994). In some cases, such as standing in the entrance to a large room, the full extent of an environment can be viewed from a single vantage point and one moves, with the eyes, from location to location. This is akin to the *range space* of a sensory device, for instance implemented as *QVGraphs* by Bhatt, Schultz, and Freksa (this volume) or *Local Maps* as implemented by Chown (this volume). These **gaze tours** differ from route-based ones, and these differences are also reflected in descriptions based on them (Levelt, 1982, 1989; Ehrich and Koster, 1983; Shanon, 1984). Ullmer-Ehrich (1982) identified three defining differences between route and gaze tours. First, gaze and route tours use different spatial reference frames (Levinson, 1996a). Second, for route tours, the origin varies with described movement whereas with gaze tours the origin is fixed. Third, route tours relate locations to the addressee's imagined location while gaze tours relate locations to other, previously described, locations. Although not specifically identified by Ullmer-Ehrich, other differences include the extent of described and implied motion within the environment and associations with different scales of space (Montello, 1993). Gaze tours have been associated with vista or figural space while route descriptions are associated with environmental space (Tenbrink et al., 2011). The importance of this additional point comes when considering embodied perspectives on mental representations (e.g. Kosslyn, 1980; Paivio, 1990; Barsalou, 1999, 2005; Zwaan, 2004). Although different, gaze tours and route tours do share the sense of relating information from the ground level, as if within the environment, both use egocentric spatial terms, and both have groundings in perception and action (Barclay and Galton, this volume; Chown, this volume; Wang et al., 2012). Further, topological networks can be used as an underlying structure when modelling both types of tours (Chown, this volume).

A third common way of experiencing an environment is from a vantage point above it. Such a vantage point can be gained from atop a nearby hill or by flying over the environment in a plane or helicopter. Maps provide this vantage point, commonly referred to as **survey perspective**, symbolically. Survey perspective descriptions relate location information by systematic scanning from this above-environment view. Just as route descriptions and gaze tours differ, but also overlap in features, so do survey and gaze tour descriptions. Survey descriptions and gaze tours both relate locations to other known locations and both use a single vantage point. Yet, the vantage points themselves differ between these approaches; whereas a gaze tour might use a doorway vantage point that is level with the described environment, a survey description uses a bird's-eye vantage point, external to the environment. As such, gaze tour descriptions appear to blend aspects of route and survey perspectives, but at the same time have unique characteristics. Further, as with the other description types, topological networks have been used when modelling survey knowledge (Chown, this volume). See Denis and Fernandez (this volume) for additional discussion of description perspectives.

Thus, there are three common ways of experiencing environments—walking through them, standing in one place in them, and viewing them from above. These common experiences help structure the way spatial information is related through language; that is, spatial descriptions share correspondence with our real-world experiences. That descriptions can relate spatial information in different ways is a moot point if people do not successfully extract accurate spatial information from varied linguistic experiences. Thus an important research topic is whether spatial descriptions can effectively relate spatial information, regardless of their structure. It is clear that language can effectively convey spatial information about relatively complex environments, and that the mental representations that readers form from various spatial descriptions afford accurate spatial inferences (Perrig and Kintsch, 1985; Taylor and Tversky, 1992b; Denis and Zimmer, 1992; Ferguson and Hegarty, 1994; Avraamides et al., 2004; Lee and Tversky, 2005; Noordzij and Postma, 2005; Brunyé and Taylor, 2008b; Avraamides and Kelly, 2010). Less clear is how different description types affect the spatial mental representations that arise from linearized spatial relations. Given that gaze tour descriptions have different overlapping properties with route and survey descriptions, it can be difficult to disentangle their unique contribution. As such, much of the research addressing this question has focused on comparing route and survey descriptions. Because route and survey descriptions differ substantially, we have gained some insights into mental representations based on such descriptions. Further, these insights have been used as a basis to address whether mental representations learned from different description types bear resemblance to those learned from the analogue experiences on which these description types are based (i.e. navigation and maps; Taylor and Tversky, 1992b; Chabanne et al., 2003; Peruch et al., 2006). Avraamides, Mello, and Greenauer (this volume) additionally touch on these points.

2.3 Developing and using spatial mental models from descriptions

Research exploring mental representations based on extended spatial descriptions has probed how the structural framework of different description types impacts mental representation development and use. From a functional point of view, a critical underlying question is the extent to which the resultant mental representation affords perspective switching and spatial inferences generation, skills critical to spatial problem solving. Understanding this is important in that the format and perspective available when one learns an environment may not match how it is eventually used (Taylor et al., 1999; Brunyé and Taylor, 2009), whether this use is intentional or not. Returning to the opening scenario, the description you received to find your way home took a route perspective, describing the turns you should travel and the features you will see along the paths back to your apartment. However, what if along the way you find ambiguity between what was described and what you see? For example, the description and accompanying gesture suggested that you would find just one path going roughly straight. Instead you find that the path, which you consider a main path because it's gravel rather than grass, turns and a secondary path continues in the direction you understood you need to travel. At this point of ambiguity, you need to problem solve. Being able to mentally simulate the environment, as opposed to physically exploring it, allows you to more efficiently consider multiple options when problem solving. One option through mental simulation is switching to a survey perspective on the environment. The survey perspective, because of the overview it provides, may help you decide which path will guide you to your destination. Similarly, a survey perspective may help you reorient if you end up making the wrong path choice. But, taking a survey perspective in these problem-solving situations requires mentally conjuring up a spatial representation with features different from those imparted from the learned description.

Our research suggests that spatial description perspective plays a role in developing mental representations that afford inferences across perspectives. Brunyé and Taylor (2008a) examined how spatial mental representations develop with increased environment experience. Participants read route or survey descriptions of the town and convention centre environments used by Taylor and Tversky (1992b) and then drew sketch maps and verified statements about the environments. Half of the participants read each description once, and half thrice. Results suggested that readers needed more time with route, compared with survey, descriptions to achieve the same perspective flexibility. More specifically, after reading a description one time, participants had higher accuracy and lower response times when verifying spatial inferences following survey relative to route descriptions. They also had more accurate sketch maps after reading a survey description one time, compared to reading a route description once. But, the difference based on description perspective was not apparent with better-learned environments (i.e. descriptions read three times). A second experiment showed convergent results when people could control how much they read. In this study, participants had to read the descriptions three times, but in a self-paced manner. They spent more time reading route descriptions on the second and third reading cycles than they did survey descriptions, a finding confirmed by Brunyé et al. (2008). This study supports the notion that participants could develop perspective-flexible mental representations, but required more time to do so with route descriptions. This finding corroborates other work that also suggests relative difficulty developing mental models with route relative to survey descriptions (Lee and Tversky, 2005; Noordzij and Postma, 2005; Noordzij et al., 2006).

A corollary of having less perspective flexibility after reading route descriptions is having greater perspective specificity. Brunyé and Taylor (2008a) found some evidence of this corollary with route descriptions. While participants made gains in spatial inferences in both perspectives, readers of route descriptions could inference more accurately from a route perspective after having read the description three times. Brunyé, Rapp, and Taylor (2008) also supported this corollary. In their study, participants read descriptions of real-world neighbourhoods (modelled after Pittsburgh and Detroit, USA and previously used by Schneider and Taylor (1999)), half taking a route and half a survey perspective. After each description, they verified statement accuracy. Of the spatial statements verified, the first four took the same perspective as the description and the fifth switched perspective. This fifth spatial statement was of primary interest. In addition to supporting Brunyé and Taylor's (2008a) finding that route descriptions need more reading time, this study showed that participants developed perspective-flexible representations following survey descriptions, but less flexible representations following route descriptions. Their second study showed that perspective-specificity could also be reinforced for survey descriptions. If participants studied a map of the environment prior to reading a survey description, they showed decreased perspective flexibility. The survey representation presented by the map reinforced the description perspective. The map prior to reading, in contrast, increased perspective flexibility for route descriptions.

It should be noted that not all research supports this notion of perspective flexibility. Some research suggests that mental representations maintain the learned perspective, either completely or to some extent (Bosco et al., 1996; Perrig and Kintsch, 1985). Shelton and McNamara (2004) found a perspective switching cost and orientation specificity in scene recognition. Tlauka and Wilson (2006) also showed orientation specificity linked to perspective. In some cases, the perspective maintenance was specific to particular individuals or aspects of the environments/descriptions. Perrig and Kintsch (1985) found this with females. Schneider and Taylor (1999) found perspective-specific properties linked to integration in memory. Other work suggests that representations change with time and additional exposure to the spatial description. Lee and Tversky (2005) showed that, during their construction, mental representations showed perspective sensitivity. While reading, participants took longer to read sentences that switched perspective, but these perspective effects diminished when the description had been learned further. Brunyé and Taylor (2008a) showed that readers needed more time with route, compared with survey, descriptions to achieve the same level of perspective flexibility, but that with time it was achieved. The development of perspective flexibility may also reflect individual needs, such as those discussed in reference to blind and sighted individuals by Gallay, Denis, and Auvray in this volume. Thus, results suggesting perspective specificity or sensitivity may not rule out developing representations that will eventually support perspective flexibility. In sum, we contend that these latter findings support perspective flexibility, but indicate that it develops with experience.

These findings with spatial descriptions corroborate research that compares learning from navigation versus maps. In particular, previous work has suggested that increased experience with environments solidifies spatial mental models (Appleyard, 1970; Ladd, 1970; Golledge and Spector, 1978; Kuipers, 1978; Thorndyke and Hayes-Roth, 1982; Sholl, 1987; Lee and Tversky, 2005). Further, increased experience with navigation leads to mental representations more similar to those gained by studying maps (Golledge and Spector, 1978; Thorndyke and Hayes-Roth, 1982). All of this work coincides with the seminal model of spatial knowledge development proposed by Siegel and White (1975). They propose that during navigation people sequentially solidify spatial information in a developing mental representation, progressing from representing landmarks to landmark interrelationships to more integrated mental models of the environment. Although the sequential nature of this progression has been questioned (e.g. McDonald and Pelligrino, 1993; Ruddle et al., 1997; Ishikawa and Montello, 2006) the types of spatial information have generally been accepted. Brunyé and Taylor (2008a), with spatial descriptions, suggest that even with limited experience developing models are not inextricably bound to the learning perspective, thus suggesting that mental model development begins earlier rather than later.

Evidence that mental representations may not be bound to the learned perspective comes from two other types of findings. First, when people describe environments they do not necessarily maintain the learned perspective and will sometimes switch perspective (Taylor and Tversky, 1992a) and may not linguistically signal the perspective switches, which sometimes occur mid-sentence (Taylor and Tversky, 1992a; Filipi and Wales, 2004). Second, the perspective represented in memory may more closely match one's goal for learning the environment than the perspective learned (Taylor et al., 1999; Brunyé and Taylor, 2009).

What might account for the representational differences between route and survey descriptions? Brunyé and Taylor (2008a) suggest four possible mechanisms. First, route descriptions may have greater working memory demands. The increased working memory requirements could have multiple sources, including being a by-product of structure differences. Survey descriptions structure spatial information hierarchically (breadth then depth). Route descriptions present information serially and as such may require more online information maintenance within working memory during both reading and updating relative to a principle reference vector (e.g. Shelton and McNamara, 2004). Second, and strongly related to the first mechanisms, route and survey descriptions differ in the extent to which a configural framework is related directly. The same configural information that survey descriptions convey directly must be inferred from route descriptions. This would yield a greater cognitive load when reading route descriptions, a contention supported by discourse processing research that has suggested that spatial inferences may be formed both during learning (online) and at retrieval (i.e. Graesser, Singer, and Trabasso, 1994) and presumably require a high degree of cognitive resources (i.e. Estevez and Calvo, 2000; Linderholm and van den Broek, 2002; Rapp et al., 2006).

Route descriptions also seem to promote more active egocentric imagery, given that they describe an ego walking (or driving) through an environment. Thus, it makes sense that they would evoke perception and movement based imagery. Such imagery, however, may induce additional cognitive load at the expense of higher-level abstraction, at least with limited experience (i.e. De Beni et al., 2005; Deyzac et al., 2006). This possibility will be further discussed within the framework of embodied cognition later in this chapter (e.g. Glenberg and Kaschak, 2002; Barsalou, 2005; Zwaan and Taylor, 2006). Finally, route descriptions involve temporal sequence information (e.g. turn left *after* passing under the bridge) that may vie for some of the same visuospatial resources as the spatial information (e.g. Zwaan and Radvansky, 1998). In the rest of the chapter, we discuss the empirical support for these contributing factors.

2.4 Working memory demands with spatial descriptions

Spatial descriptions provide an interesting medium for investigating how proposed working memory mechanisms (Baddeley, 1992) might work in concert. Spatial descriptions use language to describe a spatial scene. Thus, of interest is the roles played by the articulatory rehearsal loop (verbal subsystem), the visuospatial sketchpad (visuospatial subsystem), and the central executive, when reading spatial descriptions. Of additional interest is how the working memory subsystems interact during processing. Neuroimaging work has suggested frontal-lobe specialization for representations requiring integration of verbal and spatial information (Prabhakaran et al., 2000). This suggests different processes for the interaction above and beyond the individual contributions. Further, do these roles change as a function of the spatial description's perspective? Such questions can be effectively examined using secondary working memory interference tasks.

In general, these interference task studies have people read a spatial text while also performing a secondary task designed to specifically interfere with visuospatial or articulatory working memory processes. Performance decrements on subsequent memory tasks are interpreted as indicating involvement of the targeted working memory subsystem. Comparing route descriptions and non-spatial, procedural instructions (how to make wine), De Beni et al. (2005) showed that while the verbal component of working memory was predictably involved with both types of descriptions, only the spatial description used the visuospatial sketchpad. This finding has been further supported by other research (Deyzac, Logie, and Denis, 2006; Pazzaglia et al., 2007). Brunyé and Taylor (2008b) extended this work in several ways. First, we compared working memory subsystem involvement between the two spatial description perspectives. Second, we examined working memory involvement both at encoding, while reading the descriptions, and at retrieval, while completing memory tasks. Third, we specifically examined central executive involvement, looking at two different hypothesized central executive functions by using two different secondary tasks. Random generation engages the central executive because of its requirements to inhibit repetition and avoid falling into perceptible patterns (i.e. Duncan et al., 1996; Baddeley et al., 1998; Miyake et al., 2001). Auditory sequence monitoring requires continuous information updating of sequential information, a proposed central executive function (Rabinowitz et al., 1982; Miyake and Shah, 1999; Brunyé et al., 2006).

The results strongly suggested that description perspective impacted working memory involvement. First, findings suggested greater visuospatial sketchpad and central executive involvement for route relative to survey descriptions. The visuospatial secondary task impeded spatial inferences, particularly those from the unlearned perspective, to a greater extent with route descriptions. Further, the effect of the visuospatial secondary task exceeded that of the articulatory task for the route description. For the central executive, random generation impeded inferencing in all respects, that is, regardless of description and test statement perspective. In contrast, the sequence monitoring secondary task only affected inferencing with route description learning and more so when making inferences from a route perspective. This finding strongly implicates the need to process temporal information in route descriptions. Our studies examining working memory involvement at retrieval, in this case secondary working memory tasks when verifying statements, showed that the visuospatial secondary task consistently and selectively interfered with inferencing, particularly inferencing time. The central executive tasks interfered with perspective switching, but not inferencing from the learned perspective. These findings suggest that additional inferencing, beyond that done during learning, occurs during retrieval and that inferences that switch perspectives engage both visuospatial and central executive processes.

What are the implications of increased working memory involvement with route compared to survey descriptions? The higher working memory maintenance may detract from cognitive resources necessary for higher-order integration and consolidation (i.e. Sweller, 1988). Perspective flexibility, in turn, relies on this higher-order integration and consolidation. A second implication comes from potentially identifying more specific roles for working memory subsystems. In particular, the differential roles of the working memory subsystems identified in relation to spatial perspective suggest that spatial mental models recruit articulatory (verbal) mechanisms for direct information retrieval, and recruit visuospatial and central executive mechanisms for inferences (Brunyé and Taylor, 2008b). This view is supported by neuroimaging research. Prabhakaran et al. (2000) suggest that the prefrontal cortex's role in integrating multiple information types in working memory may underlie its unique contribution to high-level cognition requiring flexible mental representations. Work from other labs has further supported differential working memory involvement as a function of description perspective. Deyzac, Logie, and Denis (2006) also showed that the verbal aspect of working memory played a primary role only in processing landmarks from a survey perspective. This suggests that this landmark information could be accessed directly if learned from a survey perspective, but not if learned from a route perspective. In summary, route descriptions enlist working memory resources above and beyond those used for survey descriptions. The need for these additional resources helps explain why route descriptions require additional processing time to yield the same inferencing facility.

2.5 Configural information

A survey perspective relates configural spatial information directly while this information must be inferred from a route perspective (e.g. Bilge and Taylor, 2010). To relate this configural information, survey descriptions often have a hierarchical structure that serves to divide an environment into spatial areas (Noordzij and

Postma, 2005). This difference between description perspectives has implications for differential working memory demands as a function of description perspective. As discussed in the previous section, route descriptions appear to have greater visuo-spatial sketchpad and central executive requirements (Brunyé and Taylor, 2008b). A contributing factor to these increased working memory demands may be that readers engage resources towards building global spatial relations into their mental representation. Inferences require significant cognitive resources (i.e. Estevez and Calvo, 2000; Linderholm and van den Broek, 2002; Rapp et al., 2006). While discourse processing research has debated the extent to which people make spatial inferences while reading (e.g. Zwaan and Van Oostendorp, 1993), it seems safe to assume that processing goals for spatial descriptions would encourage the active tracking and representation of spatial relationships (Morrow, 1994; Rinck et al., 1997). This assumption is grounded in the fact that people can build accurate mental models from spatial descriptions; ones that enable accurate spatial inferences (Perrig and Kintsch, 1985; Denis and Zimmer, 1992; Taylor and Tversky, 1992b; Ferguson and Hegarty, 1994; Avraamides et al., 2004; Lee and Tversky, 2005; Noordzij and Postma, 2005; Loomis et al., 2007; Brunyé and Taylor, 2008b; Avraamides and Kelly, 2010).

Current research in our lab is exploring how specifically defining the environment boundaries, one aspect of its overall configuration, in route descriptions may reduce the inferencing burden. Some of our earlier results (e.g. Brunyé et al., 2008; Brunyé and Taylor, 2008a) might be partially attributed to the fact that route descriptions often proceed through an environment without necessarily defining the environmental boundaries. Survey descriptions, in contrast, typically define overall environmental structure early on in the description (e.g. *The town is shaped like a vertically elongated rectangle*). Within a route description, understanding an environment's overall structure by travelling the perimeter might provide readers with configural information that ultimately aids perspective flexibility. Our recent work is examining whether providing configural information early in route descriptions (through a global tour around the environment) will aid perspective flexibility. If this is true, then earlier results might be attributable not only to perspective differences, but also the informational dissimilarities that restrict early configural knowledge development with route descriptions.

The availability (or not) of configural information may play another role tied to spatial perspective. In addition to increasing cognitive demands for route descriptions because of increased inferencing, configural information can ease cognitive load for survey descriptions. People naturally structure memory hierarchically (Stevens and Coupe, 1978), mentally forming boundaries even when they are perceptually absent (McNamara et al., 1989). Spatial hierarchies can be based on boundaries, such as roads (McNamara et al., 1984). People use non-spatial information to form these hierarchies as well, basing them on functional similarity (Hirtle and Jonides, 1985), perceptual similarity (Hirtle and Kallman, 1988), semantic similarity (Hirtle and Mascolo, 1986; Thomas et al., 2012), temporal order (Parmentier et al., 2006), and similarities based on social variables, such as race (Maddox et al., 2008). When organized hierarchically, information about particular spatial relations can be inferred from relations between hierarchical groupings, thus reducing cognitive load. Being able to generalize to spatial categories leaves open cognitive resources for other processing, including inferences (Brunyé et al., 2007). This may contribute to results showing easier information access when learned from a survey perspective (Chabanne et al., 2003). It should be noted, however, that relying on spatial categories can sometimes also lead to memory distortions (e.g. Stevens and Coupe, 1978; McNamara et al., 1984; McNamara et al., 1989; Maddox et al., 2008).

2.6 Mentally simulating extended spatial discourse

Understanding how spatial perspective might differentially affect development and use of resultant mental representations requires insight into the nature of the representations. Two complementary theoretical perspectives have been proposed to explain the nature of spatial mental models. First is the amodal symbolic position, wherein spatial memory involves networks of amodal symbols, such as a vast connectionist network, that code for spatial relationships (e.g. Stevens and Coupe, 1978; McNamara et al., 1989; McNamara, 1991). The second, more recent view suggests instead that spatial representations involve modal, analogue, and sometimes image-like components (Paivio, 1990; Glenberg, 1997; Barsalou, 1999). This view, often referred to as embodied or grounded cognition, would seem to make sense for spatial descriptions. Spatial language usually conveys information that is or will become perceptible and actionable. Take our opening scenario. The spatial description is designed to guide you on paths and through intersections, passing salient environmental features (e.g. windmills, bridges, turns) that will help you find your way back home (Tenbrink and Winter, 2009). This embodied approach has been applied more generally to concept understanding (Paivio, 1990; Glenberg, 1997; Barsalou, 1999) and language processing (for a review see Fischer and Zwaan, 2008).

With respect to reading comprehension, several theorists have proposed that it involves the activation of perceptual symbols (i.e. Kosslyn, 1980; Paivio, 1990; Barsalou, 1999, 2005; Zwaan, 2004) and motoric representations (Lakoff, 1987; McNeill, 1992; Glenberg, 1997; Glenberg and Kaschak, 2002; Zwaan and Taylor, 2006) while readers or listeners mentally simulate what is being described. For spatial descriptions, this would mean that the mental representations that support thinking about spaces are supported by the same cognitive (Zwaan, 2004; Barsalou, 2005) and neural (Ruby and Decety, 2001; Hauk et al., 2004; Pulvermüller, 2005; Tettamanti et al., 2005) mechanisms that support direct perception and action in those spaces. Moving within an environment would involve several modalities and these would also be part of mental representations, including vision (e.g. Zwaan et al., 2002; Yaxley and Zwaan, 2007), motor movement (e.g. Glenberg and Kaschak, 2002), and sound (e.g. Kaschak et al., 2006; Kurby et al., 2009; Brunyé et al., 2010b; Brunyé et al., 2010a). Relevant to making inferences about the environment, proponents of this embodied view argue that these multimodal mental representations form the basis for successful inference generation (Zwaan, 1999; Fincher-Kiefer, 2001; Brunyé and Taylor, 2008b) and prepare comprehenders for subsequent perception and action (Barsalou, 1999).

Recent research has considered how perceptual and motor-based representations might specifically underlie spatial description processing. Evidence for these embodied processes comes from an examination of how readers spontaneously activate multimodal representations when reading spatial descriptions. Not only do people simulate the actions and motions described for a character, but do so with fictive motion as well (Matlock, 2004). A sentence with fictive motion uses an action verb, but does not actually describe motion, such as 'The path *runs* straight until it dead-ends.' The nature of route and survey descriptions suggests the differential need of such modal representations. Route descriptions would more heavily recruit active three-dimensional mental imagery associated with a within-environment perspective (Brunyé and Taylor, 2008b; Farmer et al., 1986; Miyake et al., 2001; Pazzaglia et al., 2007) and motoric connections to orientation updating (e.g. Rieser and Pick, 2007). Thus, route descriptions would more likely involve mental simulations involving perceptual and motor imagery.

To a large extent, spatial description research has only provided indirect evidence for use of mental simulations when reading route descriptions. Indirect evidence has been supplemented by phenomenological reports. When reading route descriptions, people claim to imagine themselves moving through the environment and using mnemonic imagery strategies (e.g. 'I imagined seeing cows in the field as I ran down the path'). Such reports can include movement through the described environment, visual characteristics of described landmarks, and reports of a sense of presence. Reports after reading survey descriptions lack the same perceptual and motoric qualities. Current research in our laboratory hopes to disentangle the relative contributions of visual and motor imagery in comprehending and representing different perspective spatial descriptions.

One recent study from our lab, however, directly implicates perceptual and motor imagery involvement in spatial description comprehension (Brunyé et al., 2010b). This work involved two experiments comparing the extent to which readers integrate motor-related sounds into spatial representations of route and survey descriptions. Experiment 1 had participants read route and survey descriptions at their own pace, one sentence at a time. While reading, they listened to one of two sounds: footsteps on gravel or a metronome. The pace of the sound either started slow and switched to fast, akin to walking versus running and operationalized as 1 versus 3 sounds per second, or started fast and switched to slow. After reading, participants verified spatial statements, some requiring inferences. Dependent measures included reading time and spatial inferencing from memory. In a second study, participants read a route or survey description while listening to either a fast or a slow sound (footsteps or metronome) and then drew a map of the described area and estimated the scale of their map.

Results of this study showed differential embodiment of the sounds into the spatial representations. Reading times provided the first piece of evidence. Participants read both perspective descriptions faster with fast compared to slow metronome sounds. In other words, reading rates followed the rate of the metronome sound. However, the test of embodiment hinged on the effect of the footstep sounds. Changing the speed of footstep sounds only affected route description reading times. The difference between the footstep and metronome sounds suggests that the inherently sensorimotor characteristics of the footstep sounds guided mental simulations of movement through the environment, thus affecting reading times during route, but not survey, description processing. Memory task performance showed similar differential embodiment based on perspective. The footstep sounds appeared to reinforce route-perspective information, thereby leading to a reduced ability to switch perspectives to make inferences. Participants reading a route description while listening to footsteps had more difficulty making survey-perspective inferences. In other words, readers showed perspective specificity when both the spatial description and the presented sounds enforced a ground-level perspective. Further, footsteps with a route description led to better within-perspective inferencing, an additional suggestion that listening to footsteps while reading a route description encouraged thinking about the environment from a ground-level perspective. This finding also suggests that results showing perspective-specificity after limited route description exposure may be based on greater ground-level mental simulation (Shelton and McNamara, 2004; Brunyé et al., 2008; Brunyé and Taylor, 2008b). Neither footsteps with a survey description nor metronome sounds with either perspective produced this extent of perspective specificity.

To conclude, route descriptions would appear to evoke greater mental simulation and embodiment than survey descriptions. Route descriptions relate the motor and perceptual information one would experience when travelling through the environment. The increased cognitive processes needed to create this mental simulation, however, appear to contribute to the need for more learning time with route descriptions.

2.7 Temporal information

Learning spatial information necessarily has a temporal component. This is true regardless of whether the information is learned from navigation, through maps, or from descriptions. The temporal component, however, often goes unnoticed as it is strongly correlated with spatial proximity (Clayton and Habibi, 1991; Curiel and

Radvansky, 1998; Naylor-Emlen and Taylor, 2009). Spatial descriptions can have temporal contributions based on their nature and content. Spatial descriptions reveal information over time, a necessity of the linear nature of language. But, route descriptions contain additional temporal information in what is described, in particular instructions on where to go in what order and what to do at various decision points (Daniel and Denis, 2004; Shi and Tenbrink, 2009). With survey descriptions, the spatial relationships described often follow a given-new ordering (Haviland and Clark, 1974). Beyond following this referential convention, locations could be described in alternate orders without much disruption. The implication of this is that the order in which information is presented in a description has greater import for route compared to survey descriptions. This is not to say that the temporal order in which survey descriptions present information is not encoded. Taylor and Tversky (1992a) found that the order that participants drew landmarks on sketch maps matched the order they had been related through descriptions, regardless of perspective.

How temporal information might differentially affect spatial description reading also ties into working memory involvement. Much of the research on serial order memory has focused on the phonological loop of Baddeley's working memory model (e.g. Burgess and Hitch, 1999). The phonological loop's involvement centres on building activation strengths across serially connected items (Page and Norris, 1998). Other work, however, has suggested central executive involvement for serial order (Jones et al., 1995). Jones et al. (1995) compared serial memory for spatial position to serial recall of verbal material. The results suggested that serial spatial recall has similar limitations as seen with verbal material. This similarity across verbal and spatial content suggests central executive involvement. This suggestion fits with our findings specifically examining different working memory functions with spatial description processing. As noted in our discussion above on the role of working memory, the sequence monitoring secondary task had specific consequences for the route perspective, both learning and inferences (Brunyé and Taylor, 2008b). Another domain that also requires memory for serial order, that of procedural instructions, shows similar findings. Brunyé et al. (2006) found broad-based memory decrements when participants engaged in a secondary sequence monitoring and updating task (Miyake et al., 2000; Miyake and Shah, 1999; Rabinowitz et al., 1982) while learning procedures.

In conclusion, some of the additional demands required for processing route relative to survey descriptions may arise from the fact that route descriptions have more temporal content. Our findings suggest that the increased temporal content, in turn, requires additional central executive resources during processing.

2.8 Conclusions

Language provides a viable means for learning environments. From descriptions readers can form accurate spatial mental models (Perrig and Kintsch, 1985; Denis and Zimmer, 1992; Taylor and Tversky, 1992b; Ferguson and Hegarty, 1994; Avraamides et al., 2004; Lee and Tversky, 2005; Noordzij and Postma, 2005; Loomis et al., 2007; Brunyé and Taylor, 2008b; Avraamides and Kelly, 2010). Spatial perspectives, route, survey, and gaze, structure how space can be linearized into language. Comparing learning from and using representations based on these different perspectives has helped us explicate the more general cognitive underpinnings of spatial language comprehension.

In this chapter, we have reviewed four possible explanations for mental representation differences based on route and survey descriptions. We have focused on one specific difference that has interesting implications for spatial problem solving. Namely, we have focused on the finding that learning from route descriptions requires additional experience and cognitive resources to reach the same level of inferencing prowess (Lee and Tversky, 2005; Noordzij and Postma, 2005; Noordzij et al., 2006; Brunyé et al., 2008; Brunyé and Taylor, 2008a). Note that while this finding has generally been described as a failing of route descriptions, it can also be interpreted as an advantage of survey descriptions. The explanations, as evidenced through our discussion, are not independent from one another. Looking at each pair-wise combination of the four, one will find overlaps. Yet taken together, they cover what empirical data has revealed. Together, these explanations provide insight into spatial description comprehension more specifically and contribute to spatial cognition research more generally.

These explanations also have implications for navigation systems. In-vehicle, commercial navigation systems include route-based spatial descriptions. They may provide a single route instruction, such as 'in 200 metres, turn left'. Or they may provide two to three consecutive instructions, if actions need to be engaged in relatively rapid succession, such as 'take the exit right and then stay in the left lane to turn left at the next intersection'. The route-descriptions of navigation systems are much more temporally disconnected than typical oral or written route descriptions, but also are embedded or coincident with what a driver has the potential to perceive when the description is delivered. In other words, the driver can see the right-hand exit when instructed to 'take the exit right'. Applications of our explanations to in-vehicle assistance systems should consider how to maximize the tie between the route-instruction and perceptually available information (while not visually distracting the driver). Additional research in this regard could inform in-vehicle assistance system design as well as further contributing to our general understanding of spatial description processing.

3

Spatial representations for described and perceived locations*

MARIOS AVRAAMIDES, CATHERINE MELLO, AND NATHAN GREENAUER

3.1 Introduction

Although the visual system is for most people the primary means of acquiring and representing spatial information in memory, the other sensory modalities often supplement or replace vision in many situations. For example, people can establish spatial representations by hearing sounds emanating from different locations, by touching objects, and by moving from one location to another. In addition to these direct modes of encoding, information about space is often acquired through symbolic media such as pictures, maps, diagrams, and language. Language in particular is frequently used in daily life to communicate spatial information about both our immediate surroundings (e.g. when a colleague says 'hand me that book on your left') and remote environments (e.g. when we instruct a student to 'go to the lab and switch on the computer that is next to the TV'). Nowadays, language is used to guide our navigation in unfamiliar environments (e.g. when used in GPS-based navigational systems) but it has also long served to immerse us in familiar, unfamiliar, or fictitious settings, allowing us to build vivid representations that contain spatial information (e.g. when reading descriptions of distant places in novels).

In this chapter, we evaluate whether, and to what extent, sensory and linguistic stimuli support the formation of functionally equivalent spatial representations. We begin by reviewing what is currently known about the organizational structure of spatial memories derived from vision. In the context of this work, we review the conditions known to give rise to similar or different representational structures based on what is known about the unique characteristics of spatial language. Finally,

* Support was provided by research grants OSSMA-206912 from the European Research Commission and KOINO/0308/17(BE) from the Cyprus Research Promotion Foundation.

we describe and contrast several accounts of functional equivalence in terms of their ability to integrate these findings.

3.2 Models of spatial memory derived from vision

Walking to the front door of our house after we have switched off all the lights or recalling where in our living room we placed our keys, are two of the many tasks that we carry out on a daily basis that rely on spatial memory. In the first example, walking towards the door in complete darkness entails maintaining and using, in real time, information about locations in our immediate surroundings (e.g. potential obstacles) and processing them in relation to our own changing position (a process referred to as *spatial updating*). Such tasks involving reasoning about immediate locations are commonly referred to in the literature as *online* (Waller and Hodgson, 2006; Avraamides and Kelly, 2008). Recalling where we left our keys could also involve online processing, as we might be searching for them while still in our living room. However, if we have already left for work, thinking about the keys we left at home is no longer an online task. In this case, we must recall information about previously visited locations that are part of a remote environment with which we are no longer actively engaged. Tasks involving reasoning about remote spatial locations are referred to as *offline* (Avraamides and Kelly, 2008).

Kelly, Avraamides, and Loomis (2007) contrasted spatial reasoning in online and offline contexts by examining how well people localize previously-viewed objects under various circumstances. Participants studied layouts of objects presented around them in a virtual environment and then pointed to them from various imagined perspectives. Following learning, the objects were removed and participants in an *immediate* testing condition were instructed to turn 90° to their left or right while monitoring where the objects were located. Another group of participants were assigned to a *remote* testing condition. Following learning, these participants were taken to a different room before being instructed to turn 90° to their left or right and to imagine the objects around them. After allowing participants as much time as needed to imagine the objects based on the new direction they were facing, participants' memory was evaluated by asking them to point to each object's location from several different imagined orientations (i.e. imagining views from several different orientations without turning). Results from this study revealed that in the immediate testing condition participants performed better when pointing to objects from imagined perspectives corresponding either to the learning or testing orientation than from any other orientation. In contrast, in the remote testing condition, participants performed best when mentally adopting the learning perspective; participants' performance at the perspective corresponding to their facing direction at test was no better than at any other non-experienced perspective. Thus, the findings of Kelly and colleagues (2007) revealed that when reasoning about a remote environment the orientation of one's body exerts little influence on performance. Reasoning about one's immediate environment, however, appears sensitive to previous experience as well as awareness of one's current orientation.

Avraamides and Kelly (2008) subsequently elaborated on their earlier findings by proposing that participants encode spatial layouts in memory on the basis of both an online representation that codes self-to-object relations in a sensorimotor framework (see also Rieser et al., 1986; Wang and Spelke, 2000; De Vega and Rodrigo, 2001; Waller and Hodgson, 2006) and an offline representation that maintains knowledge of object-to-object relations (Wang and Spelke, 2000; Sholl, 2001; Mou et al., 2004; Waller and Hodgson, 2006). When encoding a layout, particularly from a central vantage point (as in Kelly et al., 2007), the reference direction of the offline representation formed is, in the absence of instruction or conflicting environmental cues (e.g. an intrinsic structure of the layout), aligned with the initial learning perspective (Greenauer and Waller, 2008; Shelton and McNamara, 2001).[1] Thus, judgements aligned with the learning perspective regardless of one's current physical orientation and location (i.e. environment) are indicative of reliance on an offline representation. Conversely, the presence of an advantage for judgements aligned with one's currently embodied perspective, rather than a previously experienced perspective, indicates reliance on an updated, online framework. According to this model, offline representations preserve long-term knowledge of spatial relations that remains valid and accessible even after an observer moves to a remote location. The transient and situational nature of online representations, however, renders them unavailable when no longer in the encoded environment (but see Rieser et al., 1994; Kelly et al., 2007; May, 2007, for reestablishment of online codes through visualization and self-motion).

Although several distinct two-system models have been proposed, the distinction between online and offline representations has been both ubiquitous and widely supported by behavioural studies (e.g. Wang and Spelke, 2000; Mou et al., 2004; Waller and Hodgson, 2006; see also Burgess, 2006 for a review). In general, online representations are conceptualized as transient and dynamic, used for the monitoring of self-to-object relations in real-time, considered automatic or obligatory, and limited in capacity. Offline representations, on the other hand, refer to relatively abstract and enduring knowledge structures that maintain information about object-to-object relations in an environment. In the context of these distinctions, we now consider how the characteristics of language may support and influence memory when spatial information is not directly perceived.

[1] Although egocentric experience can exert a strong influence in determining how spatial relations are represented in the offline system, several studies have demonstrated that non-egocentric cues can also be strong determinants in organizing enduring knowledge structures (e.g. McNamara et al., 2003; Mou and McNamara, 2002). Thus, a preferred orientation in memory could be misaligned with one's experience. To the extent that the processes and behavioural outcomes discussed subsequently will be true for any preferred orientation in an offline representation, we focus on the simple case of one's previous experience in order to simplify these discussions.

3.3 Spatial memories for linguistically acquired environments

3.3.1 Key distinctions from vision

Whereas most current models describing the organization of both short-term and long-term spatial memories are based on methodologies employing primarily visual stimuli, many researchers have argued that spatial representations derived from different sources are functionally equivalent (Bryant, 1997; Loomis et al., 2002; Loomis et al., 2007). That is, if differences in learning are equated so that the fidelity of the resulting representations is equivalent and biases are controlled for (e.g. it is well known that encoding 3-D sounds leads to underestimation of distances; Ashmead et al., 1995), then these representations can be used to carry out any spatial task with similar behavioural outcomes.

Functional equivalence is not unexpected with regards to different sensory modalities, inasmuch as these have evolved to respond to correlated sources. However, language differs in several important ways from the spatial senses, and might be expected to result in a qualitatively and quantitatively different representational form. For instance, while vision allows for the near-simultaneous encoding of metric information, language is inherently serial. Thus, a reader or a listener must integrate several pieces of information that are presented at different points in time in order to form a single coherent representation, with clear implications for working memory load (a point further elaborated on in Taylor and Brunyé, this volume). Visual learning, conversely, allows for the sampling (and re-sampling) of any available information in any order, as well as the ability to focus on global or local properties of spatial relations. Consequently, the serial nature of spatial language may be most problematic in the formation of configural knowledge, such as that involved in judging inter-object relationships. Note, however, that this quality of language is not necessarily a liability, as it parallels certain spatial behaviours it is intended to support, such as route navigation, and can thus facilitate recall of information associated with it (e.g. landmarks along a route, Denis and Fernandez, this volume). Furthermore, the large-scale spaces encountered during route learning are themselves often experienced sequentially, such that descriptions may bear considerable similarity to perception in such cases.

Establishing an understanding of spatial relations based on language may be further complicated by uncertainty about the precise locations and relationships involved when metric information is not provided. For example, knowing that something is 'to your left' provides only categorical knowledge about location. Moreover, as pointed out by Taylor and Brunyé (this volume), route descriptions in particular often contain references to time or ordinal relationships. This has clear implications for considerations of functional equivalence of representations formed from linguistic and perceptual stimuli. Specifically, although the temporal and physical dimensions of space are experienced *together* by a situated observer, and thus provide redundant information, a listener or reader generally has to infer spatial relations from temporal or ordinal statements, with additional working memory costs. Importantly, even when metric information is incorporated into a description, the measurement supplied (e.g. metres) remains an abstract, arbitrary construct that must be converted into behaviourally meaningful units by the recipient. In this sense, spatial information conveyed through language is never accessed directly, but is mediated by associations between linguistic labels and spatial concepts such as direction or distance. Similarly, whereas vision and other sensory modalities allow an observer to directly perceive the relationship between points in space and their physical movement, for instance by experiencing optic flow while walking, linguistic spatial information is typically divorced from the body and its actions. Chown (this volume) has suggested that retinal position and the movements of the eyes, head, and body of an observer as they move in an environment may in fact be the 'metric' by which object locations and directions are represented in the cerebral cortex. If correct, establishing these coordinates based on categorical and often imprecise linguistic statements would require substantial conversion and inference (a point we return to in our discussion).

Finally, in contrast to vision, the act of describing spatial layout information requires that a reference frame be selected to fully specify locations in terms of direction or distance. Although language is flexible in the reference frames it permits to be used to describe a given configuration (for a popular classification, see Levinson, 1996b; Tenbrink, 2011), the selection of a reference frame often determines what information is readily available to the recipient of such a communication, and what information must be inferred. For example, consider the phrases 'the mug is 5 cm to my left' and 'the computer is 30 cm in front of me'. Both statements communicate the linear and angular relationships existing between the targets (the mug and computer) and a *reference object* (here, the speaker) relative to a *reference direction* or perspective (here, implied to be the speaker's facing direction). From this example, it is clear that a listener who wishes to know the relationship between the mug and computer would need to perform a series of spatial transformations in order to infer this relationship. In contrast, although vision provides its own reference frame (i.e. retinal coordinates), objects that are perceived visually can be at once related to the observer, as well as to other persons, objects, or environmental features, with none of these potential reference frames being obligatory or superseding others. It is a well-established fact that space perceived through direct experience or visual depictions can be described in numerous ways depending on observers and task requirements (for some examples of this diversity, see Denis and Fernandez; Taylor and Brunyé; Seifert and Tenbrink, this volume) precisely because of the richness of the information afforded. Due to the reference frames implied or explicitly conveyed in such communications, it is as yet unclear whether a described space could likewise be imagined or recreated from multiple perspectives.

Based on these differences between language and vision, one could expect that spatial representations derived from language (1) are more difficult to construct and maintain, (2) may not be treated as a physical reality, and (3) may be more restricted in terms of reference frames available for encoding and retrieval. Although the first prediction has generally been supported by research, recent findings show that the second and third predictions do not necessarily hold true. Nevertheless, despite differences that may exist between vision and language, if the representations derived from these stimuli can be used to carry out a spatial task with similar behavioural outcomes, these representations are regarded as functionally equivalent.

3.3.2 Challenges to the formation and updating of linguistic spatial information

Previous research has indeed verified that spatial representations are generally more difficult to construct from language than vision. In one study, Klatzky et al. (2002) asked participants to learn the locations of three or five targets from vision, 3-D sound, or spatial language. In the vision condition, participants turned their heads within a virtual environment to observe the names of objects that appeared at specified head orientations. In the 3-D sound condition, participants' head orientation remained stationary while locations were indicated by tones emanating from loudspeakers arrayed around the room. In the language condition, locations were provided by verbal labels based on clock positions (e.g. 'ball, 3 o'clock') played back to the stationary participant via headphones. Following learning, participants made a series of judgements in which they pointed to object locations until a performance criterion was met. Results revealed that in the demanding 5-object condition, language produced slower learning than either vision or 3-D sound, suggesting that spatial representations based on non-perceptual stimuli are comparatively more difficult to construct. Similarly, in a study by Avraamides et al. (2004), participants learned the locations of four objects either visually or through spatial sentences in the form of 'there is a ball at 8 o'clock, 4 feet away from you'. Learning involved alternating exposure to the objects and pointing tests until participants could point to all objects accurately. In one experiment, visual objects were presented sequentially, with all previous objects removed each time an object was introduced. The authors observed that participants took more trials to reach the learning criterion after encoding locations from language than vision. Thus, the inherently serial nature of linguistic stimuli does not appear to be the only factor underlying the difficulty of acquiring spatial information from language—a point further developed in Section 3.4.3.

The challenges associated with encoding described spaces do not, however, preclude the resulting memories from being used in ways analogous to their visual counterparts. A number of studies have examined spatial updating for objects encoded through language and vision, and have provided converging evidence for functional equivalence. For example, a study by Loomis et al. (2002) examined updating of a single target that was either presented as a sound from a loudspeaker positioned at different locations or a verbal utterance of the form '4 o'clock, 9 feet'. After encoding, participants, both blindfolded sighted and congenitally blind, were asked to walk either directly or indirectly towards the target. For indirect walking trials, the experimenter guided participants to a new position that was 2.74 m in front of them before being asked to walk straight to the target. Therefore, in order to execute the task correctly in the language condition, participants had to first convert the verbal utterance into awareness of a real location and then update their position relative to the target while moving to the new position. The authors compared the separation of the terminal points of the walking trajectories in direct and indirect trials. Results revealed only small separation between direct and indirect paths, regardless of input modality, suggesting that participants successfully updated the position of the target during their movement to the new standpoint under both conditions. Additionally, separations were only slightly greater in language than audition.

A follow-up study by Klatzky et al. (2003; Experiment 2) used a similar procedure, but asked participants to update multiple targets (3 or 5) and included a vision condition. After learning target locations to a criterion, participants walked directly or indirectly to them. As in other comparable studies (e.g. Klatzky et al., 2002), learning target locations through spatial language was found to be more difficult than with vision (but not spatialized sound). Additionally, indirect walking was associated with greater updating noise (relative to a direct path) in the linguistic condition. However, differences between modalities in updating performance were negligible overall, indicating that linguistically described targets were updated relatively efficiently despite their costly acquisition and relative imprecision.

In summary, owing in part to its inherently serial and imprecise (if not categorical) nature, language is associated with a slower, more effortful construction of spatial representations than vision. However, despite this initial encoding challenge, linguistically derived spatial representations nevertheless lend themselves to relatively efficient updating during locomotion. Thus, rather than indicating qualitatively distinct representational forms, differences in online processing between modalities appear to be mainly quantitative in nature, manifesting in the speed and precision of responses, and in the comprehensiveness of encoding and the subsequent noise accrual over motion.

3.3.3 Online and offline processing of described spaces

The previous section presents evidence that linguistically acquired targets can be easily updated with physical movements, however it has been suggested that this may be the exception rather than the rule. Wang (2004) asked participants to learn the location of visual targets (laboratory objects) and either describe a familiar setting (objects from their kitchen, Experiment 1) or listen to a description of other objects (Experiment 2) as though these unseen objects were also currently arrayed around them. Participants subsequently rotated with respect to either the real or the imagined layout before indicating the position of all objects from their updated egocentric orientation. When participants had rotated with respect to the described objects, they were equally efficient when pointing to targets within this attended layout and within the visual layout. In contrast, rotating with respect to the visual layout did not lead to efficient updating of the targets within the described layout. Previous research by Wang and colleagues (Wang and Brockmole, 2003a, 2003b) had already demonstrated that visually perceived spaces are no longer updated automatically when they cease to be relevant to immediate, situated interaction by an observer (i.e. when they become remote environments). Based on the results with linguistically described spaces, Wang (2004) argued that unlike spatial information derived from vision, described layouts are not treated as physical realities because they are neither potential obstacles nor goals for locomotion. Thus, described spaces are not updated *automatically*—online—like visual targets.

Although described spaces may generally not be treated as real, this does not mean that they cannot be, particularly when doing so is situationally appropriate. Indeed, by asking participants to take described objects into account while moving, Avraamides et al. (2004) provided findings indicating that, under some circumstances, described spaces can be treated as physical realities and thus need not be relegated to offline processing. For example, in one experiment participants were required to move to a new standpoint after they had learned the locations of several surrounding objects via language. Importantly, prior to learning, participants were informed that they would move before being tested. In contrast to previous experiments where no such movement instructions were included (see the previous section), no differences were found between the vision and language conditions during learning. One possible explanation for these results is that the movement requirement (or anticipation of movement) had encouraged participants to transform the verbal descriptions into spatial coordinates, thereby creating a representation identical (or at least similar in form) to that derived from vision. Under other circumstances, however, the authors speculated that participants may instead maintain the surface structure of the linguistic stimuli in memory until required to produce a response, rather than actively constructing a spatial representation during encoding.

Even though physical motion can induce or facilitate the conversion of linguistic stimuli into a functional, spatial representation as in the previous example, other studies show that it may not necessarily lead to updating. For example, Avraamides (2003) asked participants to read narratives that described a protagonist in an unfamiliar environment (e.g. a museum) with objects described at locations around him or her. After the initial description of the environment, the protagonist was rotated to a new orientation. Participants were asked to indicate the directions of described objects from the protagonist's updated orientation. In one experiment, participants carried out the task from a fixed orientation that was aligned with the initial orientation of the protagonist (i.e. they had to mentally rotate to align their real orientation with the updated protagonist orientation). In a second experiment, participants physically rotated so that their orientation matched that of the protagonist at all times. Regardless of the mode of rotation, participants performed better when adopting the initial orientation of the protagonist than any of the novel orientations. This suggests that participants did not update protagonist-to-object relations with either imagined or physical movement. In a third experiment, however, where coloured discs serving as perceptual placeholders for described objects were placed around the participant, performance for initial and novel orientations was equivalent, but only in the physical movement condition. Thus, although physical movement did not by itself induce spatial updating, the presence of physical markers in the learning environments was sufficient to prompt participants to process the narrative in an online fashion.

The lack of updating following physical movement by Avraamides (2003) contrasts with the results of Rieser et al. (1994), who demonstrated that imagined updating in remote environments could be facilitated by executing the corresponding physical trajectories. It should be pointed out, however, that Avraamides (2003) did not instruct participants to interpret their own physical rotation as a rotation within the imagined environments. Thus, one possible explanation for these apparently discrepant findings is that updating within described environments may not occur spontaneously but, rather, result from instructions or perceptual cues (e.g. the coloured discs) highlighting the relationship between linguistic targets and self-motion. This speculation has been supported by a recent study conducted in our laboratory (Pantelides, 2010). In two experiments, participants learned a virtual layout of objects by physically rotating to face each object in turn. Object presentation was contingent on the direction faced and was effected visually (i.e. the object appeared) or verbally (i.e. by 'you are facing the...' statements). Regardless of whether the object was grounded in a rich, structured visual environment (Experiment 1; compare to Avraamides (2003) Experiment 3) or paired with only optic flow (Experiment 2), described layouts were updated as efficiently as their visual counterparts. This was true even when the fidelity of the representation was initially (i.e. prior to updating) superior for visual targets (Experiment 1).

Based on these findings, we suggest that the fact that listeners do not always treat described spaces as physical realities spontaneously may explain differences in spatial performance relative to vision. Coupled with processing demands present at learning, such as those imposed by the serial (and often imprecise, categorical) nature of language, this tendency may make participants unlikely to immediately convert descriptions into a functional spatial representation. In effect, the features of linguistic communication may encourage participants to resort to nonspatial strategies, which in turn are less suited to support fast, efficient, and accurate behaviour.

However, when participants are informed of how described targets coexist with directly perceived features of the environment (e.g. Avraamides, 2003, Experiment 3; Pantelides, 2010, Experiment 1) or made aware of the relationship between these descriptions and their own movements (as in Avraamides et al., 2004, Experiment 3; Pantelides, 2010, Experiment 2), they appear quite capable of representing and retrieving spatial information online, with an efficiency comparable to that observed in visual learning situations.

Inasmuch as linguistic descriptions can serve not only to outline features of remote environments with little direct relevance to the listener (e.g. in narration), but also to indicate behaviourally important objects present in the immediate environment (e.g. when warning someone of an obstacle lying outside of their field of view), offline (nonspatial or spatial) and online processing are both circumstantially valid and useful strategies. This many-faceted nature of spatial language should thus be taken into account in any comparison with the spatial senses. Indeed, described spaces may only be processed in a manner functionally equivalent to perceived spaces when the knowledge conveyed is intended to serve a similar purpose (e.g. to guide locomotion), with other modes of linguistic processing (i.e. verbal or abstract, offline coding) being otherwise preferred.

3.3.4 Reference frames in the communication and encoding of space

Although a reference frame is implicit in many linguistic communications of spatial information, a number of findings suggest that the resulting verbally acquired spatial memories are not necessarily tied to relations directly given or the perspective used to specify them. For example, Taylor and Tversky (1992a) asked participants to read descriptions of environments written from either a route or a survey perspective. Subsequently, participants were asked to verify statements that contained spatial information that was either explicitly mentioned in the descriptions (verbatim) or statements involving spatial relations that could be inferred (inference). Results revealed that although participants who read descriptions thrice responded faster to verbatim than inference statements, route and survey inference statements were verified equally fast regardless of learning condition. This result supports the notion that spatial representations derived from language can be flexibly organized in a reference frame that differs from that used in the description. Note, however, that after a single reading, survey descriptions yielded faster inferences than route descriptions, implying that perspective flexibility may take more time or effort to achieve as a function of the type of description supplied. Brunyé and Taylor (2008a; see also Taylor and Brunyé, this volume) provide further support for the notion that not all linguistic descriptions are created equal in terms of reference frame flexibility.

Further evidence for the non-definitive nature of reference frames given in verbal descriptions comes from Avraamides et al. (2004, Experiment 3), who, as previously described, asked participants to make allocentric judgements (i.e. indicate the relative direction of two objects) after they had encoded all objects by vision or by egocentric spatial language. Although deriving object-to-object relationships based on self-to-object statements is arguably less straightforward than when objects can be perceived directly, participants were equally accurate and fast in their allocentric pointing judgements in both encoding conditions.

The reverse reference frame conversion, from allocentric statements to egocentric spatial awareness, has also been documented. Avraamides and Kelly (2010), replicated the findings by Kelly et al. (2007) with regards to online and offline processing in immediate and remote environments. Verbal descriptions were provided in allocentric terms, using objects present in the laboratory as referents (e.g. 'a ball is placed on the chair that is next to the TV'). As with the experiments conducted on visual targets (Kelly et al., 2007), participants were tested in the same or an adjacent room. An influence of participants' physical perspective at test was observed in the immediate but not in the remote testing condition. Thus, participants were not only able to acquire an object's egocentric locations based on allocentric statements, but also to incorporate these targets in a dynamic, updatable memory framework tied to their sense of position.

In addition to overriding a reference frame provided in a description for the purposes of updating, readers or listeners also appear capable of integrating multiple perspectives to form representations from novel vantage points. After reading scene descriptions written from a single observer's point of view, participants typically employ the protagonist's perspective when recalling details of the scene (e.g. Franklin and Tversky, 1990). However, when participants read a description encompassing multiple views and are subsequently asked questions from the viewpoint of multiple observers, participants will integrate the perspective of multiple observers at recall (Franklin et al., 1992). Recent scene recognition studies have demonstrated that such integration may occur on the basis of visual perception. Indeed, participants trained to recognize a scene from multiple views have been shown to exhibit superior recognition performance when probed from a novel perspective central to the trained views than from previously studied perspectives (Friedman et al., 2011; Waller et al., 2009).

The reviewed evidence suggests that despite the apparent rigidity of descriptions, the spatial representations based on linguistic stimuli are remarkably flexible in the behaviours they support. A speaker's adoption of a route or survey perspective when giving directions does not preclude a listener from accessing spatial information from a different perspective. Similarly, object-to-object relationships can be obtained from egocentric descriptions and body-centred, dynamic representations can emerge from allocentric statements. Finally, people can integrate information acquired from multiple described perspectives rather than being limited to any individual reference frame through which it was conveyed.

3.4. Implications for models of functional equivalence

3.4.1 Overview

In summary, a review of existing research suggests that spatial representations derived from language are functionally equivalent to those constructed from vision, although forming spatial representations from language requires in most cases extended learning to attain the fidelity of representations acquired visually. Furthermore, acquisition may be more taxing on cognitive resources due, for instance, to language's inherent imprecision or serial presentation. These difficulties inherent to linguistic stimuli can promote the use of verbal rehearsal strategies (Avraamides et al., 2004) rather than spatial encoding. Such strategies may in fact overshadow an important feature of linguistically-derived spatial representations, namely their ability to support online, situated interaction or the use of reference frames and relationships differing from those provided. In fact, as Loomis and colleagues (2007) argued, when encoding difficulties are controlled for, representations from language and vision may be functionally equivalent. Based on the reviewed research demonstrating that aspects of spatial processing (e.g. online updating) are intrinsically tied to the body and the physical senses, it should not be surprising that updating does not necessarily occur spontaneously with language. However, we would add that a listener's intentions and available strategies must also be taken into account before drawing any conclusions about functional equivalence.

The present chapter reviewed evidence supporting the conjecture that representations derived from language are functionally equivalent to those constructed from perceptual modalities. Overall, this evidence is consistent with the notion that representations derived from different sources can be organized using the same reference frames, and can thus support spatial actions in similar ways. Although functional equivalence between sensory modalities has been demonstrated repeatedly in empirical work (e.g. Tlauka et al., 2005; Giudice et al., 2007; Tlauka et al., 2009), the representational and computational underpinnings of equivalent performance have yet to be ascertained, particularly when non-perceptual information sources such as language are considered. In this final section, we review three classes of hypotheses that reflect the current conceptualization of this question, and highlight the empirical support for, as well as challenges to, each account.

3.4.2 Separate-but-equal modalities

One representational account of functional equivalence, referred to as the *separate-but-equal* hypothesis, suggests that similar behavioural outcomes can derive from distinct, modality-specific representations. The idea that multiple modality-specific representations are established is compatible with the findings of Yamamoto and Shelton (2005). These authors showed that when participants learned the same layout based on two successive experiences, one visual and the other proprioceptive, they maintained two separate representations, each organized around its own reference direction. Functional equivalence could, in theory, arise in the absence of any form of cross-modal communication. The physical senses, for instance, are exposed to modality-specific signals that co-vary because they stem from the same objects or features of the environment. The actions and thoughts individual sensory modalities support are likewise correlated. Therefore, separate modality-specific systems could likely evolve and develop to respond to environmental contingencies in a similar manner, without necessarily interacting with one another. Although this rationale for separate-but-equal spatial systems is logically tenable, it does not represent a cognitively efficient means of representing space. Additionally, studies using vision and touch (Giudice et al., 2009; Kelly et al., 2010; Kelly and Avraamides, 2011) have demonstrated that spatial information *can* transfer across modalities, resulting in a single, unified representation.

3.4.3 Recoding and modality-specific simulations

As a second possibility, the *common-recoding* hypothesis suggests that information from different modalities is recoded into a common format, with vision being considered a strong candidate. Based on studies showing visual-image effects (e.g. scanning time as a function of distance; Denis and Zimmer, 1992; Denis and Cocude, 1997; see Kosslyn et al., 1978) and the benefits of imagery in verbal description comprehension (Denis and Denhière, 1990; Gyselinck et al., 2007), it is indeed plausible that modality-specific imagery may underlie some aspect of spatial language processing. However, visual processing cannot be the sole medium for the construction of spatial representations, as congenitally blind participants have been shown to perform spatial tasks as well as sighted participants (Loomis et al., 2002).

Rather than single-modality recoding, a more general concept of embodiment has recently been put forward and may shed light on the processing of spatial statements (e.g. Barsalou, 1999; for a review, see Taylor and Brunyé, this volume). According to this view, reading or listening to a description recruits the sensory and motor resources of the body to simulate the experience of the environment, route, or actions being conveyed. For instance, the position and orientation of the body influence the ease of acquiring and retrieving spatial information acquired from language (e.g. Spatial Framework Theory; Franklin and Tversky, 1990; Bryant et al., 1992; Avraamides and Kelly, 2010) and in the act of communicating spatial information, spontaneous gestures reflect the perspective adopted by the narrator (e.g., Emmorey et al., 2000). Note, however, that although the grounding of spatial descriptions in simulated actions and sensations may facilitate understanding, this process can also be a source of interference and additional costs. Embodying imagined perspectives may help to access new vantage points through self-motion, but creates interference in cases where body orientation is incompatible with the spatial judgement to be made (Avraamides and Kelly, 2010). Likewise, listening to slow or fast footsteps not only affected the speed at which participants in Brunyé, Mahoney, and Taylor (2010b) read route descriptions, but was also associated with greater difficulty in adopting survey perspectives. In other words, grounding one's access to imagined or recalled spatial information in a simulated egocentric perspective may limit one's ability to approach this information flexibly or more abstractly. Furthermore, as Denis and Fernandez (this volume) point out, simulations or images formed from descriptions are not necessarily equivalent: for instance, there are tradeoffs in working memory and recall costs when envisioning either a route or a survey perspective. Based on these recent findings,[2] the common recoding hypothesis needs to be reformulated to allow for recoding to occur into different modalities as a more holistic simulated experience with variable outcomes as a function of task demands.

3.4.4 Amodal convergence of modalities

A third possibility is that stimuli from different sources converge into a more abstract and *amodal* representation (Bryant, 1997). According to this hypothesis, spatial information is initially processed by modality-specific systems whose outputs converge into a *spatial representational system* (SRS). The SRS functions to represent spatial information in a format that is independent of the input modality. As Bryant argued, this amodal representation is then used to support the execution of spatial tasks. A similar account was later proposed by Avraamides et al. (2004; see also Loomis et al., 2007). According to this account, linguistic and sensory processing of information results, respectively, in the encoding of the verbal message and the creation of a spatial percept. This initial encoding is then converted into a spatial representation that may be common for language and perception. At this stage, according to Loomis et al. (2002), equivalent spatial representations are created from all spatial modalities and are used to support spatial tasks. It should be noted that the accounts of Bryant (1997) and Loomis et al. (2002; see also Cattaneo et al., 2008), in contrast to earlier amodal models (e.g. McNamara, 1991), involve both modality-specific and modality-independent processing. In essence, the resulting spatial representation could be considered as *supramodal* rather than amodal. Such supramodal accounts are therefore compatible with the findings from electrophysiological studies showing that both common and distinct areas of the brain are activated (Mellet et al., 2000; Mellet et al., 2002) or that both common and distinct EEG components are obtained (Tlauka et al., 2009) when retrieving spatial information encoded from different modalities.

[2] Note that the authors of the studies cited here do not necessarily espouse a common-recoding view of functional equivalence. Their findings are summarized to introduce potential modality-specific recoding media.

3.5 Conclusion

Future studies are expected to shed light on the nature of spatial representations derived from various stimuli. Understanding how spatial representations are obtained from other modalities is important not only for broadening our theoretical understanding, but also in order to create and improve applications designed to support spatial behaviour. For instance, consider the implementation of sensory substitution in navigation assistance, a topic discussed at length by Gallay, Denis, and Auvray (this volume; see also Golledge et al., 1998). These authors point out that blind individuals tend to prefer to receive route planning information verbally, rather than through more direct means (e.g. tactile maps). Likewise, many seek out linguistic directions for global navigation as these are considered both natural and rich in information. However, it is important to consider whether some of the cognitive costs associated with this format, as we have outlined in this chapter, may outweigh such benefits in some cases. For instance, if there are indeed costs associated with converting verbal statements into actionable information (as implied by the common recoding hypothesis), information supplied perceptually (i.e. directly) may promote faster reaction and limit the impact of concurrent cognitive load. Thus, successful design of navigation aids relies heavily on our understanding of how spatial representations derived from different stimuli are alike or different and, more importantly, how our various ways of interfacing with space interact and communicate with each other. Demonstrations of equivalence or near-equivalence in aspects of performance, regardless of input, have been critical to applied and fundamental research. However, it is crucial to recognize that these studies are only steps towards a better conceptualization of the cognitive and physiological principles that underlie equivalence in behaviour, to the benefit of all practitioners in the field.

4

The processing of landmarks in route directions

MICHEL DENIS AND GILLES FERNANDEZ

This chapter is dedicated with gratitude to the memory of Gary L. Allen (1950–2007), whose insights have opened invaluable perspectives to many students of spatial cognition.

4.1 Spatial discourse as a way of assisting navigation

The processing of spatial discourse has attracted the attention of many researchers in cognitive science since the mid-1990s (e.g. Bloom et al., 1996; Denis, 1996; Levinson, 2003; Hickmann and Robert, 2006; Dasen and Mishra, 2010; Evans and Chilton, 2010). Language is generally recognized as being able to express spatial thought (see Bhatt, Schultz, and Freksa, this volume) and as being a source of the schematization of space by human beings (see Chown, this volume). However, it is well-established that spatial discourse takes a variety of forms, and these warrant the attention and effort required to distinguish between them. The distinctions can be based on structural characteristics, but also on the *cognitive processes* that each form of discourse entails. In this chapter, we focus on the processes by which a speaker or writer refers to the objects present in an environment when their discourse is intended to assist a person whose immediate or subsequent task is to navigate in this environment. Note that the form of spatial discourse is well known to depend on the type of behaviour it is expected to elicit in the spatial domain (e.g. Bateman et al., 2007; Taylor and Brunyé, this volume).

The first type of spatial discourse to interest cognitive psychologists was the description of arrays of objects without any relationship to a navigational task. These descriptions typically refer to a set of objects arranged in a spatial configuration, as viewed from an external perspective. They are composed of sets of statements describing the relative positions of static objects in a two- or three-dimensional environment. The task that follows the processing of such descriptions typically consists either of recalling them verbally or of constructing diagrams of the objects described, but also of drawing inferences about the unstated relationships between designated objects (e.g. Ehrlich and Johnson-Laird, 1982; Mani and Johnson-Laird, 1982; Ehrich and Koster, 1983; Bryant et al., 1992; Deyzac et al., 2006; Tenbrink et al., 2011).

The present chapter focuses on spatial descriptions composed of more extended verbal materials describing complex large-scale environments (e.g. Linde and Labov, 1975; Noordzij and Postma, 2005; Gyselinck et al., 2007). Not only do they potentially refer to environments that call for various forms of displacement—a feature that is likely to impact discourse content—but also, in contrast to the classic 'table-top' situations where the configuration described can be apprehended as a whole from a single vantage point, these environments cannot be perceived holistically, but require the successive descriptions of several scenes that the reader or listener has to integrate to construct a unified representation.

Barbara Tversky (1991) has documented the distinction between two perspectives taken on such environments by a writer or speaker (and consequently by their reader or listener), which result in two distinct forms of spatial discourse, namely route and survey descriptions. *Route descriptions* adopt the perspective of a person moving through an environment and experiencing a succession of viewpoints, whereas *survey descriptions* adopt a bird's-eye viewpoint of the entire environment. Both descriptions inevitably refer to objects present in the environment, but the choice of a perspective has an impact on the structure of the resulting spatial discourse, as well as on the spatial framework and the spatial terminology used. In particular, person-centred language is typical of route descriptions ('When you turn right at the church, you will find the post office'), while survey descriptions make larger use of canonical directional terms ('The post office lies to the east of the church'). A number of studies based on the verification of inferential statements have provided evidence supporting the view that both types of descriptions can result in similar, perspective-free internal representations (cf. Taylor and Tversky, 1992b, 1996; Schneider and Taylor, 1999; but see Brunyé and Taylor, 2008a).

Both route and survey descriptions essentially provide readers or listeners with information about the structure of a configuration or environment without necessarily being intended to assist in a navigational task. The type of spatial discourse we will be considering here introduces a further feature, by explicitly providing the person who reads or listens to such discourse with *directions*. In other words, an environment is described by a writer or speaker in order to provide the reader or listener with a *procedure* of how to proceed through the environment. For example: 'When you get out of the subway station, you will be facing a movie theatre. Proceed to the right and take a short subway passage. You reach a bank. Turn right after the bank. The museum is at the end of the street.' This form of discourse belongs to the broader category of *procedural discourse* (cf. Dixon, 1987; Glenberg and Robertson, 1999). In general, in addition to their instructional components, route directions include descriptive components intended to situate landmarks relative to the moving person or to other landmarks. Based on the description of structured visual scenes, instructions are delivered to make users execute an appropriate series of moves so that they eventually reach a specified goal.

The present chapter is primarily devoted to human cognition and language as involved in wayfinding and the representation of spatial information by humans. While focusing on empirical work, it further aims at establishing relations to formal and applied research, and points to the value of taking empirical findings into account in the design of navigational assistance systems.

4.2 Models of direction-giving discourse

Route directions are a common form of daily communication. However, so far, the processes involved in their comprehension and memorization have not received as much attention from the cognitive or linguistic point of view as the *production* of such directions.

Early ideas about the generation of route directions were developed by Wolfgang Klein (1982) from a linguistic perspective, with a focus on local deixis, as a device intended to express the spatial relationships between objects in a situational context. In particular, Klein (1983) highlighted the difficulty experienced by young children in correctly manipulating deictic expressions. In cognitive geography, a model of the mechanisms underlying direction-giving was proposed by Helen Couclelis (1996). This model relied on the notion that an imagery-based representation of the environment is translated into language, but that the process is mediated by a common underlying internal representation, an abstract 'mental model', which is neither linguistic nor map-like. Couclelis postulated a succession of stages in the direction-giving process, at the core of which is the construction of a transient representation and the transformation of the information it contains into the form of a verbal output. A psychological approach was also developed by Golding, Graesser, and Hauselt (1996) as an application of a general cognitive model of question-answering to the specific situation of direction-giving.

An articulated psychological model of the processes involved in the generation of route directions was proposed by Gary Allen (1997). Allen's conceptual framework was described in an article which provided directions for a systematic investigation of what he called 'route communication episodes'. This approach resulted in a taxonomy of the speech acts involved in direction-giving, including a detailed classification of the verbal devices used to define places, refer to landmarks, or convey the movements to be executed through the environment. The most detailed part of this proposal was about the nominals used to refer to environmental features, such as landmarks (which are typically used as subgoals along the route), pathways (which are 'channels of movement': streets, etc.), and choice points (such as crossroads, which afford options with regard to pathways). Another important part of the analysis developed by Allen concerned what he called 'delimiters', that is, the statements used to designate distances or directions, or to specify places or objects that serve as points of origin or destinations. On the basis of these components, Allen delineated a set of rules for effective direction-giving, which were summarized in what he then called the CORK (Comprehension of Route Knowledge) conventions.

In 2000, Allen provided an in-depth analysis of the conventions used for conveying wayfinding information, and he drew up a set of recommended principle-based practices, the effectiveness of which for communicating route knowledge was tested in field studies involving students on a campus. The first principle pertained to the temporal ordering of the statements: People who receive route directions in an order reflecting the spatio-temporal reality perform better in a wayfinding task than those receiving directions in a sequence that is not consistent with the natural order. The second principle was concerned with the reduction of uncertainty at choice points along the route: Performance is better when directions are specified in greater detail (involving references to more landmarks) at choice points. The third principle dealt with the preferable use of spatial designations with which listeners are more familiar. For instance, there is a consistent tendency for men to privilege metric distances and compass direction information, whereas women prefer directions emphasizing environmental features (visual landmarks). Allen (2003) went on to investigate the role of gestures that accompany verbal route directions, especially deictic gestures, which map directly onto the surrounding environment.

At about the same time as Allen introduced his CORK model, we independently proposed an approach that focused on identifying the components of route directions that are thought to be crucial in assisting wayfinding (Denis, 1997). We started by postulating three cognitive macro-operations that must be involved in route-direction situations, namely: (a) activating an internal representation of the territory in which a displacement will take place; (b) planning a route connecting the starting point to the destination in the relevant subspace of the activated representation; and (c) formulating the procedure that the user will have to execute to advance along the route and eventually reach the destination. Based on the analysis of an abundant corpus of route directions collected in both campus-like and urban environments, we developed a classification of the various kinds of statements generated by describers, making a particular distinction between references to landmarks and prescriptions of actions.

Route directions were seen as a series of prescriptions of actions to be taken along a succession of segments, each new action being triggered when specific landmarks are encountered. Despite the wide differences between the respondents' individual descriptions, we developed a method that allowed us to build descriptions that delivered the essentials of a navigational episode (or 'skeletal descriptions'). A skeletal description typically contains the minimal set of landmarks and instructions needed to ensure safe and accurate navigation (cf. Daniel and Denis, 1998). An interesting extension of this approach was developed by Tversky and Lee (1998, 1999), with the intention of comparing verbal route directions and route maps. These studies confirmed that the same type of information is represented in the descriptions and depictions of routes, suggesting that the same underlying mental representations may be used to generate both.

There are many indices that contribute to the contrast between individual descriptions that are rated as 'good' or 'poor' (such as measures of their similarity or overlap with skeletal descriptions, or their richness in landmarks). Examples of measurements of the quality of route directions in terms of the number of elements or judges' ratings have been proposed (e.g. Fontaine and Denis, 1999; Lovelace et al., 1999). However, the ultimate test of the value of route directions as aids to wayfinding is, of course, how well travellers perform in actual spatial settings. The effectiveness of skeletal descriptions and of actual descriptions rated as good by independent judges (in contrast to the inadequate assistance provided by the descriptions rated as poor) was attested in a variety of environments, from complex urban environments involving an intricate network of streets (such as the city of Venice; see Denis et al., 1999) to more open spaces (like university campuses; see Daniel et al., 2003). People who are brought to such places for the first time and invited to use various sets of route directions display contrasting patterns of behaviour, as reflected by directional errors at crossroads, pauses during locomotion (attesting to the travellers' hesitations at choice points), and requests for assistance. Navigation times are longer for people who have been given poor descriptions than for those who have read good descriptions or skeletal descriptions. It is also well-established that people who say they prefer the survey to the route perspective in navigation make more errors when using poor descriptions than people more inclined to rely on visual memories of landmarks (Denis et al., 1999).

4.3 The functional significance of landmarks in route directions

Three main functions are typically assigned to landmarks in direction-giving. The first is to signal locations where specific actions are to be accomplished ('Then you will see a fountain; go round it and then turn right') or to be terminated ('Keep going until you come to a merry-go-round'). Another function consists of locating a remote, less visible landmark ('You will see a church; to the right of the church is a memorial; just to the right of the memorial is a path; take this path'). The third function is related to the need to confirm to a traveller who is carrying out a lengthy action that he or she is still on the right track in spite of the absence of any reorientation ('Walk for 400 metres along the same street; you will successively pass a theatre, a post-office, and a big florist's shop before turning right into Liberty Street').

There are considerable individual differences as regards the frequency with which people refer to landmarks in route directions. At one extreme are those describers who only refer to the ground substratum on which walking takes place (streets, squares, bridges, etc.). At the other extreme are those who tend to provide too much landmark information, thus running the risk of overloading the working memory capacities of the people they are addressing with unnecessary references to irrelevant information. There is no way of determining the optimum number of landmarks to use in a description. However, if some amount of landmark information is typically needed, the most critical and useful ingredients of route directions are landmark/action pairs. There is evidence that the most efficient route directions are those which closely connect actions to landmarks, that is, which intimately articulate the prescriptive and the descriptive parts of route directions. This is one of the main features of skeletal descriptions, which typically convey the essentials of directions by providing plain indications of 'what to do where'. Furthermore, when route describers are required to be concise, they usually shorten their descriptions by reducing the number of landmarks they mention, mainly by eliminating those they judge not to be crucial for assisting wayfinding (Daniel and Denis, 2004).

Although landmarks are usually distributed all along a described route, a higher density of references to landmarks is found near the starting and destination points. In between, places where reorientation is called for elicit more frequent mentions of landmarks. More frequent mentions are also found at other critical places, such as intersections along extended segments that do not involve any reorientation, or wide-open spaces to be traversed without any direction change. In other words, places where a direction change is required, but also those where a mistaken reorientation might well occur, are similarly treated as critical nodes, thus calling for more elaborate descriptions (Denis, 1997; Michon and Denis, 2001). These findings confirm the intuition that landmarks play a genuine cognitive role in route directions by signalling the portions of a route calling for special attention.

Not only do describers spontaneously include references to landmarks in spatial discourse, but also people receiving route directions experience cognitive difficulties and complain when they are confronted with instructions that are mere lists of street names and left/right turns. Landmarks are spontaneously introduced in discourse by most describers, and they are expected by the people who receive these descriptions. The particular cognitive status of landmarks can thus be seen as residing in the fact that they help a traveller create an *advance visual model* of the critical parts of an environment, as seen from a route perspective, which prepares the traveller to behave appropriately in ambiguous contexts requiring crucial decisions (Denis et al., 2006). This view is consonant with the hypothesis that the representations derived from language are functionally equivalent to those resulting from direct experience, even though some qualitative differences remain possible (see Avraamides, Mello, and Greenauer, this volume).

While the relevance of landmarks and their value for direction-giving are well documented, little attention seems to have been paid to comparing the use of landmark versus street information in route directions. In a study where people performed a wayfinding task with lists of written instructions referring to either landmarks or street names of the neighbourhood traversed, we found that performance (based on determination of the frequency of directional errors, pauses, and checking) was better when people had processed landmarks. Furthermore, as well as being less frequent, pauses and checking episodes were significantly shorter for people who processed landmark information than for the other participants (Tom and Denis, 2003). In another study, the longer times taken to read street than landmark information were taken as reflecting the difficulty of processing street-related items and integrating them online into a mental spatial representation of the route. The recall and recognition of landmark information (e.g. a market) were better than those of street information (e.g. Market Street), a finding which once more confirmed that landmarks are more useful than street names for guidance purposes (Tom and Denis, 2004).

Landmarks mentioned in route directions can be any kind of object in an environment that functions as a point of reference. Landmarks are not 'intrinsically' landmarks. They are environmental features that are *selected* and *used* by describers, and intended to serve as reference points for the addressees. Any object in the environment may gain the status of being a landmark, provided that this function has been intentionally assigned to it by a describer. Note also that people moving along a route lined with various landmarks are able to treat various subsets of them differently. In particular, there is a critical distinction between landmarks situated at navigationally relevant locations (such as intersections, where a crucial decision has to be made), and those at turning points that do not involve a choice between several alternative routes (when a street makes a right angle). The first sort of landmark is better memorized than the others (as attested by recognition tests), and they elicit specific cerebral activity (in the parahippocampal gyrus, as shown by neuroimaging studies) (Janzen and van Turennout, 2004; Janzen et al., 2007).

The significance of landmarks in route directions warrants in-depth analysis of their functional value and, in particular, their memorability. They cannot be investigated without considering the relationships between their processing and the mental imagery potentially associated with their evocation. In tasks involving the production of route directions, Vanetti and Allen (1988) showed that participants with high visuospatial abilities produced a more efficient spatial discourse than low visuospatial imagers. Our own analysis of the production of route instructions revealed that the directions generated by high visuospatial imagers contained more frequent references to landmarks than those of low imagers. This finding is consistent with the expectation that if people are more likely to retrieve visuospatial information from their memories, they are more likely to refer to such information in their own verbal productions (Denis, 1997).

4.4 Visuospatial imagery and the processing of route directions

Beyond the data on the cognitive processes underpinning the production of route directions, much remains to be found out about how this type of discourse is processed and memorized. An integral part of this question is the issue of the nature of the representations that remain available to people once they have processed a package of directions intended to assist their navigation in an unfamiliar environment.

One research question pertains to the role of mental imagery in the processing of spatial texts or discourse. In our present approach, we identify a need to account for the mechanisms by which people who process route directions use their imagery capacities to encode landmark and action information. Could mental imagery be a process that facilitates the encoding of navigational information? Could this mental activity be implemented as a cognitive anticipation of the successive scenes that will make up the navigational episode to be executed? Visuospatial imagery accompanying the processing of route directions is presumed to be a factor in navigational performance by providing representations wherein information is more efficiently preserved from decay and thus more available in long-term memory.

Another research question pertains to the distinction introduced above between route versus survey perspectives in spatial texts (Tversky, 1991; Klatzky, 1998). Both perspectives are candidates for the representation of spatial information in memory, with distinct implications as regards the processes implemented in the construction of such representations. According to the first perspective, the representations constructed from a set of instructions consisting of the successive views that the traveller is about to encounter, plus the actions associated with each of them. If this type of route-like processing is implemented, the resulting representation is likely to preserve the (sequentially structured) route perspective inherent in navigational instructions. The alternative option is that people may prefer to use the information available in route directions to construct a map-like view of the environment to be traversed. If this type of coding were to be implemented, it would result in a survey representation, with the implication that distinct resources have to be devoted to such coding.

Note that another possibility is that the implementation of such strategies could depend on individual cognitive characteristics. In particular, the propensity with which people construct visuospatial images while reading or listening to verbal materials might well affect the likelihood that a strategy calling for imagery will be implemented. There is good evidence that high imagers encode descriptions of spatial configurations more quickly than low imagers (Denis and Denhière, 1990; Denis and de Vega, 1993). Furthermore, it has been shown that when people reading a spatial description are invited to implement a strategy of visualization, the efficiency of imagery is to a large extent thwarted if they are simultaneously performing a concurrent task tapping their visuospatial resources (Gyselinck et al., 2007).

Given the relevance of imagery in the processing of material referring to the visual appearance of an environment, it is appropriate to rely on measures of individual proficiency in image production and manipulation to see if individual differences are likely to affect the processing and memory of spatial text, in the absence of any explicit incitation to use imagery. It is well-established that people with high visuospatial abilities have special abilities for processing descriptive prose (cf. Vanetti and Allen, 1988; Denis and Denhière, 1990; Denis and Cocude, 1997; Sadoski and Paivio, 2001). If visuospatial imagery is spontaneously implemented during the processing of route directions, high imagers should be expected to outperform their counterparts in the processing and memory of route directions, especially if these are highly saturated with visual landmarks.

Empirical answers to these questions have been provided by a series of experiments conducted in our laboratory (see Fernandez, 2000). When the sentences that compose a set of route directions separate references to landmarks from prescriptions of actions, the former take longer to process than the latter, which reflects the fact that special attention is given to the landmarks that compose the visual scenes encountered along the route. Further tests show that the recall of landmarks is better than that of actions, indicating that the differential processing of the two components of route directions results in differential memorization. Furthermore, when landmark descriptions are enhanced with details, their memorization is further increased. Lastly, participants classified as high imagers process route directions more quickly than low imagers. This pattern confirms the well-established fact that imagery favours the processing of various forms of spatial discourse, and also suggests that people with a greater capacity to construct visuospatial images are in a better position to construct visuospatial representations based on the descriptive part of route directions. When recalling information, high and low imagers reach comparable levels of performance, but to attain this level, low imagers have to devote more time and effort than high imagers to constructing visuospatial representations that incorporate relevant pieces of spatial information.

The demonstration that mental imagery may be profitably involved in constructing the mental representation of an itinerary relies on an approach based on individual differences, as measured by standard visuospatial tests. However, we need to be more specific about the content of imagery activity in this context, and to bear in mind the distinction already made between route and survey perspectives. Route imagery maintains an egocentric perspective on the imagined environment. It matches the perspective adopted during the forthcoming navigational experience. Alternatively, imagery can consist of constructing visuospatial representations of the environment traversed from a bird's-eye view. Here, step by step, the reader constructs an allocentric (map-like) representation of the environment organized around a system of coordinates.

There is no doubt that route imagery is more compatible than survey imagery with the implicit perspective associated with route directions in general (see also Taylor and Brunyé, this volume). Route imagery does not require any transformation of perspective of the representation under construction, and is therefore expected to be especially suitable for the processing of route directions. However, it is likely that this form of imagery has a cost resulting from the need to form visual images of each step of the itinerary. Processing times in fact tend to be longer with route than with survey instructions. It also takes longer to process landmark than action sentences. With regard to landmarks more specifically, it takes longer to process them in the route than in the survey condition. In terms of recall, landmark recall turns out to be poorest when a survey perspective has been adopted.

Route imagery instructions are particularly compatible with the perspective imposed by standard route directions (namely, visualizing scenes from an egocentric perspective, and compiling a succession of such views). Although costly in terms of processing, route imagery remains closely connected to the cognitive operations required by the processing of route directions. The route perspective is inherently present in any set of route directions, even in the absence of explicit instructions. The structure of the representations implemented from route instructions matches the sequential structure imposed by the processing of any set of route directions. Another interesting outcome is that action recall is especially favoured by route instructions.

The findings summarized above give us the following picture. On the one hand, the transformation of sequential information conveyed by route directions into map-like representations is achieved at a relatively low cost, but it does not help recall. On the other hand, sticking to the sequential structure of the information and strengthening this information by mental imagery has a cost, but also has a beneficial effect on recall. The most effective strategy is undoubtedly that which parallels the linear, egocentric structure of the progression prescribed by the route directions. Despite its cost, route imagery outperforms the other strategy in terms of both landmark and action recall.

4.5 Landmarks in route directions: when psychology meets computer science and robotics

Based on the literature and the findings of the experiments reported in the previous section, two main conclusions can be advanced. (1) The representations derived from processing route directions are essentially organized around a set of visual landmarks, each of which triggers an associated action at crucial points along the route. (2) Although there are various ways of using visual imagery to create representations of spatial information, the cognitive activity that accompanies the processing of route directions with the greatest productivity consists of visualizing successive scenes of the environment to be navigated, just as they will appear to the traveller from a route perspective.

All of this sheds light on human cognition, and more specifically the representation of navigational spaces and language-based wayfinding. These conclusions also have repercussions beyond psychology, in the area of computer science applied to the processing of geographical information, and the design of computer-based navigation aids (see Seifert and Tenbrink, this volume). Here, we discuss selected examples of work that is extending behavioural research to computer-oriented applications and human–machine interactions.

The use of landmarks as identifying features in an environment is essential in navigation, and provides people with a way to locate themselves and establish goals. Sorrows and Hirtle (1999) proposed a comprehensive framework in which landmarks are seen as tools for navigation, serving the purpose of identifying points along a route where navigational decisions are to be made. In contrast to the sense of disorientation created by the lack of landmarks and the uniform appearance of labyrinths or mazes, landmarks contribute to the cognitive differentiation of complex environments by offering distinct, memorable cues along a route. Sorrows and Hirtle listed the main characteristics of landmarks, including their singularity (or contrast with the surroundings), prominence, and accessibility. Three categories of landmarks were considered, namely, visual, cognitive (or semantic), and structural. The theory was applied to human navigation in real and electronic spaces, and the same framework was extended to robot navigation (Hirtle, 2008). This approach was a source of inspiration for scientists interested in navigation involving artificial agents, whether these agents are providers or users of route directions.

A particular concern has motivated an important line of research intended to clarify how landmarks are extracted from an environment, and in particular whether this process can be automated to some extent. Raubal and Winter (2002) noted that navigation services communicate route instructions that are typically based on geometric data from the street network, which is frequently the only data set available. The challenge for people designing navigation systems is to enrich wayfinding instructions with local landmarks, just as ordinary people do in everyday communication. Raubal and Winter proposed a formal approach to extracting features from an environment, based on measurements of their saliency as landmarks. The assumption was that route directions enriched by local landmarks would be easier to understand than those that are only direction- or distance-based. Measurements of façade area, shape, colour, and visibility (from the egocentric perspective of the moving person) are characteristics that can be used to create a contrast between a building and its surroundings. Statistical measures can be applied to find out whether these values differ significantly from those of adjacent buildings. The building with the highest measure of landmark saliency at a decision point will be automatically selected and a verbal instruction will be created by the navigation service. Additional factors that are cognitively relevant for the selection of salient features may include their advance visibility for a person approaching a decision point (Winter, 2003). A computational model of the extraction of salient landmarks has been implemented by Nothegger, Winter, and Raubal (2004), and cross-checked using landmark selection by human participants. Evidence has been reported that the automatic selection of landmarks correlated closely with human choices at decision points. The system was further extended by Klippel and Winter (2005) using a formal language for the specification of conceptual route knowledge (see also Hansen et al., 2006).

Starting from the evidence that landmarks are not part of most route guidance information systems, alternative methods were proposed to automatically derive landmarks from existing spatial databases, using knowledge-discovery processes and data mining methods. One such approach was developed by Elias (2003), who compared various methods (involving either supervised or unsupervised learning) for extracting landmarks. In the same spirit, a web mining approach was proposed by Tezuka and Tanaka (2005). More recently, Sadeghian and Kantardzic (2008) expressed the concern that the design of automatic landmark detection systems calls for greater efforts to determine the attributes of objects that are analysed in order to select landmarks. In particular, the approach might include not only the static attributes of objects, but also attributes that represent how people have interacted with these objects. Furthermore, efforts to extend the types of objects that can be considered as candidate landmarks to objects other than buildings may prove to be important if the systems are to be applied to non-urban environments (see Burnett et al., 2001).

The design of navigation services must also take the issue of spatial reference systems into account, and make decisions about what type of knowledge is most suitable for the users' navigational needs. In this respect, it is important to identify the respective values of route and survey knowledge, and to acknowledge their features that will meet the users' cognitive expectations. Of course, in the generally accepted hierarchy of the three levels of spatial knowledge (landmarks, routes, and survey representations), it is survey knowledge that is thought to have the highest cognitive value by abstracting knowledge from specific sequences and integrating information collected from different sources into a single model. Nevertheless, there are good reasons to think that the best strategy for assisting navigation is to use the type of representation that corresponds to the route perspective. For this reason, models of computer-based assistance must take into account what we know about cognitive preferences from psychology experiments. Even though people can easily combine perspectives or switch from one to the other in spatial descriptions, we can only recommend that navigational assistance systems stick to the route perspective.

The general idea, here, is that although human-generated directions are less than ideal in some cases, it is a sound approach to include some of their characteristics in computer-based systems. For instance, when delivering route directions, a typical 'friendly' method consists of speakers adapting their instructions to what they know or assume to be the spatial knowledge of their addressees. In contrast, navigation systems frequently assume no prior knowledge (or a minimal amount of such knowledge). One way of personalizing navigational assistance consists of determining the knowledge possessed by addressees, and thus provide them with adaptive (individualized) route directions. A formalization of this problem and the development of an algorithm for the dialogue-driven generation of route directions have been provided by Richter, Tomko, and Winter (2008). Along similar lines, Tenbrink and Winter (2009) have addressed the issue of the variable spatial granularity of route directions. Some sections of directions, especially those intended to assist travelling through environments of varying complexity, call for fine granularity. Human describers show considerable flexibility in switching from coarse to fine granularity or vice versa, and it is a challenge for system designers to include this capacity in automatically generated route directions. Similarly, just as people readily differentiate between the easy and complex steps in an itinerary, and adjust their descriptions in consequence, the next generation of automated route guidance systems can be expected to develop ways of paying special attention to the 'tricky parts' of routes, for instance when the geometry of the street network violates the expectations of the navigating person. Valuable suggestions for the automated identification of such difficult parts and the automated adaptation of descriptions have been proposed by Hirtle, Richter, Srinivas, and Firth (2010).

Despite the differences between the sensory and behavioural capacities of humans and robots, to a large extent the navigational issues in both areas appear to be similar (see Barclay and Galton, this volume). Although robotics is a research domain in its own right, with a scope extending beyond the issues discussed in the present chapter, the possibility of guiding artificial agents by natural language instructions has been a challenge for many years in cognitive science (cf. Werner et al., 1997). Advances in this domain were heralded by the computational models of spatial cognition developed in artificial intelligence in the mid-1990s (e.g. Herzog and Wazinski, 1994; Chown et al., 1995). Of course, in order for any effective interaction to occur between a human being and a robot, a common representational ground is a prerequisite, which calls for efforts in the robotics community to construct functional systems of representation in artificial agents. Most of the work reported attests that it is possible to execute navigational tasks based on natural language, provided that commands are formulated in a concise way, and that the typical uncertainties and indeterminacies of natural linguistic information are overcome. Illustrations of this can be found in several command systems for mobile artefacts (e.g., Lauria et al., 2002; Tellex and Roy, 2006; Elmogy et al., 2009; see also Coventry et al., 2009b; Shi and Tenbrink, 2009).

Computer science and robotics have confirmed that they can make a significant contribution to extending the scope of the cognitive science of spatial cognition, in particular in the domain of navigational instructions and wayfinding assistance. The repertoire of the human cognitive system has been shown to include efficient methods for orientation and navigation, as well as for communicating information to this effect. The cognitive approach has been the starting point for developing artificial systems endowed with similar capacities. The value of extending these concerns from natural to artificial systems has been that it has forced scientists from both communities to make explicit the algorithms and representations manipulated by the systems, and to make them converge on common conceptualizations.

Part 2

Computational models

5

Selection of reference objects for locative expressions: the importance of knowledge and perception

MICHAEL BARCLAY AND ANTONY GALTON

5.1 The importance of locative expressions

If your husband or wife, late for a meeting, asks you where their car keys are, giving the answer 'In the house somewhere' is unlikely to win approval. On the other hand, giving the answer 'By the telephone' is more likely to lead to domestic harmony. The generation of a useful locative expression, and in particular the choice of an appropriate reference object ('the telephone' as opposed to 'the house' in this instance) is an important factor in everyday communication. The purpose of this chapter is to investigate the characteristics of objects which determine their suitability as references in locative expressions. It seems that, compared to choosing a reference frame or assigning a preposition, the choice of reference object is relatively under-researched, but it is an important instance of humans translating judgement about the world into language.

It is necessary to stress that giving the answer 'Your keys are the ones with the small red fob' to the question 'Where are my keys?' will also not gain approval; your wife or husband will already know which his or her keys are and wants to know *where* they are. The issue of reference choice for a locative expression is not the same issue as that of generating referring expressions. In referring expression generation (see e.g. Dale and Reiter, 1995; Duwe et al., 2002) the target is disambiguated from a group of similar objects by adding qualifiers to the target, so for example 'the big red dog with the collar' might serve to specify a particular dog in a group of small animals. None of the group may require locating as such and clearly the expression has not helped the listener to locate the target.

Spatial location can also be used to disambiguate, as in 'the small white dog next to the big red dog'. This is a hybrid expression serving to disambiguate and possibly, but not necessarily, locate the target. Inasmuch as the location element is not concerned solely with disambiguation, it would fall within the scope of this study, but the question addressed by this study would be why was the 'big red dog' chosen to help locate the 'small white dog', rather than some other object, not whether the referent was effectively disambiguated by the 'big red dog'. Spatial location in disambiguation is addressed in work by Tenbrink (2005) and Varges (2004) among others, although the factors affecting the choice of reference, with regard to its locating a target object, are not investigated.

The examples of the keys and the dog given so far might be termed simple locative expressions, containing a single reference and preposition. More complex expressions are used when a single reference is inadequate to locate the target object. These complex locative expressions often contain hierarchical reference objects, as in 'The keys are on the desk under the window in the common room'. This study is restricted to simple locative expressions, although extensions of the model to cover complex expressions are discussed in Section 5.8. The characteristics of the different references in a complex locative expression are likely to be similar (as each reference in turn becomes the target for the next reference in the sequence), and the process for determining the suitability of even a single reference turns out to be highly complex.

5.2 Formation of locative expressions

From consideration of the process of forming a locative expression some necessary parameters of an experimental platform for investigating reference object choice can be derived. The process of forming a locative expression, given a target object, requires the speaker to make three decisions:

1) Decide on an appropriate reference object.
2) Select a reference frame (given the reference object).
3) Assign a suitable preposition (given the reference object and reference frame).

The assumption made at the outset of this study is that, if the purpose of the expression is genuinely locative, the first step in the process is, as listed above, the choice of a reference object. Herskovits (1998) proposes that the process of generating spatial language is one of progressive schematization of, or perhaps filtering of irrelevant information from, the scene being described. This idea is expanded by Chown (this volume), who also discusses the cognitive processes underlying it. Given the information bandwidth reduction from a real-world scene input to the human visual system, to the vocal output of a locative expression (also noted as a requirement by Chown, this volume), schematization seems reasonable. This bandwidth reduction is the opposite of the process of deriving a visual representation of a scene from a linguistic description as discussed by Bhatt, Schultz, and Freksa (this volume) in which metric assumptions must be made to add information to the largely qualitative linguistic descriptions. The following sequence of steps seems to accord with the schematization process and delivers simplification at each step:

1) The objects in the scene are 'recognized', a step which allows us to replace a complex visual representation of an object with a name and its necessary geometric characteristics.
2) Choosing a suitable reference object effectively removes extraneous objects from the scene, leaving only the target, the reference, and possibly a few key distractor objects that may influence the following decisions.
3) A reference frame is selected; this process is often, but not always, trivial (see Carlson-Radvansky and Radvansky, 1996; and Carlson-Radvansky and Irwin, 1994 for some more complex cases). The result, however, is that contending reference frames are removed from consideration, leaving the chosen reference frame to fix the axes for the next step.
4) The geometry of the remaining objects together with the topological relationships between them are combined in the assignment of an appropriate preposition.

If this sequence is accepted, it can be seen that, while it is valid to research preposition assignment and reference frame choice in experimental situations (scenes) containing very few objects, research into reference object choice will require more complex scenes containing multiple objects. In addition to this, to provide information about the different characteristics of objects and their influence on reference suitability, the objects in the scenes should be as varied as those encountered in the real world. Moreover, the scenes should be three-dimensional representations to allow a realistic range of geometric and topological relationships to be portrayed.

The necessary complexity of the scenes, and the large number of possible influences on reference object selection, suggests that a machine learning approach would be useful in determining which influences are important and how they interact in the reference choice task. This chapter uses such an approach.

5.3 Interpreting a locative expression

The three primary influences on reference object suitability can be derived from the two necessary steps a listener must take on hearing a locative expression, with the addition of some representation of communication cost. Presented with a locative expression and the task of finding the target object the listener must:

1) locate the reference object;
2) search for the target object in the region constrained by combining the reference object location with the spatial preposition.

On the assumption that the speaker intends their communication to be effective, or at least is trying to cooperate with the listener, it will follow that the speaker will have

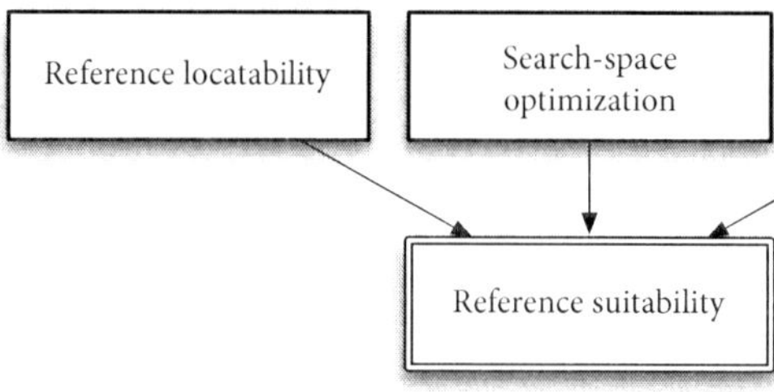

FIGURE 5.1 Three primary influences on reference suitability

chosen the reference object to be easily locatable; and also that, in conjunction with the preposition, the reference will suitably constrain the region in which the listener must search for the located object. These must be the main criteria for an object to be a suitable reference.

There is some evidence for this cooperation with (or consideration for) the listener in spatial communication (see Mainwaring et al., 2003) in the adoption of reference frames that reduce the mental effort required by the listener. Also, in most way-finding literature (see Section 5.4) the starting assumption is that the speaker is attempting to assist a listener in a navigation task.

Note that Carlson-Radvansky (1996) interposes a third step, 'A reference frame is superimposed on the reference object', between the two steps above. Not all prepositions require a reference frame and so this step, when required, is implicitly included as a sub-task in determining the region associated with the preposition in the second step listed above.

Communication cost will be an important consideration if a potential reference is ambiguous (requires some descriptive qualification to be properly locatable) or if the time taken for the communication is comparable to the time the listener will take to locate the object. While in many cases, for a human speaker, this may not seem a major consideration, it will be necessary to constrain a computational model from over-specifying a locative expression, particularly in the case of complex expressions with hierarchical references.

These three primary influences on reference choice are shown in Figure 5.1. The purpose of the work reported here is to determine which characteristics of reference objects affect reference locatability, search space optimization, and hence reference suitability, in near real-world settings.

5.4 Related work on reference object selection

Machine learning of spatial language (for language generation as opposed to interpretation) has received only sporadic attention over the past few years. Machine learning of spatial preposition use has been attempted by Regier (1996), Lockwood et al. (2005) and Coventry et al. (2005) among others, but these systems have 'pre-selected' references and all deal with two-dimensional scenes.

Machine learning of reference selection takes place to an extent in the 'Describer' system (Roy, 2002). This uses a two-dimensional scene set with abstract shapes and limited influences on reference object choice. It has a wide scope, tackling reference choice, target disambiguation and preposition assignment simultaneously. The system tended to choose references that were easily identifiable but very poor at locating the target, and this is probably the best illustration of why reference choice for disambiguation (referring expression generation) and reference choice for locative expression generation must be considered as different tasks.

The 'VITRA' system (Herzog and Wazinski, 1994) is an ambitious scene description system including all elements of spatial language, but it does not include machine learning. As part of the VITRA project Gapp (1995) proposes a model for reference object selection and some of the variables used, and issues raised, in the paper are addressed elsewhere in this chapter. The model is based on a Euclidian distance measure between vectors of variables: the lower the distance between a vector for a candidate reference and the vector for an ideal reference, the more suitable a reference is considered. The model is not learned and does not allow for complex dependencies between variables. No results are given for the model.

Elements of a comprehensive system for interpreting spatial language in a three-dimensional virtual environment are described in Kelleher and van Genabith (2004) and Kelleher and Costello (2009). The work incorporates computational models for spatial preposition use and referring expression generation in a context including visual and dialogue memory. Of particular interest is the derivation of a composite variable for calculating the 'visual salience' of a reference. This is managed by projecting the three-dimensional scene onto a two-dimensional pixelated plane and then integrating the pixels relating to the visible parts of an object multiplied by their distance from the centre of focus. The measure is used by Kelleher for reference disambiguation but is clearly a candidate attribute for reference selection in locative expressions (on its own or combined with other factors). Variants of the measure are investigated in this study and are described in Section 5.6.

Within the extensive literature on landmarks, work on selection of landmarks for wayfinding relates directly to the general problem of reference object selection, augmenting the work from linguistics. Wayfinding landmarks do not usually locate objects so much as places or decision points; however, the processes involved are analogous, and some key studies on landmark selection are discussed here. Note that this discussion focuses on landmark (as a subset of reference object) characteristics; a full discussion of the use of individual landmarks, and sequences of landmarks, in wayfinding can be found in Denis and Fernandez (this volume).

Nothegger et al. (2004) develop an algebraic model for the salience of features in their work on the use of buildings as landmarks. As such it is a machine model of reference choice, although in a very limited field, and it is similar in structure to that of Gapp (1995). Elias and Brenner (2004) describe a system for automatic landmark selection that uses machine learning to a certain degree. Information from a geographical information system is fed into a decision tree which is used to decide which of the buildings in the vicinity of a route decision point are 'unique'. This 'uniqueness' is used as a substitute for 'is a good landmark candidate' as no training data were available as to which buildings in the data set were considered good landmarks. All 'unique' candidates from this process have their visibility assessed and a candidate with high visibility is then chosen as the appropriate landmark. The system appears to be more comprehensive than many others and is possibly the closest to that used in this study, however no results are given from using the system.

Aspects affecting landmark choice such as historical or cultural significance are discussed in more detail in Barclay (2010). In this study the objects presented have no individual or specific significance (there is no Nelson's column, for example), although they may have categorical significance. As an example, houses may be considered more significant than trees. In the work presented here, however, we concentrate on objectively determinable factors (e.g. geometry or topology), and our empirical results show that even in the absence of historical or cultural salience, a fair approximation to human performance can be obtained.

Various psycho-linguistic investigations into reference object characteristics have been performed which identify potential characteristics for reference objects. The current experiments provide some confirmation for the results of these linguistic investigations with data more representative of the real world than any previously available.

Miller and Johnson-Laird (1976) note that the scale of the reference and target objects is important in selection of a reference, as for example: 'It would be unusual to say that the ashtray is by the town-hall'. Talmy (2000) lists attributes of reference objects relative to the target, and the elements of his list relevant to reference selection are as follows, where the reference is likely to be: more permanently located, larger, geometrically more complex in its treatment, and more immediately perceivable than the target. Although there is agreement that the size of the objects as well as the distance between them and their mobility will affect their suitability as references, the different possible ways of representing the size of objects and the distance between them is not investigated in depth. Ascertaining which measures are relevant becomes important when machine models are being constructed as well as being informative of the mental processes used by humans in forming locative expressions. A range of measures for object size and geometric extension are considered in this study and these are described in Section 5.6.

Visibility is discussed by Gapp (1995) and is inherent in the visual salience calculation of Kelleher and Costello (2009). Although it is tempting to say that an obscured object would be a poor reference, it may not be that straightforward. If the case of the cloth covering the table is considered, the obscured table may still be the better reference than the cloth. Talmy's term 'perceivable' is probably better than the term 'visible' although 'visible' is used in this chapter. Variables relating to visibility are described in Section 5.6.

Recent experimental work by Carlson and Hill (2008) indicates that the geometric placement of a reference is a more important influence than a conceptual link between target and reference, and that proximity and joint location on a cardinal axis (e.g. target directly above or directly to the left of reference) are preferred. The experiments were carried out using two-dimensional object representations on a two-dimensional grid. The angular relationships between the target and reference are not considered here as the focus of this study is on the characteristics of the reference object itself. Earlier work by Plumert et al. (1995) focuses on hierarchies of reference objects in compound locative phrases but also finds a relationship between description complexity and search time for the target object, that is to say if an object is reasonably easy to locate the speaker will reduce their description accordingly. Although a very useful study in locative expression formation, it does not deal with the characteristics of reference objects.

In the case of urban navigation, Burnett et al. (2001) derive the following characteristics of good landmarks from interviews with subjects who have chosen particular landmarks in an experimental setting: permanence, visibility, usefulness of location, uniqueness and brevity of description. They also note that landmark characteristics can be traded against each other; a more visible landmark may not need to be so usefully located as one less easy to discern, for instance.

The factors which contribute to 'visual and cognitive salience' in urban wayfinding are investigated by Raubal and Winter (2002) and Nothegger et al. (2004), who test automatically selected landmarks against those selected by humans. Cognitive salience, including cultural or historic significance, is in practice related to the issue of prior knowledge of the specific landmark by the listener which is not relevant in this study. Visual salience combines the aspects of visibility (discussed above) and identifiability. Identifiability is clearly related to uniqueness or (lack of) ambiguity. It is an important issue and how it affects reference choice is discussed in Barclay and Galton (2008b), but it is not covered in this chapter.

Sorrows and Hirtle (1999), along with landmark characteristics already mentioned, include prototypicality as an important characteristic of landmarks. Prototypicality is an important factor as without specific knowledge of a landmark or reference, categorical knowledge is required. A church that looks like a supermarket, or otherwise departs from its expected visual category, is a problematic reference. In this study the issue of prototypicality is not central, in a sense it is considered part of the object recognition sub-system and in this study the objects are already recognized. For the comfort of the human subjects all objects in the scenes are designed to be recognizable members of their categories, although this makes the study, at least in part, specifically English. Cross-cultural studies with mosques instead of churches and French bar-tabacs instead of pubs will need to be the subject of future work.

In earlier work on reference object choice (Barclay and Galton, 2008b) a comprehensive influence model for reference choice was developed, the primary purpose of which was to organize the different characteristics for references described above into a functional model. Different researchers, as described, had arrived at variously overlapping lists of characteristics but no attempt had been made to organize them in a conceptual framework. As noted, not all the characteristics of references identified have been included in this study, which mainly deals with the geometric characteristics of the reference objects.

5.5 A near real-world data set for locative expression generation

For complete coverage of the reference object selection problem, human subjects would describe real-world locations and the computer model would be trained and tested through analysing stereoscopic images of the same scenes. This is not yet possible or desirable for a variety of reasons. First it expands the problem of generating spatial language to include the error-prone task of object recognition in potentially highly cluttered environments. It also increases the computational load unnecessarily for the investigation at hand and makes the collection of training and test data from human subjects far more costly. The strategy adopted in the current research is to use schematized images generated from constructed or real-world scenes such as that shown in Figure 5.2. The number of objects appearing in the scene is reduced to give a reasonable computational load for machine analysis of the scene and also to reduce visual clutter and time needed for the human subjects describing the scenes in the openGL representation at the reduced resolution of a computer screen. The construction of the scenes is further described in Barclay and Galton (2008a). The average number of top level objects (those which can be selected as reference objects) in a scene is just over 27; the objects are typically composed of multiple named parts (e.g. a table has a top and four legs) although this information is not used in the current experiments. Overall the scenes need to be representative of reality with all the major candidates for reference objects in place and with their significant geometric features intact.

The machine learning system is presented with the openGL vertices of the objects making up the scene and the corresponding names of the objects. Variables calculated from this vertex representation directly or by ray-casting are described in Section 5.6.

The scene corpus currently consists of 133 scenes, most of which have four target objects, giving a total of 529 test cases. In machine learning terms there are in the region of three positive and perhaps ten non-redundant negative reference examples for each test case giving nearly 7000 test and training examples for reference object selection.

FIGURE 5.2 A scene from the test data set together with the photograph from which it was derived. The target objects are shown in the constructed scene

The corpus contains 14 series of between 7 and 10 scenes taken from a particular viewpoint. This allows the machine learning system to establish which objects in a scene are fixed and which are, to some extent, mobile. About half of the scenes in the corpus are derived directly from photographs and about half are manipulated or constructed to ensure coverage of particular spatial relationships that might not occur in even a large sequence of scenes based on photographs. This includes for instance the idea of 'among', e.g. a group of objects distributed among the members of another group.

The scenes cover a range of scales from table-top, through room-scale to street and vista-scale. Over 140 different object types are named in the scenes. Some sense of the range of scenes and the diversity of objects within them can be gained from inspection of Figures 5.2 and 5.7. The scenes are defined solely by the objects within them and the objects are constructed from a simple set of three-dimensional geometric primitives (cones, prisms, spheroid sections, etc).

In a machine learning approach to problems such as this, enough data must be provided to enable the training of the system as well as to provide enough test data to establish significant results. The learning process is 'supervised', so the training and test data consist of the scenes as described above along with, for each test case (target object), a list of reference objects considered as 'good' (by implication the other objects in the scene are not good references). Appropriate prepositions are also provided, to enable future work with the data set, although they are not used in the experiments presented here. As, on average, it takes 30 seconds to annotate a test case with a good reference object and preposition, to annotate the entire test data set would take nearly 5 hours non-stop work. Within the resources available to the project, annotation of the entire test set by a sufficient sample of volunteers (ideally at least 20) was not possible. Informal experiments suggested that annotation of 30 test cases by a volunteer in a single session was leading to complaints and possibly to errors from lack of concentration. Each volunteer was therefore asked to annotate

20 test cases. Given a group of 80 volunteers, it would be possible to annotate each test case twice or 80 of the test cases (just over 15 per cent) 20 times. The latter approach was chosen, as useful information is obtained on the distribution of reference choices by a group of subjects. For each of the 529 test cases the first author has provided his opinion as to suitable reference objects (up to a maximum of three), providing the remainder of the annotation. The results from the 80 cases annotated by multiple subjects are used to ensure that the first author's use of spatial language is not completely idiosyncratic, or biased in some detectable way.

A group of 40 volunteers from among Exeter university research staff and students and a further 40 from the authors' acquaintance each provided opinions on groups of 20 test cases. Of the volunteers three were acquainted with the research in more than outline form. The instructions presented to the volunteers gave an outline of the research and the reasons for the volunteer's participation.

For each of the 20 cases the subject was shown a scene such as that in Figure 5.2 and asked the question 'Where is the <target>?'. They then had to choose a reference object and preposition from drop down lists to complete a simple locative phrase of the form 'The <target> is <preposition> the <reference>'. The target object and selected reference were highlighted in the scene being shown. It is important to note that the subject's choice of reference object and preposition were independent. Any one of 24 of the static English spatial prepositions could be chosen along with each candidate reference object, giving about 650 possible different descriptions for each case. The subjects were not led to the choice of a particular reference by the presence of a preposition supplied by the experimental environment.

The scenes were presented in random order on a laptop computer, and the environment was not controlled in any other way. An instruction page was provided but no practice examples. There was no time limit imposed and it was possible to review and amend answers. The scenes given to the subjects were chosen by the following method. Two scenes were chosen at random from each series (without replacement) to ensure coverage of the various scene scales. From each scene chosen one of the four test cases was chosen at random. If the test case was not suitable for presentation to a human subject, the next test case in sequence was used. (The only reason for unsuitability is complete or near complete occlusion of the target object, making it difficult for a human subject to identify.)

The subjects were also asked to indicate their gender and age (within bands) and to state whether English was their native language. It should be noted that this was not designed to be a psycho-linguistic experiment in its own right. Although some useful information can be derived from it, the purpose of this exercise is to provide validation data for the machine learning test set. Further information on the results of this exercise including the preposition list and preposition usage can be found in Barclay (2010).

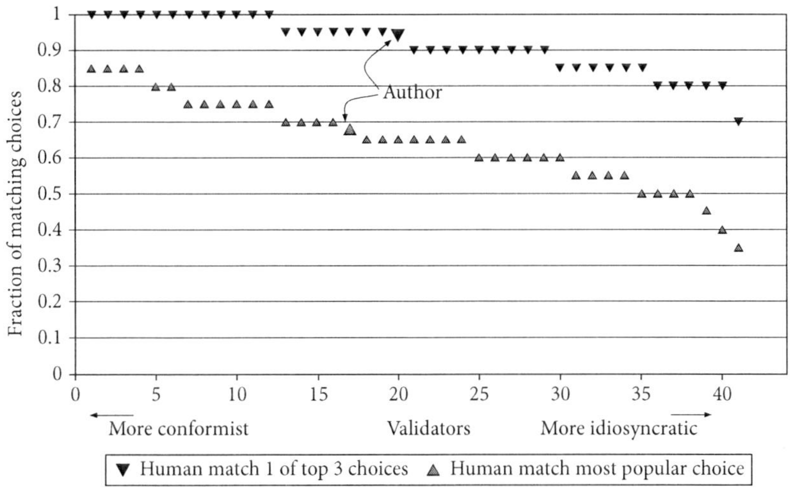

FIGURE 5.3 Conformity of individual validation subjects to the group choice of reference. The cases of matching the most popular choice of the group as a whole, or one of the three most popular choices of the group are shown. The two cases are separately ranked

Figure 5.3 shows, for each of the 40 subjects in the validation exercise, how often they agreed with the consensus of the group as a whole as to the choice of reference object. The lower trace shows how often each individual agreed with the most popular choice of the group as a whole, and the top trace shows how often each individual chose one of the three most popular reference choices of the group as a whole. There are, in most of the scenes, several objects that make effective references and the average number of different references chosen by the group of humans is approximately 4.5. This figure will include some idiosyncratic and possibly some erroneous choices, so will be somewhat higher than the actual figure for the number of effective references in a scene. For this reason the chosen measure for the machine models is whether they can match one of the three most popular reference choices of the human validators, which allows for a realistic number of effective references while not including ineffective references. As can be seen this measure is achieved by the median human 90 per cent of the time. Of course if there is only one reference chosen by the group of humans in a given scene (that is, total agreement) the machine models must choose this reference to 'match human performance'. Also a reference must be chosen by more than one human to constitute a valid reference, and a machine model matching a reference chosen by a single human would not be considered to be correct.

From Figure 5.3 it can also be seen that the first author, who has provided some of the reference choices for the test data set, appears not to be making idiosyncratic choices and thus training the machine models on the author's choices of reference is not likely to produce biased models.

As noted there appear to be a few cases where the human subjects have given erroneous answers, or answers that they may not have intended, owing to misinterpretation of the experimental process. No attempt has been made to remove these from the data set. The statistical learning processes used for the machine models will disregard these cases; however, they will be reducing the conformity among the human subjects in a manner that may not be present outside of the experimental setting. That is, the median conformity of 90 per cent quoted is more likely to be a low, than a high, estimate.

5.6 Descriptive variables for reference objects

Details of the variables used in the machine models, and how they are derived from the scenes in the test data set, are given here. The variables are discretized although many are continuous in nature (for instance the distance between a target and a candidate reference object). Combining continuous variables typically requires some assumption to be made about the form of the joint probability distribution, with weighting values then being learned. The nature of the variables in this study suggested that making any such assumption would be unsafe. The discretization process also introduces limitations and potential errors in the model and these are described in Barclay (2010). Typically the bins are organized to give a roughly uniform distribution of objects in each bin, within the constraint of the near-logarithmic bin spacings. The number of bins is set to the minimum possible before performance degradation is evident. The abbreviations of the variable names (given in parentheses in the list below) have been included to make the graph legends a manageable size.

5.6.1 *Classification variable*

The classification or ranking variable is referred to as 'Reference suitability' (refSuitability), and this is effectively the output of the machine learning system. (This is to say that this is the output variable and the variables described in the following subsections are the input variables.) During training the value is given by the assessment of the human subjects in the validation exercise, or by the author. Up to three objects in a given test case are decided to be suitable. Although some may be thought to be more suitable than others no account is made for this during training of the machine model, each being given the same weight. During testing, the value of this variable is determined for each candidate reference object by the machine learning system, and the reference with the highest suitability value is returned as the model's choice of reference.

5.6.2 *Variables related to the distance between objects*

The distance measure used is the proximal distance (ProxDist). This is the distance between the closest points on the candidate reference object and target object, derived from an exhaustive search of the distance between the facets making up the surface of each object. Other experiments have determined that the distance between object centroids is a significantly less good predictor of reference suitability. The distance measure proposed by Gapp (1995), between the centroid of the target object and the closest point to the target on the candidate reference object, performs less well than the proximal distance, though not significantly so. Five logarithmically spaced bins are used for discretization of the distance measure. The bin boundaries are: $0.02S$, $0.05S$, $0.1S$, $0.2S$ where S is the length of the diagonal of the scene bounding box (in metres). The scaling of the distance measure by the size of the scene enables scenes of very different scales (from table top to street scale) to be conveniently combined in a single model. Although this seems intuitive there is an implicit assumption that humans scale distance with respect to overall distances in a scene rather than (say) with respect to significant object sizes. This assumption has not yet been explicitly tested.

5.6.3 Variables related to object size

The following variables related to object size are used:

1) Bounding box volume (BbVolRef). This is the volume of the bounding box of the object. The bounding box is minimal and does not vary with the object's orientation with respect to the x, y, z axes of the overall scene. On this measure a fruit-bowl will be bigger than the apple inside it. Eight values are used with bin boundaries as follows: $0.025k$, $0.1k$, $0.4k$, $1.6k$, $6.4k$, $25.6k$, $124.0k$, where k is $\frac{S^2}{16{,}000}$, S is the scene bounding box diagonal as before, and the constant 16,000 simply gives more or less manageable numbers for the bin boundaries. As an illustration, an object with a volume of about a quarter of a litre (a teacup say) is about on the lowest bin boundary in a room-scale scene measuring 4 metres by 4 metres by 2.5 metres. The more intuitive measure relating the volume of an object to the volume of a scene returned different bin distributions of objects between large (street scale) and small (table-top scale) scenes and distorted the results. Why this should be so and what it means for the processing of different-scale scenes by humans is not yet understood.
2) Material volume (MaterialVolRef). This is the volume of 'material' in the object. On this measure a thin walled fruit-bowl might be smaller than the apple inside it. The bin organization is the same as for bounding box volume.
3) Convex hull volume (HullVolRef, HullVolTarget). For both the target and candidate reference objects, this is the volume of the convex hull of the object. The convex hull of an object can best be visualized by imagining it covered in plastic film, so that volumes between protruding points are filled in. The convex hull volume will be smaller than the bounding box volume and larger than or equal to the material volume. On this measure a fruit-bowl will usually be bigger than the apple inside it. The bin organization is the same as for bounding box volume.

4) Ratio of target to reference convex hull volume (HullVolRatio). This is the target convex hull volume divided by the candidate reference convex hull volume. Eight values are used with bin boundaries as follows: 0.0001, 0.001, 0.01, 0.1, 1, 10, 100. This measure allows the relevant object sizes to be used without recourse to scaling by the scene dimension.
5) Maximum dimension (MaxDimRef). This is the maximum dimension of the object's minimum bounding box. As with bounding box volume it is independent of orientation. Eight values are used with bin boundaries as follows: $0.005S$, $0.01S$, $0.025S$, $0.05S$, $0.1S$, $0.25S$, $0.5S$ where S is the length of the diagonal of the scene bounding box (metre).
6) Minimum dimension (MinDimRef). This is the minimum dimension of the object's minimum bounding box. The bin organization is the same as for Maximum dimension.
7) Height (HeightRef). This is the highest point on the object, not the vertical dimension of the object. Gapp (1995) suggests that height might be a 'characteristic' dimension for some objects (people for instance) and this measure embodies this as well as allowing for a shorter man on a table to be a better reference than a tall man on the floor, which seems intuitive. The bin organization is the same as for maximum dimension.
8) Ratio of minimum to maximum dimension (RefDimRatio). This is simply the minimum dimension of the candidate reference's bounding box divided by the maximum dimension. Five values are used with bin boundaries as follows: 0.01, 0.033, 0.1, 0.33.

5.6.4 *Variables derived from ray casting*

In the current study 10, 000 rays are 'cast' into each scene at random. Each ray is effectively a vector from the camera position with a random component in the x and y directions (the 'z' direction is into the scene from the camera position).

The point at which the ray intersects the first object in the scene along its length progressing from the camera position is recorded. If the ray does not intersect any objects in the scene it is not used, but is not replaced, so for a sparse scene there will be fewer than 10, 000 intersections.

1) Viewability (Viewability) This is the number of times an object is intersected by a ray. With this measure large but mostly obscured objects and objects only partially in the field of view will have low viewability. The eight bins are approximately logarithmically spaced with boundaries 5, 10, 25, 50, 100, 250, 500. This measure incorporates object occlusion by other objects, as well as the distance of an object from the viewpoint, and variations in apparent object size due to orientation with respect to the viewpoint (seeing a long thin object end on, say).

2) Sight-line salience measure (salienceSL). This is an analogue rather than a direct equivalent of Kelleher's salience measure (Kelleher and Costello, 2009) because in this study the centroid of the target object is not always in the centre of the field of view (in fact very rarely, since the target can be anywhere in the scene). So Kelleher's weighting of pixels (or, in this study, ray intersections) by the fraction of their distance between the centre and the edge of the field of view is not applicable. Instead, using $\overrightarrow{VH}$ as the vector from the viewpoint to the intersection point between the sight-line and an object ('hit-point'), and $\overrightarrow{VT}$ as the vector from the viewpoint to the target centroid then

$$hitAngle = \arccos \frac{\overrightarrow{VH}}{\| \overrightarrow{VH} \|} \cdot \frac{\overrightarrow{VT}}{\| \overrightarrow{VT} \|} \tag{1}$$

defining *F*(*hitAngle*) as

$$F(hitAngle) = \begin{cases} 1 & \text{if } \sin(hitAngle) < \sin(viewAngle) \\ 0 & \text{otherwise} \end{cases} \tag{2}$$

which is necessary as target objects are not always in the centre of the scene. The 'viewAngle' is the openGL view angle (i.e. the projection angle). The sight-line salience (*salienceSL*) is then

$$salienceSL(Obj) = \sum_{hitpoint \in Obj} F(hitAngle)\left(1 - \frac{\sin(hitAngle)}{\sin(viewAngle)}\right) \tag{3}$$

simply summing over all intersections between rays and the given object (without, hopefully, needing a complex formal definition for $hitpoint \in Obj$). The measure incorporates object viewability with a measure of distance from target to reference.

3) Proximal salience (proxSalience) An issue with Kelleher's salience measure is that it is a measure derived from the projection of a three-dimensional scene onto a two-dimensional surface, whereas humans have a depth perception faculty and an ability to judge distances in a three-dimensional space. A version of the salience measure which combines a candidate reference object's viewability and the distance between the candidate reference and target objects in three dimensions is defined by

$$F(hitDistance) = \begin{cases} 1 & \text{if } \| \overrightarrow{HT} \| < sceneBBdiag/2 \\ 0 & \text{otherwise} \end{cases} \tag{4}$$

where $\overrightarrow{HT}$ is the vector from the hitpoint to the closest point on the target. This is necessary as target objects are not always in the centre of the scene. The three-dimensional proximal salience (*proxSalience*) is then

$$proxSalience(Obj) = \sum_{hitpoint \in Obj} F(hitDist)\left(1 - \frac{2 \cdot \| \overrightarrow{HT} \|}{sceneBBdiag}\right) \quad (5)$$

4) Proximal salience squared (proxSalienceSqr) A problem with compound variables such as the salience variables described is that there is an implicit assumption of the nature of the combining function. To assess whether this might be a significant issue another measure can be defined as follows:

$$proxSalienceSqr(Obj) = \sum_{hitpoint \in Obj} F(hitDist)\left(1 - \left(\frac{2 \cdot \| \overrightarrow{HT} \|}{sceneBBdiag}\right)^2\right) \quad (6)$$

where the distance is squared so that the value of the measure will fall off more quickly as the distance between the objects increases.

5.7 Simple models of reference object suitability

A Bayesian network approach is taken for the machine learning system as this seems to fit well with the structure of the problem. It is possible that a support vector machine approach might give better absolute results, in a machine learning sense, but there are other advantages to using Bayesian networks. In particular it is easier to illustrate, using a Bayesian network, which influences, and combinations of influences, are important in human choice of reference objects. The ability to cluster variables around 'hidden variables' which can then be mapped on to concepts related to the overall problem is the key reason for choosing Bayesian networks as the machine learning structure; this will be relevant in future work, but not that presented here.

The results in this chapter are derived from the restricted form of network shown in Figure 5.4, in which the full joint probability table for the variables used is available. That is to say that, for each combination of values of all variables, there is a corresponding entry in the joint probability table. No assumptions need to be made about variable independence. This limits the number of variables that can be used, since results deteriorate as the size of the joint probability table approaches the

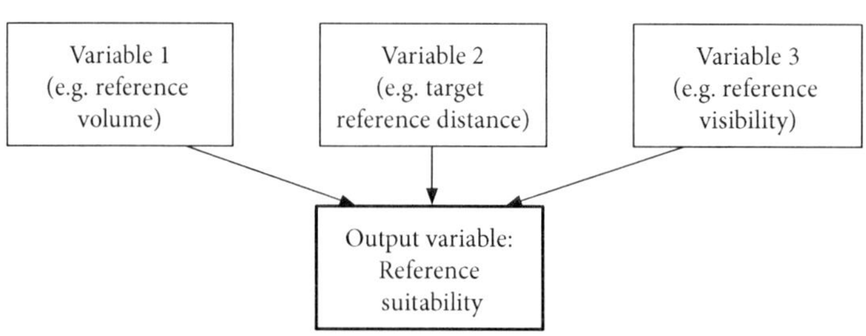

FIGURE 5.4 Restricted form of Bayesian networks used in this chapter

number of training cases available. Combinations of up to four variables only are used here, keeping the probability table to approximately 1000 entries and avoiding significant performance degradation. In work using more complex models (to appear) more general Bayesian networks are used.

Ten-fold cross validation is employed to maximize use of the training/test data set. The data set is traversed ten times and each time a different 10 per cent of the data set is reserved for testing. The networks are trained on the remaining 90 per cent of the data set. This also yields partial result values which are used to assess the statistical significance of the difference between results from different networks. Wilcoxon's signed rank test (Wilcoxon, 1945) is used in this study as no assumptions need to be made about the distribution of the result values. The use of Wilcoxon's signed rank test can lead to an apparent discrepancy between the figures given for significance and the performance values given in the graphs. It is possible that even if model 'B' out-performs model 'A', model 'A' may be significantly better than a third model 'C', while 'B' is not.

In interpreting what is meant by a statistically significant result some care must be taken. What is actually true is the following:

> There is an apparent significant difference between two machine models in matching the judgements of a group of humans as to the suitability of reference objects, which were made for schematically represented scenes, displayed on a computer screen.

In the following sections a shorthand is used which could be read as:

> This model is significantly better than others for determining reference suitability.

Whether the scenes are real enough is discussed in Section 5.8. It should be noted that, while the machine models are matching considerably fewer group choices of reference than the median human in the group, it is probably reasonable to suggest that an increase in the number of matches represents an improvement. As the median human matches about 90 per cent of one of the top three group choices of reference this is always the case here.

Figure 5.5 shows results from three types of model. The first (top) set of bars shows results using only a single variable relating to the size or shape of the candidate reference objects. The best of these, the volume measures and the ray-cast measure of viewability, correctly predict about 25 per cent of reference choices on their own. It is hardly surprising that this is the case as, wherever a reference is situated in the scene (however far from the target), it is equally likely to be a good reference according to these models. This illustrates the important point that poor performance of a variable in a poor model does not necessarily mean that the variable is unimportant. The random baseline figure is derived from running the model with random numbers in the joint probability table. Given that the task is to match one of three human chosen references out of an average of 27 candidates in a scene this figure is as expected.

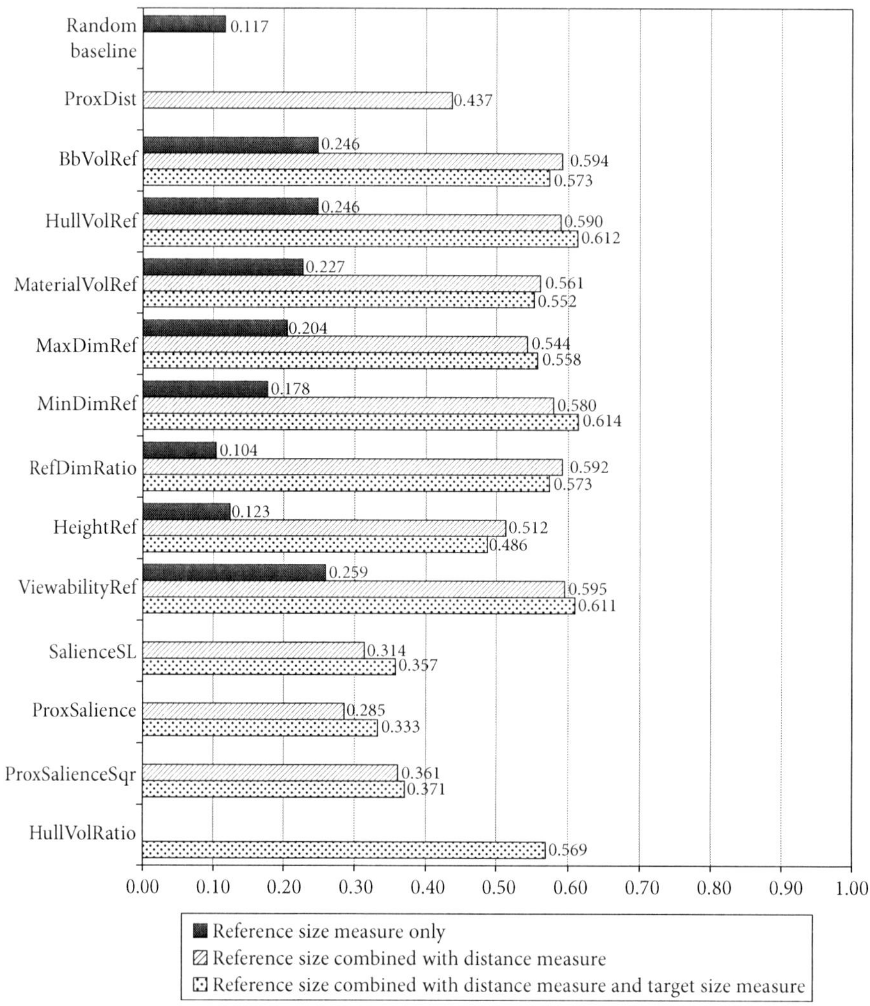

FIGURE 5.5 Fraction of machine choices matching one of the top three human choices of reference when the indicated variables are included in the model. Where applicable the distance measure is ProxDist and the target size measure is HullVolTarget

The second (middle) set of bars shows results from models combining the reference size or shape measure with a distance measure. Also shown is the result for a model using the distance measure on its own and three composite variables derived from ray-casting, which contain distance and size information. The distance measure is better on its own than any of the single size or shape measures. Inspection of the model shows that it is the closest objects which are favoured and so this result could be paraphrased as 'choosing the closest object in the scene as a reference will concur with human reference choice 44 per cent of the time'. The most surprising result is that the composite variables all perform significantly worse than the distance measure on its own (at the 0.005 level $W = 45$, $N = 9$). There is clearly some relationship between size and distance that humans are using that is not reflected in the way size and distance are combined in these measures. All of the size measures, used in combination with distance, are significantly better than distance alone (at the 0.005 level $W = 45$, $N = 9$ or better, except refHeight which is at the 0.025 level $W = 38$, $N = 10$). Hence it could be said that 'choosing an appropriately sized object in reasonable proximity to the target as a reference, will concur with human reference choice about 61 per cent of the time'. The fact that the minimum dimension of the reference predicts reference suitability so well might seem surprising, but it should be remembered that an object with a large maximum dimension could be one-dimensional, having both an inappropriate shape and volume for a good reference object. This is less likely to be true for an object with a large(ish) minimum dimension.

The third (bottom) set of bars shows results from models which have the convex hull volume of the target object included as well as the distance and reference object size measures. Also shown is the model combining the reference to target volume ratio and the distance measure. There is no significant difference in any of the individual models from adding a measure of target object size, which is a little surprising. This indicates that the absolute measure of reference size, related to its locatability, is more important than its size relative to the target, related to its ability to constrain the search for the target. Some caution needs to be exercised in interpreting this result; the assumption is being made that the data set contains sufficient examples of very small objects for which there are references that are too large to be useful. In most scenes the convex hull volume ratio between the largest and smallest objects is greater than 100 000 : 1 (for example a desk at $1.7m^3$ and a pen at $0.000006m^3$), which is getting close to the limit of visibility in the experimental setting of a standard personal computer screen, and seems suitably extreme. The composite measures are not significantly improved by addition of a target size measure. There is some indication that the intuitively attractive measure of convex hull volume is a better predictor of reference suitability than the material volume of an object.

Figure 5.6 contains results from further models which combine the information elements in different ways and, most importantly, look at reference geometric extension (or shape factor) as an influence in addition to, rather than as an alternative to, reference size. The first (top) result simply replaces the absolute target volume measure with the measure relating reference and target volumes in a model containing the reference volume and distance. This representation performs worse than the absolute target volume measure although not significantly so. It suggests that the relatively good performance of the volume ratio in combination with distance measure (Figure 5.5) is due to the volume ratio being well correlated with reference volume.

The first three results in the second group in Figure 5.6 add different measures of reference geometric extension to the model containing reference and target convex

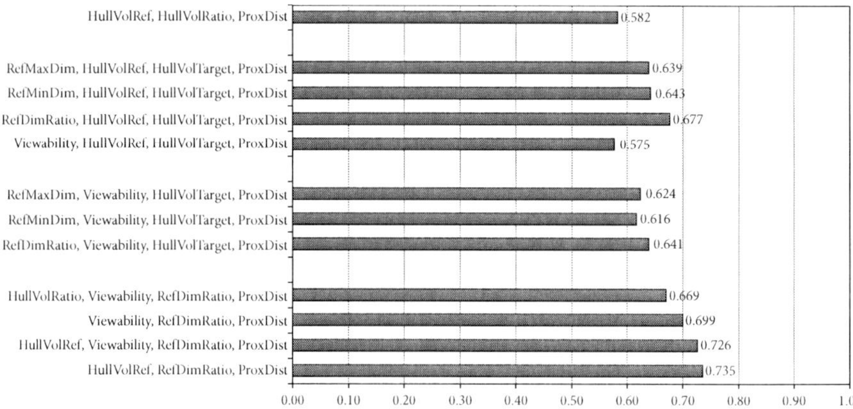

FIGURE 5.6 Fraction of machine choices matching one of the top three human choices of reference when the indicated variables are included in the model

hull volumes and the distance measure. All of the models show an improvement over the model containing only object volume and distance, and the result for the model using the refDimRatio measure is significant at the 0.025 level ($W = 41, N = 10$). The next result adds the viewability measure instead of a geometric extension measure and this produces a worse result, although not significantly so. It suggests, though, that viewability does not add useful information to a model containing a measure of reference volume and is, of itself, not as good a predictor of reference suitability. The third group of three results combine distance, target volume, a geometric extension measure, and viewability (as a measure of reference size). These show the same pattern with respect to the geometric extension measure as the similar models using reference convex hull volume, although not performing as well in each case. The final four results show models that combine a reference size measure, distance, and reference geometric extension (refDimRatio), but omit any direct measure of target size. Perhaps surprisingly, these models perform better than the corresponding models including the target size measure. The model containing HullVolRef along with ProxDist and RefDimRatio performs significantly better (at the 0.025 level, $W = 39, N = 10$) than the model which also includes HullVolTarget. As before, the viewability measure does not seem to add useful information.

5.8 Knowledge and perception in reference object choice

5.8.1 *Influencing factors and variable representations*

The best model for reference choice presented here matches the choice of a group of human subjects 73.5 per cent of the time, as compared to the median human who will match the group choice 90 per cent of the time (see Figure 5.3). These are fairly simple models, however, and omit consideration of various factors which other commentators and researchers have suggested may be important, including object mobility, salience or ambiguity, topological relationships between target and reference objects, and angular relationships between reference and target. More complex models have been investigated using some of these factors, and this work suggests that the performance of machine models can be taken to the point where they match human reference choice in at least 80 per cent of cases, still some way short of the median human. How best to represent these variables is still under investigation and results will be published in due course.

Some cases where the machine did not match one of the human choices of reference are shown in Figure 5.7 and the issues arising from these are discussed where appropriate below. It is felt that the machine choices illustrated are not wildly inept and indeed seem less so than some of the outliers from the human selections.

It has been possible to learn a variety of things about individual influences on reference choice and the variables that best model them and these are discussed in the following paragraphs:

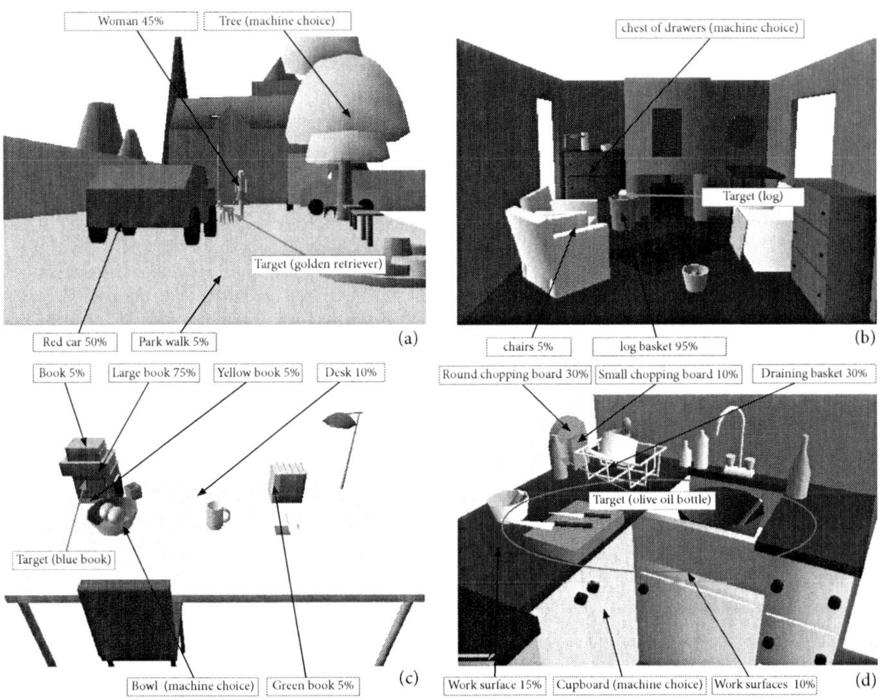

FIGURE 5.7 Cases where the best machine model chooses differently to the group of human validators. The percentage of humans making each reference choice is shown

1) Perhaps not surprisingly, the idea that the reference object should be 'appropriately located' or 'near' (see Burnett et al. 2001, in the landmark context) the target is confirmed. This is the single most influential factor in determining reference suitability, contributing to the ability of the reference to define a small area in which the listener must search for the target. Although the nature of the distance measure is not investigated in detail in this chapter, a measure of distance has to be included to bring the models to a level of performance at which it is possible to make meaningful judgements about other factors.
2) The absolute size of the reference is clearly important, and this again is as expected, since the listener must clearly be able to locate the reference in a scene. The best measure for reference size appears to be convex hull volume and it seems that the visible area of the reference does not play an important part in the process. This is surprising and is counter to the expectations of some commentators (e.g. Gapp, 1995). The explanation could lie in the fact that in this exercise the speakers were not given any information regarding the position of the listener, who could potentially have a different viewpoint from which a reference is not obscured. It could also be that the shared knowledge (between speaker and listener) of the likely physical size of an object is the important characteristic rather than a perceived projection of an object in one particular direction.
3) In this study the size of the target, either in absolute terms or expressed relative to a candidate reference, does not help selection of a suitable reference. This is surprising but may be related to the restricted nature of the exercise, in which there is no option to adopt a compound locative phrase (using more than one reference). In this case, for small targets, the most easily locatable of what would ideally be a sequence of references tends to be given. This is to say that, given a choice between two sub-optimal references, the more easily locatable (larger) is given, in preference to the one that better locates the target but is itself harder to find.
4) Variables related to geometric extension are clearly important in selection of reference objects and the variable which best expresses this is the ratio of the maximum to the minimum dimension of the reference. No other explanation for this is evident except that it relates to the goodness of the reference in defining (along with distance) the space in which the listener has to search for the target. This characteristic of reference objects may equate to Talmy's 'Geometrically more complex in treatment', and it is clearly necessary, in addition to some representation of the overall size of the reference object, in an effective model for reference choice. Whether the geometric extension used is perceived in the scene or derived from prior object knowledge is not clear. If the case of a man standing in front of a long train is considered, however, it can be seen that he may be adequately referenced by the train whereas a man

standing beside it may not. In this case the orientation of the extension of the train relative to the position of the target could only be as perceived in the particular scene. To say that a man is 'on a sidewalk' does not effectively locate him (in the direction along the sidewalk) if the sidewalk is long. In this case it could be envisaged that the sidewalk would be discounted as a reference due to prior knowledge of its likely extension. Variables that capture this orientation of extension are under investigation.

5) Variables that assume simple relationships between size and distance representations perform badly. This suggests that simple computational models which implicitly include similar relationships, such as the Euclidian distance model proposed by Gapp (1995), will not be satisfactory. This does not mean that the salience measure proposed by Kelleher and Costello (2009) will not be appropriate for reference resolution, which is a different task, and it should perhaps be further investigated.

5.8.2 *Possible shortcomings of the platform and models*

To assess whether the scenes were sufficiently complex to be representative of the reference choice task in the real world, the correlation coefficient between the likelihood of a correct match and the number of objects in a scene was calculated, and is 0.02, with the number of objects in a scene ranging from 11 to 42, averaging 27 as noted. There is no correlation between the number of objects in the scene, a measure of its complexity, and the results (in terms of matching human reference choices) from the machine models. This suggests that the process of ignoring some of the 'clutter' in real-world scenes (thereby reducing the number of objects) will not have affected the results.

The representation of the objects themselves, in terms of detail or realism, does not materially affect the machine models. So long as no significant geometric features are missing and the size of the objects is accurate, the machine models will return the same result irrespective of the level of realism. No evidence from the results bears on the question of whether the object representations affect the human subjects; so we are left with anecdotal support in that none of the human observers commented on the crudeness of the object representations or the difficulty of interpreting the scenes. Some types of objects are either absent from, or badly represented in the scenes, and these include non-rigid or conformable objects, such as fabrics, and penetrable or partially space-filling objects such as foliage. This leads to the omission of some spatial relationships (for instance 'the tablecloth is over the table') and possibly some difficult reference choice cases. Overall it is not expected that object representation has materially affected the results and it should also be noted that in this respect the data set used in this study is at least as good as in other contemporary studies (e.g. see Byron et al., 2009).

There seems to be no way to assess whether the corpus as a whole is sufficiently diverse and no target for 'sufficient diversity' can be easily defined. In the limit a highly diverse scene set would contain scenes strange to many humans (who may not have been to the Sahara desert, say) but on the other hand it is true that the human ability to describe scenes is not dependent on having prior experience of them. As noted, there is a range of scene scales that might be characterized by large indoor spaces such as an open plan office or an airport check-in hall, or small outdoor spaces such as a suburban garden, that are missing from the corpus. Scenes with even larger scales and scenes with significant height extensions (hills) are also missing as they are significantly more difficult and time consuming to produce. On balance it is felt that lack of diversity is not a major limiting factor on the performance of the experimental platform as a whole.

The process of selecting a reference object, for a target in a scene on a computer screen, from a drop-down list, as described in Section 5.5, is clearly not the same as that of forming a verbal description of a real-world scene. Experiments could be performed to compare the validation results against scene descriptions in real-world scenarios, but these would be relatively resource- and time-consuming. At present reliance is placed on the fact that the vast majority of the descriptions given by the human subjects are reasonable and effective in locating the target. Also note that only using the three most popular reference choices in a scene for training tends to filter out odd references that are either mistakes, or effects of the clumsiness of the process. Future validation exercises should try to use direct selection of objects from the scene with a pointing device, primarily because this would reduce the effort required rather than because it would produce more realistic results. Use of natural language descriptions to annotate the data set is clearly more problematic, in particular because of object naming differences. A desk might be called a table or workbench, for instance, and the study has already run into a minor cultural difficulty by calling a 'dumpster' (American usage) a 'skip' (British). Natural language descriptions would either have to be cleaned up by hand or an ontology would have to be employed to reconcile object names. Even so, errors due to language use rather than selected reference object quality would almost certainly be present in the training data set.

The current models assume that humans treat scenes of all scales in the same way and it is not clear that this is the case. Initial experiments suggest that, if the data set is divided into indoor and outdoor scenes and separate models are learned for each, then the performance of the models improves by about 3 per cent. The reason for this is not entirely clear. It seems true (although this needs further research) that the distribution of sizes of objects varies with the scale of the scene, as represented by its bounding box, and this makes averaging scene-relative sizes for reference objects over a range of scene scales difficult. It may just be that the correct measure for the scale of a scene has not been identified. If the ray-cast variables had proved to be more important, this difficulty would have been removed as the view angle of the scene would be the relevant parameter, and this is effectively scale independent. Further work is required in this area to understand whether humans are making linguistic adaptations to the scale of the scenes they are describing. If so, the models currently used need modification.

Candidate reference objects are considered independently and do not interact in the machine models with other references. Although there is little experimental evidence from the test cases, there are plausible mechanisms (see Gapp, 1995) by which the choice between two references could be altered by the characteristics or positioning of a third, and this requires further investigation.

An assumption implicit in the machine models is that all objects are treated the same way by humans. That is to say that humans use the same geometric measures for all objects. It is possible that, for instance, knowing that a bowl is a container, a human would assess its size by its convex hull volume but that a chair or table might be assessed by its material volume, as this relates to its mass or perhaps 'presence' (the assumption being made that these considerations are appropriate for reference object selection). Again, it might have been thought that the ray-cast viewability variable would have been better at addressing this sort of issue, in particular for objects (such as roads) that do not have an easily definable 'volume', but are defined by an area. It is possible that one reason for this apparent failure of the viewability measure is that it is used by humans for some objects but not for others (or at least it is given more weight for some objects) and the current model does not allow for this. Different geometric treatment of different object types directly would require a full ontological background for classes, and perhaps uses, of objects.

5.8.3 Human performance in the reference choice task

From the results presented we can infer that humans are using a sophisticated model for choosing reference objects. Certainly it is more sophisticated than simply picking an object of the right size in reasonable proximity to the target.

It is probable that different people are using different model variants and that this leads to different levels of conformity to the reference choices of the group as a whole. However, taken as a group, the human models produce very similar results. This conformity does not necessarily equate to effectiveness but it can be argued that the only plausible driver for this conformity is the need for effective communication.

It is not possible to say whether humans are using the same variables organized in the same way as the best machine models, but it seems unlikely. What can be said is that there is a strong correlation between the variables used by humans and those used in the best machine models, that is they must express the same concepts and processes used to arrive at a judgement of reference suitability.

It seems even more unlikely that humans learn the reference choice task in a manner analogous to the machine models, from correlation of scene and object characteristics with reference choices of other humans. Humans have the possibility, not present in the machine models used here, of monitoring their own visual search task and selecting references that minimize this. This selection method may be augmented with ontological knowledge of the objects involved (mobility, animacy, salience) that might be acquired from observation of the objects directly, or from other humans.

The apparent (although limited) divergence of different human models may be due to the relative performance of humans in different parts of the target location task. The overall task consists, if the hypothesis model is accepted, of an unguided search for the reference object and subsequently a guided search for the target once the reference has been found. A human that performed worse in the unguided search might be expected to choose a more obvious reference even if it made the guided part of the search more difficult, as this may minimize the overall difficulty. A human with poorer visual search capability overall might choose a more 'costly' reference (involving, for instance, more frequent use of compound locative expressions) if this minimized the visual search task. Of course, background, education, gender, and cultural factors may also play a part.

5.8.4 Extension to compound locative expressions

Various possible algorithms for this can be investigated using the model as described in an iterative fashion to produce an expression containing multiple references. For instance, this could be achieved by conceptually moving the listener within the scene to a point closer to the target object and selecting an appropriate reference object and then making this the new target object, with the listener moved further towards their initial position. Plumert et al. (2001) suggest that in fact the process when giving directions happens in the reverse order. This has the advantage of allowing the listener to start the location process by moving to the region of the most apparent reference before the more detailed parts of the locative expression are uttered, thus saving some time. Both directions of expression formation are illustrated in Figure 5.8 which also highlights some of the issues involved. In particular it can be seen that if the references are being identified in descending order of apparency (that is, starting with the 'finance office' in Figure 5.8), there is no obvious way of stopping the model choosing a long series of references that are all very apparent or unambiguous. The next target is not yet defined when the reference is chosen, so a relevant search space is difficult to define. As shown, some form of cumulative cost could be obtained and used to ensure that reasonable progress was being made towards locating the eventual target.

Working outwards from the target (that is, conceptually moving the listener to a point near the target to define the first reference, the 'computer' in Figure 5.8), is a more obvious process. Targets are always defined, the last reference becoming the target at each iteration of the reference choice model. The tendency to string together too many

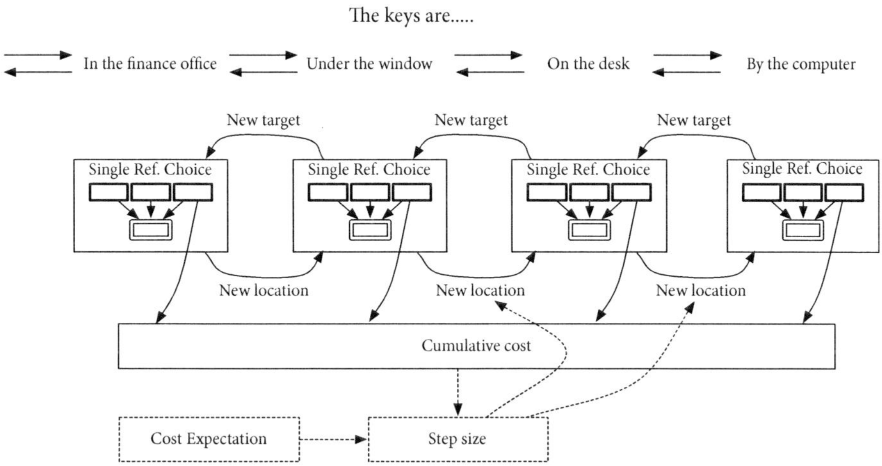

FIGURE 5.8 An extension to the model to generate hierarchical references. The model for selecting single suitable references (described here), could be applied sequentially, conceptually 'moving' the listener closer to, or further from, the target at each stage

references could be countered either by monitoring cumulative cost as with the descending reference case or by conceptually moving the listener 'away' from the target until only one or two references met some pre-determined threshold of suitability.

In many cases there are recognized steps in the process of hierarchical reference formation: buildings are divided into floors and rooms; towns are divided into districts, streets and buildings; this, once learned, simplifies the process to a large degree. On the other hand a complicating factor is that the process often doesn't proceed in a single direction, in particular when disambiguating references, or references that are parts of objects, are involved. Extending the example in Figure 5.8 illustrates this case; the description 'the keys are in Harrison building, in the finance office on the second floor, on the desk under the window, by the computer', which contains several changes of direction, is lengthy but does not sound odd.

5.9 Next steps

The results reported here confirm the work of linguistic commentators such as Miller and Johnson-Laird (1976) and Talmy (2000) in a setting nearer to the real world than previously possible and shed light on the particular attribute representations that are important in reference object selection for locative expressions. A simple machine model has been produced that, though it performs less well than humans in the reference choice task at present, has shown the potential to be extended to the point where it will be usable in a variety of applications such as computer games, training simulators, and navigation systems.

The data set and machine learning platform used in this study open up a wide range of possible further topics for research. Some of these were anticipated, and have support already in the data set, and some, to a greater or lesser extent, will need additions to the data set and further validation exercises.

An obvious extension would be to add a parallel system to learn the preposition for the simple locative phrase. Given that this would initially be a purely locative preposition assignment, it is not thought this will be a difficult task. Validation data have already been collected that will enable meaningful testing of machine learned systems. At this point the addition of some simple syntax will enable a system to produce basic locative expressions from scenes with potential applications in computer games and training simulators.

The ability of a system to produce compound locative phrases using hierarchical references and/or parts of objects or regions associated with objects (for instance 'the church is at the end of the road') would allow for far more realistic locative phrases, but is not a simple matter to achieve. Objects in scenes have all their parts available as named geometric entities but the regions that can be associated with objects and the volumes of space associated with them would need to be the subject of another machine learning exercise. This would probably be a two-step exercise in itself requiring first some translation from the geometric characteristics of an object to the regions that are appropriate and secondly an exercise analogous to learning the volume of space associated with a projective preposition. Hierarchical references can be used either because a single reference does not satisfactorily limit the search area or because an otherwise appropriate reference is ambiguous. In this second case, the addition of a referring expression generator is required. For all of these it would also be necessary to devise a system to make a judgement between the increased communication cost of using a hierarchical reference and the benefits to the listener in terms of search time.

There could be some advantage to linking the system to an ontology in the manner of Lockwood et al. (2006). This would certainly help in classifying objects and helping to learn further object characteristics such as animacy and function (container (for a bowl), support (for a table), for instance). It is not clear, however, that an existing ontology such as Cyc would have the needed information in a manner that would be immediately useful, in particular for object aggregation. For regions associated with objects it might be better, as noted, to learn these from grounded examples.

The use of machine learning techniques to investigate linguistic behaviour in near real-world environments is still new, but it has clear potential. Important psycholinguistic results can be confirmed outside of their necessary experimental settings and combined and extended into more complex models capable of delivering new insights.

6

Spatial prototypes*

ERIC CHOWN

6.1 Introduction

Like much of cognitive science in general (Newell, 1990; Staats, 1991; de Waal, 2002), the literature on spatial cognition is fractured into many subgenres. For example, there are large literatures on cognitive mapping (Golledge, 1999; Kuipers, 2000), high-level spatial representations (Johnson, 1987; Rohrer, 2005; Kuhn, 2007), the role of the hippocampus in spatial processing (O'Keefe and Nadel, 1978; McNaughton et al., 2006) and spatial perception (Farah, 1988; Pylyshyn, 2006; Gregory, 2010). What is still rare, are attempts to unify all of these areas. Many models of cognitive mapping, for example, do not deal with the problems of low-level perception. This separation of perception from higher-level processing is still common in cognitive science and is seen as the correct approach in some corners (Newell, 1990). For Newell and other believers in the **Physical Symbol Hypothesis** (Newell, 1990), perception is merely a means to activate the correct high-level symbols. This debate continues in the world of spatial cognition; for example, there are still debates on whether spatial representations are like 'pictures in the head' or more propositional as would befit the Physical Symbol Hypothesis (Farah, 1988; Thomas, 1999; Pylyshyn, 2006; Gregory, 2010). Representational theories that separate cognition and perception, and especially those founded on abstract symbols, suffer from what Harnad (1990) has termed 'the symbol grounding problem' or what is referred to in philosophy as the problem of 'intentionality' (Brentano, 1874). This relates to the issue of how the symbols in cognition are meaningfully linked to objects or things in the real world.

In addition to those who do not believe that spatial representations are in any meaningful way related to perception, there is another camp that believes that much of the schematic nature of spatial representation arises from how people talk about space. As Talmy (1983) put it in a highly influential work, 'language schematizes

* This work was supported by the National Science Foundation (NSF) under grant No. 1017983. The author gratefully acknowledges that support.

space'. The quote is provocative and variations are still pervasive (Deutscher, 2005). Further, even if one does not agree with the quote, it does raise an interesting question, namely how exactly is space schematized?

This chapter presents a theory of how space is schematized that is in accord with a growing literature on the importance of embodiment in cognition and especially in spatial processing (Johnson, 1987; Thelen, 1995; Lakoff and Johnson, 1999; Rohrer, 2005). The representations presented are what Clark (1989) has termed 'semantically transparent', which is to say the representations arise naturally from perception and how they are used rather than being, for example, artificially transformed into semantically meaningless symbols or propositions. The Denis and Fernandez chapter of this volume contains numerous examples of how this plays out when people give directions, especially using what the authors term 'route imagery instructions' which they define as 'visualizing scenes from an egocentric perspective, and compiling a succession of such views'. As we shall see, this 'succession of views' maps naturally onto the structures described in this chapter.

In more concrete terms, when we perceive space it activates areas in posterior parietal cortex and the hippocampus. When we think about space, the same brain areas are active and generally in the same sequences that they were originally active. A pure symbol system, by contrast, would require an initial transformation from perception to symbols. The advantage of this is that those symbols can then be arbitrarily manipulated as a part of reasoning. There is no reason for such reasoning to activate the same brain structures as are activated during perception, as 'reasoning' is just symbol manipulation and unrelated to perception. Among the potential drawbacks of such an approach is that the result of that reasoning is still just a set of symbols, which in turn need to be transformed back into a form that is meaningful to the body (e.g. the motor system). There is a long line of research on topics ranging from imagination (Perky, 1910) to mental rotation (Shepherd and Metzler, 1971), to neuropsychological evidence (Farah, 1988) that calls such theories into question. In their chapter Denis and Fernandez discuss some of the strengths and weaknesses of various approaches to this issue. The representations in this chapter, called **Spatial Prototypes**, by contrast are a direct product of how the perceptual system processes spatial information.

In this chapter it will be argued that the question of how space is schematized is not one of high-level cognition or language, rather it is a product of the structure of the visual system. Further, the need for schematization can easily be seen on computational grounds. In the first case the visual system provides one natural sort of schematization by splitting visual information into two streams referred to as the *ventral* and the *dorsal* systems (Norman, 2002). While the primary purpose of the ventral system is object recognition, the dorsal system is more oriented to spatial processing and does not contain the fine-grained detail of the ventral system. This reduction of information provides one form of schematization. Landau and Jackendoff

(1993) examined some of the impact of this split on the way that people use space in language. For example, in locational expressions objects are normally described in highly schematic terms even with regard to their spatial properties, whereas when the same objects are described as part of a category, their descriptions are rich with detail, including spatial detail. The idea is that when speaking about objects, people access information from the ventral system, while when speaking about locations they access information coming from the more schematic dorsal system.

Even with the reduction in detail in the dorsal system, there is still too much information to be useful. Visual information is coming into the brain in a continuous stream. Meanwhile storage is not infinite. This argues that there must be some sort of filter, or filters, that takes a continuous stream of spatial information and reduces it to the most useful parts. Such filters should help the cognitive system save the most important parts of perception while ignoring less important elements. Meanwhile such filters must be effective even for infants as they too are learning about space. This suggests that the filters operate automatically and are not a product of cognition. This chapter proposes such a filter based upon simple visual change. The filter itself is also grounded in the information processed in the dorsal system and is strongly correlated to the Gateway construct first proposed in the Plan model of cognitive mapping (Chown et al., 1995). The chapter discusses the implications of such a filter and also presents a model of it implemented on a mobile robot to show the power of the principle the filter is based upon. The Spatial Prototypes discussed are a direct result of the information processed in the dorsal system along with the filter based on visual change.

The nature of the schematization of navigation space has been well studied going back to Lynch's seminal work (1960) on cognitive mapping and to the developmental literature on how children conceive of space (Piaget and Inhelder, 1967; Siegel and White, 1975). A key property shared by virtually every major model of cognitive mapping (Chown, 1999), and many robot navigation systems (Beeson, 2008), for example, is some type of topological network. These networks often model the 'route map' stage of development where navigation is typically done from landmark to landmark. In addition, even models that contain more metric information, as found in the survey map developmental stage, often do so with a network structure (Chown, 1995). At the same time, other work has developed formal approaches to the relationship of topological structure (and other spatial relationships) as it relates to spatial reasoning (Kuhn, 2007; Egenhofer, 2009; Bateman et al., 2010). Taylor and Brunyé (this volume) explore some of the differences between route and survey representations, especially as they relate to perspectives and language.

The model presented in this chapter is also topological and is compatible with the literature on the schematic nature of navigation space. One contribution is that it attempts to connect such models more directly with spatial perception, extending a theory of how such schemas are formed and how they are grounded directly in perception. What is new is that it attempts to directly connect such models to domains that are not inherently spatial and to other important aspects of cognition such as episodic memory and the understanding of causality. The claim is that the cognitive system is constantly creating schematic structures of the same sort it creates during navigation, and that further, the basis for the creation of those structures is directly analogous to the processes that drive navigational schemata. In this way this chapter is a kind of bridge between the large literatures on navigation and work on spatial schemata, especially that of Johnson (1987) and those that have followed him (Rohrer, 2005). In many ways this chapter is of a kind with Rohrer's work (2005) except that, whereas Rohrer focuses on where in the brain image schemata might be created, this chapter focuses more directly on a mechanism and the implications of such a mechanism.

The starting point for the discussion of Spatial Prototypes is the split in the human visual system. The two visual systems are sometimes called the *what* and *where* systems (Ungerleider and Mishkin, 1982), and stem from the *ventral* and *dorsal* portions of the vision system respectively (Norman, 2002). In simple terms the ventral system is the primary engine of object recognition while the dorsal system is more spatial and appears to capture the relationships, whether they are spatial or temporal, between objects (and also relative to the observer).

Since the ventral pathway is the domain of object recognition, the chapter will focus on the dorsal pathway. The first questions addressed will revolve around what the most basic type of spatial representation would be and what function it would serve. The answer to these questions is predicated upon the kind of information that the dorsal system processes. Most of the analysis in the chapter is based upon functional and information processing arguments. For example, a basic functional question is posed: if one were to try and build an efficient representation from the information available in the dorsal system, what would be the simplest, most efficient, way to do it? The argument will be made that using the same principles that carve large-scale space up into smaller regions can also serve to divide event space into individual episodes. This idea was first explored in Chown (1999) with the further extension that many abstract environments, such as a piece of music, are parsed by people using the same principles that they would use to parse large-scale space. When complex environments are parsed this way they become more manageable cognitively, allowing a focus on a small subset of a larger system. There is a growing literature on event segmentation (Avrahami and Kareev, 1994; Zacks et al., 2007; Baldwin et al., 2008; Zacks et al., 2009). The model presented here is of a kind with the EST (Event Segmentation Theory) of Zacks and his colleagues (2007). EST is based primarily on the role of motion in discerning the structure of events. The focus of this chapter is on using motion as cues to build what Zacks calls 'event schema' as a basis for spatial representations.

The individual representations of event episodes, at least the spatial part, are referred to as EVENTS. The EVENT representations serve two purposes. First they form a basis for episodic memories of events, and second they provide the foundation for abstraction into the more general-purpose Spatial Prototype structures developed in this chapter. The processing described is completely automatic and relies only on the nature of what is perceived visually at any given moment, meaning that the structures are fully grounded in perception and do not require any high-level cognition or analysis. As a proof of concept, experiments run on a mobile robot are included to show that a robot can extract exactly the sorts of representations predicted by the theory.

While Spatial Prototypes are described as building blocks, strictly speaking they are not at the bottom of the spatial hierarchy proposed in this chapter. At the bottom is a structure that is more closely akin to the raw information processed by the dorsal system. This structure, called a **SCENE**, is drawn from the Plan (Chown et al., 1995) model of cognitive mapping.

6.2 Representing immediate space—SCENEs

6.2.1 The dorsal system

The analysis of SCENEs requires a brief review of the dorsal system. Norman (2002) has done a thorough analysis of the two halves of the vision system based on an extensive survey of the literature. Among the conclusions that he reaches are that (1) the dorsal system 'transforms visual information into an egocentric framework', (2) 'the brunt of motion analysis occurs in the dorsal system', (3) 'there is evidence for the involvement of the dorsal system in some type of shape or form analysis', and (4) 'the dorsal system has a very limited short-term memory'. In summary, the dorsal system tracks the positions of shapes relative to the observer, and is capable of doing so even in the (brief) absence of the object in the visual field. It is the central thesis of this chapter that just as repeated exposure to an object leads to the development of a prototypical representation of the object, so too will repeated exposure to typical spatial patterns lead to prototypical representations of those patterns born in the dorsal system. Further, the representations stemming from these patterns are available not only as a perceptual aid, but also in general cognition.

The representations stemming from the dorsal system can constitute basic building blocks for higher cognition. These building blocks are most obviously used by recombining them with information processed by the ventral system, but are also available to be used in other ways—as a central piece in numerous cognitive systems, including cognitive maps and language for example. The starting point is a simple egocentric representation born of the visual system. This representation will then be successively refined to provide more and more functionality to the cognitive system in an incremental fashion. Ideally these refinements should mirror evolutionary and developmental paths and describe a kind of hierarchy. The representations at the bottom of the hierarchy are directly perceptually driven and closely tied to the motor system, but at higher levels the representations become increasingly abstract.

6.2.2 SCENEs and Grasping

It is highly unlikely that the capacity for spatial reasoning evolved all at once, rather the development is likely to have occurred one piece at a time where each new piece contributed something new to the overall functionality of the system. The starting point for this analysis is the simple act of grasping. Grasping requires coordination between the visual system, which must identify the object and determine its relative spatial location, and the motor system, which must direct the body to do the actual grasping. Having spatial information organized in an egocentric form helps the motor system in this regard; the dorsal system can guide the motor system to the right location potentially using the motion of the hand in a kind of feedback loop. At this stage a representation, in the cognitive sense, is not necessary. It is likely that grasping can be accomplished in a pure stimulus–response system. However, such a system would necessarily rely on constantly monitoring the object to be grasped. Thus, an obvious functional extension would be to add the ability to grasp an object without constantly looking at it.

Being able to grasp an object when it is not in the visual field implies the use of some type of representation and memory. The simplest type of representation would be an analogue of the raw signal processed by the dorsal system. The dorsal system is already tied to the motor system and it is a small extension to give this nascent representation persistence even when, for example, one's eyes are closed. This addition of a limited form of memory is the first step down a path towards a more durable representation. Since the only real difference between this limited form of representation, which I will call a SCENE, and the raw form of what an earlier version of the dorsal system might have produced is persistence, it is reasonable to conjecture that this representation is also egocentric—the information comes into the dorsal system egocentrically organized and the motor system requires it to be egocentric, so there is no reason to suppose that any transformations would be done at this stage. Indeed the spatial representations described here are generally egocentric and 'semantically transparent' (Clark, 1989) in the sense that there is little or no need to transform them into other formats except at fairly abstract levels. While this does not entail a 'picture in the head' (especially since it does not include any detailed features) it can be effectively treated that way. Further such 'pictures' can be reconstructed by combining the SCENE with detailed stored representations of linked objects just as the cognitive system must combine information from the what and where systems during normal perception. The contents of a SCENE are essentially blobs, what Norman (2002) called 'shapes or forms'. This is shown in Figure 6.1. This representation is embodied in the sense that the locations of the blobs correspond to the eye positions where the objects are centred on the fovea, and of course there are strong connections to the motor system so that the objects might be grasped.

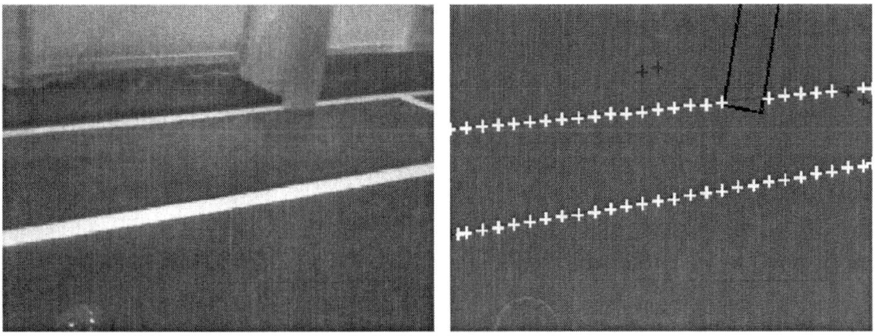

FIGURE 6.1 On the left is a fully detailed picture taken by a robot. The image contains an orange ball, a goal post and some white lines. On the right, the objects relevant to the robot have been extracted into a SCENE. This creates a representation where the details of the objects, excluding their spatial characteristics, are not present

It is at this point that the object system and the spatial system can be put back together. Cognitive organisms have many reasons to grasp objects. Most such reasons imply a desire to hold a particular artefact. This strongly suggests that the spatial system is not working blindly, but rather is engaged in grasping a particular thing that has already been identified. It represents another small step to link the blobs in a SCENE to the representations of the objects, or prototypes, created by the ventral system. It is important to note that the proposed step of merely linking to the object prototypes is small in comparison to directly combining the information processed in the ventral system with that processed in the dorsal system (see Figure 6.2). The difference is subtle but is crucial to everything that follows.

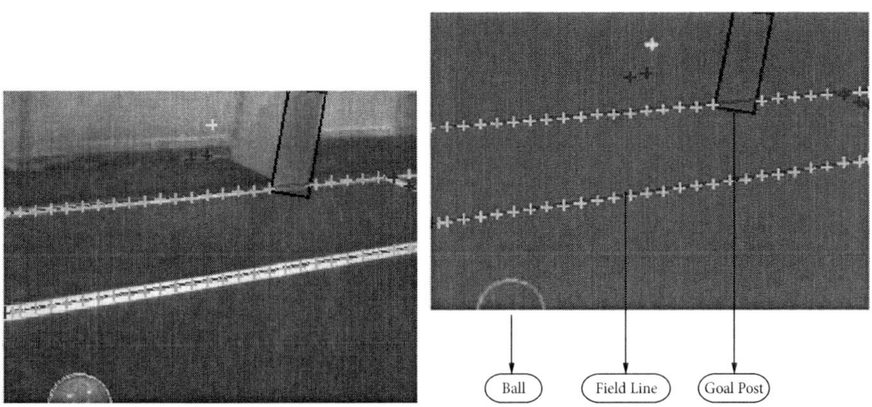

FIGURE 6.2 In the image on the left, the spatial information is combined directly with the detailed sensory information. In the representation on the right, the SCENE structure remains separate, but links to other representations of the individual objects. Those representations in turn may be quite detailed

6.2.3 *Local Maps*

Closing one's eyes is not the only reason why an object might not be in the visual field. It is also reasonable to turn one's head to look at something else. This implies that the currently viewed scene does not contain the information needed for grasping. Continuing the theme of the importance of relationship of the representation to the body, a natural extension to the SCENE representation is to include multiple SCENEs coded by head position. As Shemyakin (1962) noted, there are three basic tools for orientation: the eyes, the head, and the body. Individual SCENEs take advantage of the eyes and this new extension includes the head. The Plan model (Chown et al., 1995), which first proposed such representations, called them Local Maps (Figure 6.3). In a Local Map individual SCENEs are indexed by head positions, and blobs within the scenes are indexed by eye positions (Plan also provides for indexing by body position). The neurobiological basis of such representations comes from 'head direction cells' (Ranck, 1985; Skaggs et al., 1995), which are cells that only fire for a particular orientation of the head. It should be noted that Figure 6.3 only shows the left–right dimension of head movement, but it is highly likely that there

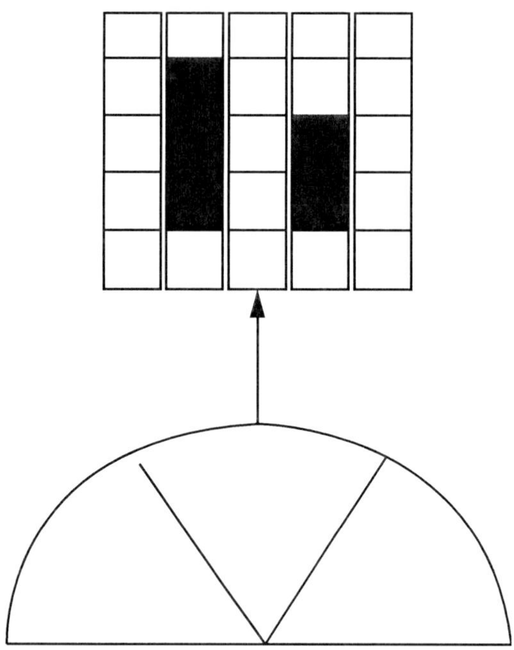

FIGURE 6.3 (Adapted from Chown, et al., 1995). Each 'slice' of the 'pie' corresponds to a SCENE that can be seen at a particular head position. The attached grid is the SCENE that is in view with the head at that position. In this case the attached SCENE corresponds to what can be seen by looking straight ahead. Both components of the representation would almost certainly be more fine-grained than what the figure shows

would also be an up–down component as well. For the sake of simplicity, the rest of the chapter will refer simply to SCENEs, with the understanding that in many cases the SCENEs are further embedded in Local Maps.

In addition to providing the ability to grasp objects that are not seen, such a representation also adds the ability to know *where-to-look* to see objects to bring them into view or to get a more appropriate view. Where-to-look in this case quite literally means where to position one's head and eyes. Such an ability is compatible with perceptual theories influenced by the Gibsonian notion of *affordances* (Gibson, 1970), such as 'perceptual activity theory' (Thomas, 1999), as the where-to-look information can effectively make the perceptual process active and feedback driven without the need for higher-level representations. What is interesting is that while Gibson effectively rejected the idea of mental representations in favour of pure perception, the affordances idea is actually useful in showing the potential power of schematic knowledge structures such as Spatial Prototypes (see Kuhn, 2007 for a formal treatment of this issue and of the power of spatial schemata in perception).

The need for the involvement of representations in perception, or even in cognition has been debated for nearly a century dating back to the Behaviorists (Watson, 1925), then to Gibson, with a more recent revival in the Behavior-Based Robotics approach taken by Rod Brooks (1987, 1991) and his followers (Arkin, 1992). Each of these groups recognized how much information was inherent in the environment and demonstrated that a remarkable amount of intelligence can be displayed without resorting to internal representations and 'thought'. In all of these models the organism or robot is completely sensory dominated—there are no internal processes that we would normally equate with cognition. Given such a limitation, it is remarkable how much of behaviour can be explained by these theories. However, such theories show their shortcomings when important information cannot be directly perceived in the environment, for example, when things are obscured or occluded, or when perceptual conditions are poor, such as during fog or at night. Even a simple representation, such as a SCENE or Local Map, can be helpful under such conditions because the organism can be cognizant of an object even when it is not directly perceived. Turning one's head away from a dangerous animal does not mean that the animal is no longer present.

Functionally, then, the SCENE representation affords more than just the ability to grasp nearby objects. It also lends itself to locating any object that is close, for example, in tracking nearby animals. This ability will, in turn, be refined in the next section.

Starting with the premise that the dorsal system provides the basis for egocentric representations of one's immediate spatial surroundings, another logical refinement is to further assume that such representations can be stored in long-term memory. Since these representations come from the dorsal system, they would only contain basic information on shapes, the relative positions of the shapes, etc. What they would not contain is the detailed information about individual objects processed by the ventral system. This further implies that the activation and retrieval of such structures might be possible with an entirely different set of objects. This is the same principle found in object recognition—we can recognize a new instance of a chair as being a chair even though the details may be different than what we have seen before. The different details in spatial terms are the individual objects in a SCENE. For example, looking at a new scene might trigger the activation of a stored representation, or SCENE, that is far different in terms of the objects present as long as the general spatial configurations of the two scenes are similar. This generic quality is a direct result of not replacing the blobs of the dorsal system directly by the prototypes of the ventral system. A major thesis of this chapter is that just as we build prototypes out of multiple experiences with an object, so too can we build Spatial Prototypes out of multiple experiences with common spatial configurations. As with the prototypes that come from the object recognition system, these Spatial Prototypes are undoubtedly hierarchical. It is likely that at the bottom of the hierarchy would be static relationships like 'close' and 'far', and configurations like 'square' or 'triangular'. As we shall see, many of them will be temporally extended as well.

Returning to the four key aspects of the dorsal system described by Norman (2002), one aspect the SCENE-based structures does not appear to capture is motion. Of course motion presents a set of problems for a memory system, not the least of which stem from the potential storage needs. This is actually similar to the core problem that drove the development of the Gateway construct in Plan—robots using egocentrically organized representations need to limit where such representations are created else they will quickly run out of memory. The Gateway concept grew from the simple realization that egocentrically organized models of space must necessarily be parsimonious with regard to storage. This holds true for the present situation as well. What this suggests is that rather than storing a continuous stream of information, the spatial system is far more likely to store *key frames*, to borrow a term from computer graphics. Thus while the dorsal system need not process static SCENEs, when memories are created from the dorsal system, it is highly likely that they will consist of a series of static SCENEs much like a motion picture consists of a series of static pictures. The next section details how the key frames might be determined.

6.2.4 Storage and retrieval

The previous subsection glossed over the details of how the spatial system is able to retrieve the correct representations. As it happens, the hippocampus appears to be a system that deals specifically with the retrieval (and possibly creation) of spatial memories (O'Keefe and Nadel, 1978; Squire, 1992). In Robotics a robot must constantly resolve what it is experiencing (i.e. its motion and what it senses) against what it knows about the world (generally in the form of an internal map as well as a model of its motion) to determine where exactly it is. This problem is called *localization*. Among the things that make localization difficult is that a robot's perception rarely exactly matches what it has stored. From this point of view, a crucial role of the hippocampus would be to resolve differences between perception and memory, in other words localization. Localization may even have been the original function of the hippocampus as the ability to navigate was crucial to our ancestor's ability to hunt (Laughlin, 1968). However, this ability to mediate between perception and memory is so powerful, and affords so much extra functionality, that it should probably not be a surprise that the hippocampus is implicated in a huge range of cognitive functioning. Accordingly the literature on the hippocampus is sprawling.

There is also a large literature on whether people maintain egocentric or allocentric representations (see for example Shelton and McNamara, 2001; McNamara, 2003). Thus far the representational scheme proposed here is squarely egocentric. However, the hippocampus may well allow one to temporarily behave as though one had an allocentric representation by accessing stored structures that are aligned somewhat differently than one's perception. For example, a robot that is using Gateways to localize might take a sensory snapshot of its current situation. This snapshot may be a good match for a stored snapshot except that the stored one was taken at a location that was slightly different and from a slightly different angle. Robots can rotate and translate these representations in order to match them up. The ability to rotate and translate a representation would be crucial if one of the roles of the hippocampus is localization. This implies then that even if the stored representations are egocentric, but can be rotated and translated by the hippocampus, then they can appear to take on virtually any alignment. In other words an egocentric set of representations could appear to be allocentric. This is in accord with the theory that posits that people normally will favour egocentric representations but will switch to an aligned frame of reference if the environment favours it (McNamara, 2003). Again the hippocampus would be crucial in mediating between the perceptual system and this other frame of reference. And again this fits the Gibsonian notion that the environment often drives perception (and in this case representation). Of course the idea that the hippocampus is then operating on a changed frame of reference means that memories should reflect those changes.

Avraamides et al. discuss these issues from a slightly different perspective in this volume. In their terms, people are sometimes processing spatial information online, such as when they are directly perceiving space, and sometimes offline, such as when they are reasoning about memories. In an online context, for example, it is difficult to make judgements based on perspectives that are not aligned with one's current perspective, but such difficulties are reduced when reasoning offline. The explanation here would be that in an online context, perception dominates the temporary spatial structures used for reasoning in the hippocampus. Changes in orientation are possible, especially when they correspond to previously experienced vantages, but such changes require cognitive effort as they require over-riding the natural imperative to process immediate spatial information egocentrically. By contrast, when spatial structures are retrieved directly from memory, there is nothing constraining the hippocampus from altering the perspective of the retrieved memory—a memory need not be aligned for the motor system, for example, because it will not normally be used to guide immediate actions. Where this theory differs from the work of Avraamides et al. (this volume) is that, while Avraamides and Kelly (2008) proposed that people essentially encode space in two ways—one which maintains online properties that relate self-to-object, and another that encodes objects to each other—the hippocampus-based theory suggests that multiple encoding strategies are not necessary. If memories are encoded from an egocentric perspective, they will tend to default to this perspective when retrieved, but if the situation dictates, the perspective can easily be dropped in favour of an apparently allocentric one. The differences in performance in an offline setting come from the fact that the hippocampus does not need to constantly mediate between two perspectives (what one sees and what one is trying to image) and can simply adopt the best perspective for the task. The viewed perspective must dominate processing normally because of the importance of linking processing with the motor system. A system that can directly link perception and action is ready to instantly respond to its environment and any perceived threats. One that requires representational transformations risks errors and slow responses.

It is also probable that the hippocampus is able to substitute different objects into what is being processed just as the object recognition system is able to cope with recognizing an object even when some of its features change. As will be seen in the next section, this makes both generalizing the structures relatively simple and gives the resultant abstractions tremendous versatility—since the representations are not tied to specific objects they can essentially act as placeholders for any object.

6.3 Representing changing space—spatial prototypes

6.3.1 Change

The foundational representations that we described in Section 6.2 provide the cognitive system with a way to find and grasp objects as well as to track other nearby objects or creatures. The SCENE construct has also been generalized to serve as a major part of a model for wayfinding and navigation in large-scale space (Chown et al., 1995; Chown, 1999). In a cognitive mapping context, stored SCENEs are necessarily static, capturing unchanging features of environments that are used in navigation (Chown and Boots, 2008). The SCENE construct can also be extended for use in dynamic settings.

The theory presented here is based upon the importance of change. To say that the dorsal system is sensitive to motion is also to say that it is sensitive to change. The Gateway construct that fostered the generation of SCENEs in Plan (Chown et al., 1995) is grounded in change as well. In Plan, as in other models of cognitive mapping there are special places in the environment that gain representational importance through either their perceptual qualities, or their use, or both. Other examples include Lynch's (1960) **nodes**, **centroids** in Traveller (Leiser and Zilbershatz, 1989), Couclelis and Golledge's **anchor points** (Couclelis et al., 1987; Golledge, 1987) and to some extent Yeap's (1988) **absolute space representations**. Gateways have a great deal in common with these and other constructs in that they capture the quality that certain locations are vital to the organization of large-scale regions of space. What separates the Gateway construct from the others is that it is grounded in explicit perceptual qualities that make them simple to identify. Indeed it is this quality that has made the Gateway construct attractive to mapping work in mobile robotics where it has been adapted to a variety of models (Kortenkamp, 1993; Chown and Boots, 2008; Beeson et al., 2010). Beeson (2008) summarizes the history of the Gateway construct in robotics and divides the models into those that are sensor-based, those that are model-based, and those that are occlusion-based. The Gateways discussed in this chapter have aspects of both sensor-based and occlusion-based models but are a generalization to domains beyond navigation.

Gateways are places in the environment where there is an occlusion followed by an opening. Gateways are relatively easy to detect because they involve a large amount of visual change or, as with a mobile robot, change in some other sensory modality. They also tend to be important because they are entrances or exits and are therefore visited frequently. These properties make Gateways ideal for the gaze tours discussed by Taylor and Brunyé (this volume). In turn, Chown et al. (1995) describe how Gateways can be abstracted into the basis for survey representations.

These Gateways can then be used to cognitively anchor and differentiate regions of space. Thus Gateways leverage a simple perceptual heuristic to parse large-scale space into smaller regions. The Spatial Prototype presented in this chapter is based upon the same principles, but it is less about change generated by the motion of an observer and more about the motion of the rest of the world. In building a map of the world, it is crucial to find invariants such as landmarks that can be used in localization. On the other hand, in event space it is the things that change that are most important to include in a representation. It might be surprising, therefore, that the same perceptual heuristic used to parse large-scale space can also serve to parse event space.

At any given time there may be change in the visual field. The change is usually due to motion, either by objects or by the observer. Meanwhile the amount of change is not constant. In some cases, as with a Gateway, the level of change can suddenly jump. This change in change—mathematically this would be the second derivative—is an important signal. Imagine a ball that rolls into the visual frame. As it enters the frame there is a great deal of change in the frame. The amount of change then levels out once the ball has fully entered the frame. The amount of change will remain relatively constant unless the ball is acted upon by an agent or another object (e.g. it hits a wall and stops), or until it rolls out of the frame. Either of these events will once again impact the level of change. The Spatial Prototype idea is founded on the theory that a large-scale shift in the level of change is a learning trigger that essentially creates a snapshot of a scene. The idea that motion is central to parsing event space is also central to EST (Zacks et al., 2007; Zacks et al., 2009).

An internal representation of an event would therefore start with a SCENE, and would consist of a series of such SCENEs each of which would be created at moments of great change (or conversely when the magnitude of change suddenly dropped off). For a person such as a young child trying to learn a model of the world, this would serve as an automatic and useful way of building cognitive structure. First, it would be relatively easy to fill in the missing pieces of the representation by interpolating between the individual SCENEs where the change is relatively constant. Second, it is highly likely that such SCENEs will capture the essential causal structure of the event. Consider a ball that has been propelled in a given direction. Once it shifts direction, a new SCENE will be created. This will happen at the moment the ball hits something. As will be shown in subsequent sections, it is then easy to make a causal link between the thing the ball hit and the act of the ball changing direction. The child need know nothing about physics or the structure of the universe because the nature of the representation presents the evidence of the causality.

6.3.2 EVENTs

We start our analysis with the simplest sorts of events that involve motion, namely an object moving across the visual field. As the object enters the visual field, there will be an increase in the amount of change in the visual field. If the observer is stationary, the amount of change will increase until the object is fully in the visual field. This transition from no change to a much higher level of change marks the start of an event. Once the object is in the visual field, if it is moving at a relatively constant velocity, then the change will be more or less constant until it begins to exit the visual field. As the object is going out of view, the amount of change will begin decreasing until the object can no longer be seen. A representation based purely upon this idea is shown in Figure 6.4(a). The shortcoming here is that there is a relatively large gap between the middle scenes. One way to fill this gap is not only to track the amount of change from moment to moment within the visual field, but also to track a second type of change, that is the change between what is currently seen and the last stored SCENE. In such a case new SCENEs will be periodically created as the item moves across the visual field. The resulting representation, called an **EVENT**, is shown in Figure 6.4(b).

To make this more concrete and to further explore how to create such representations, this process has been tested on a robot. Even the best robots could be described

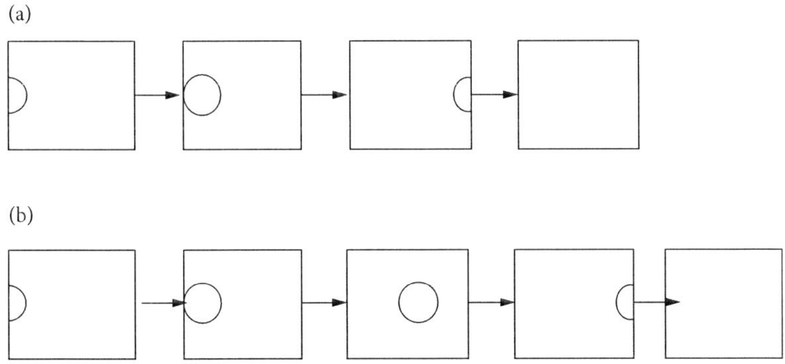

FIGURE 6.4 Each sequence is a possible representation of a ball rolling across the visual field. In (a) new SCENEs are created only at moments when the amount of change shifts suddenly. In (b) new SCENEs are also created when there is enough variation between the current visual state and the previously created scene

as visually impoverished compared to a human being and the robot used for this task has a very poor camera. The robot is the Sony Aibo and the camera is a low resolution (208 × 160) CMOS camera. The problems inherent in the Aibo's cameras make visual processing extremely difficult. The rest of the chapter includes a series of experiments done with the Aibos designed to highlight how Spatial Prototypes are created and to bridge the gap from a theoretical model to one that works in practice.

In the first experiment the robot remains still while a ball rolls across its visual field. In this, and in all of the experiments, the vision system created for Bowdoin College's RoboCup team (Chown et al., 2009) is used. In RoboCup, as in many robot applications, the first step of visual processing consists of converting a raw image into a colour segmented image. Every pixel the robot gets in an image is a three-tuple of numbers between 0 and 255 (e.g. (100, 250, 100)); colour segmentation involves converting those triples into simple colours like *green*. The converted images are then scanned for *blobs* that correspond to the objects that make up the world. These blobs are simple descriptors of possible objects, generally consisting of the object's size, colour, and location in the visual field. The process is described in detail in Chown et al. (2009). The blobs used by RoboCup teams drop visual details in a manner very similar to the dorsal system. This makes such a visual system ideal for experiments on building Spatial Prototypes.

The methodology is very simple. Colour segmentation is performed on the image and the locations of the basic objects are extracted. The creation of new SCENEs is then driven by two criteria: (1) the change in location of objects from one frame to the next, and (2) the absolute location of the objects compared to their locations in the last stored scene. If either of these numbers is greater than some experimentally determined threshold, then a new SCENE is created. It is worth noting that alternative measures of change are still being explored. For example, it is possible

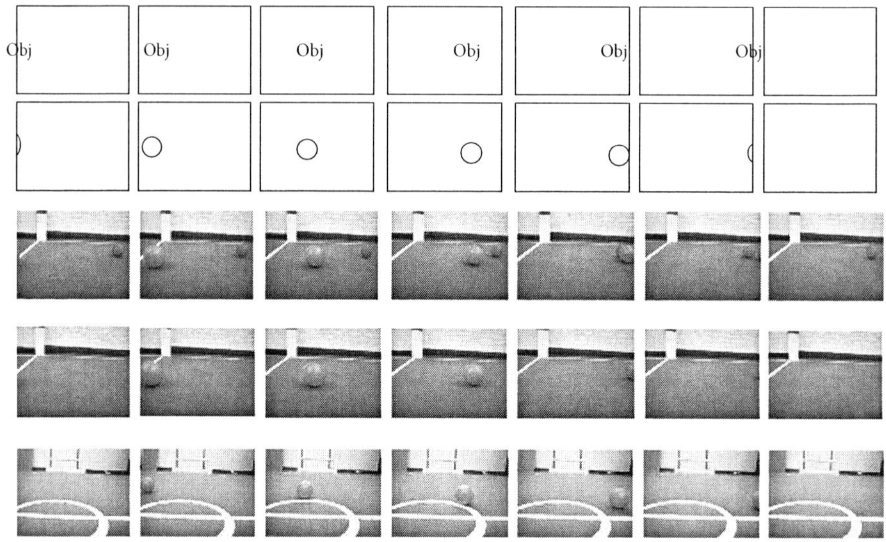

FIGURE 6.5 Each of the bottom three rows of images was extracted by a robot as it watched a ball roll across its visual field. The time course of each event was on the order of one to two seconds or approximately 30 to 60 visual frames. For these runs the robot used both change heuristics (shift in the amount of change, and difference between current frame and last extracted scene). The second row shows the result of intersecting the SCENE-based representations of the three rows. The element common to each row is the ball rolling across. In the top row this is further abstracted to just a slot that can link to any object. The top row represents the Spatial Prototype for the 'move across' EVENT. Here 'Obj' reflects the fact that any object can be instantiated into the corresponding spot

to track change in terms of pixel differences, velocity differences in the objects, and a host of other possibilities. Figure 6.5 shows the results of multiple experiments where a ball is rolled across the visual field of a robot. One of the things that should stand out is how similar the representations are despite differences in background, ball speed (the events range from 1 second up to 2 seconds long) and other factors. The top two rows show two different levels of abstraction of these three events, as will be discussed in the next section.

Given that these representations are supposed to be parsimonious representations of events, it is worth briefly examining just how much compression is done. The EVENTs shown in Figure 6.5 consisted of 30–60 frames of information. Each of these frames contains 33,280 pixels, each of which has one of 16,777,216 values. Reducing the number of frames from roughly 50 per event to an average of about 7 reduces the storage necessary to around 15 per cent of what storing the raw video would require. Of course this only reflects the savings if the raw vision frames were stored, when in fact a greatly reduced representation is actually stored.

The colour segmentation step reduces the possible number of colours to twelve. In turn the information is further reduced to blobs and their locations. The individual SCENEs comprising the EVENTs in Figure 6.5 consist of between one and four blobs depending on their individual content. This makes the effective storage necessary for these events about seven SCENEs of information, each of which contain several object locations in a manner similar to that shown in Figure 6.2. Meanwhile this tiny amount of stored information effectively captures the experience in such a way that a fairly accurate version of the event could be reconstructed.

Of course only a fraction of a person's experience occurs while the person is stock-still observing events passively. Other experiments have been run that borrow footage from Bowdoin's RoboCup team. The footage was taken by a robot as it chased an orange ball (RoboCup involves robots playing soccer and the ball the robot is chasing is the soccer ball). The footage was captured especially for RoboCup and was chosen specifically because it is anything but ideal for experiments such as these. When an Aibo runs at full speed its head bounces up and down and moves back and forth and the movement can heavily blur and distort individual frames. The results of these experiments are promising, showing similar sequences across different pieces of footage.

6.3.3 Abstracting SCENEs—Spatial Prototypes

Figure 6.5 shows how these representations might be abstracted and generalized. The result of the abstraction is a **Spatial Prototype.** Just as an object prototype strips away individual variation in detail so too must a Spatial Prototype. Whereas the representations of individual EVENTs are surely linked to specific object representations so they can form episodic memories, in an abstracted representation such specific links will not exist. The features of a Spatial Prototype can be abstract objects in general locations. This abstraction is reminiscent of *slot and filler* representations once common in artificial intelligence, giving the Spatial Prototypes a generic quality that makes them applicable regardless of the individual objects that make up the SCENEs. Together the SCENEs, the EVENTs, and the Spatial Prototypes form a hierarchy as seen in Figure 6.6. Note that at the first level up from raw perception, the representations are relatively picture-like in that they contain the information processed in the dorsal system, but at the Spatial Prototype level they are of the same basic form, but are not pictorial in any sense. If this is true, then it might be the case that representations which are not pictorial may still use the same brain structures as are used in perception, as Farah (1988) has suggested might indeed be the case.

Moments of especially large visual change will mark the beginning or end of an individual EVENT. Meanwhile new EVENTs may activate a previously stored Spatial Prototype (or Prototypes) depending on how similar they are. These existing

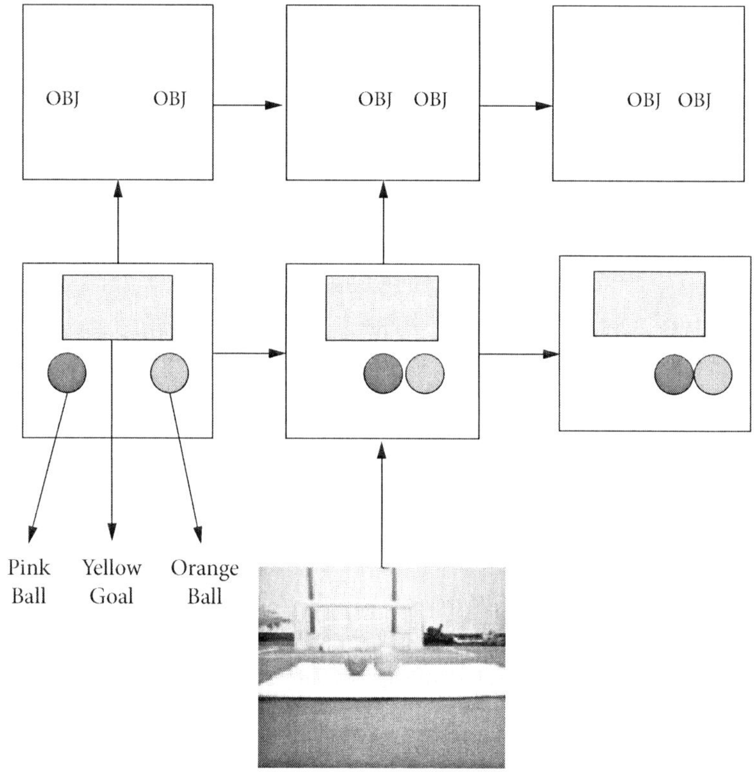

FIGURE 6.6 The bottom of the spatial hierarchy. At the lowest level is what the observer actually sees. At the next level up the dorsal system processes an abstracted SCENE minus many of the details. In turn, each SCENE is linked as part of a larger EVENT structure that serves as the episodic memory of the event. Links to the objects are only shown for the first SCENE for clarity. The top level is the Spatial Prototype layer. Here the representations are very abstract – in this case representing two objects colliding and moving off together. Notice that in this case the goal was abstracted away too since other similar collision representations would not contain one. Just as we lose details in our prototypical objects, so too can we in our prototypical spatial representations

Spatial Prototypes, when activated, can be used in a predictive fashion to fill-in possible outcomes for the EVENT. Furthermore, the Spatial Prototypes are perceptually grounded representations of space of the sort that are perfect for use in language or other tasks.

6.4 Spatial prototypes in cognition

In the Plan theory of cognitive mapping, Gateways were used to parse large-scale space into smaller regions. The cognitive benefits of this stem from the corresponding reduction in information processing required in navigation. The SCENE and EVENT structure proposed in the previous section has similar benefits as they help carve a continuous event space up into smaller, discrete chunks. It is worth noting that event space is not always inherently spatial. In this case the cognitive system may simply be reusing a mechanism beneficial in one domain, namely navigation, in a more general purpose way. As Chown (1999) has noted, this results in people describing all manner of domains in highly spatial terms. The storage benefits of a system that breaks a continuous stream of information into discrete chunks have already been discussed. In this section some added benefits are described. Primary among these is that the structures created provide natural insight into causality.

6.4.1 Complex events and causality

The nature of the relationship between events of a purely spatial character and the perception of causality were studied in a series of experiments by Michotte in the 1940s and 1950s (Michotte, 1963) and are still debated today (see, e.g., White, 2006). Michotte presented two basic varieties of experiment. In each of the base experiments observers watched a screen that had two different coloured squares, one in the centre, and one to the left. During the experiments, the square to the left would move towards the centre square and eventually 'collide' with it. In the first type of experiment the result of the 'collision' was that the first square would stop and the second would begin to move off. Of course since these were just images projected onto a screen, there was no physical collision, nor any reason to attribute any normal laws of motion or physics to the events. Nevertheless, observers would report seeing the first object *launch* the second object and hence Michotte called this the **Launching Effect**. In the second experimental type, the first square would keep moving when it 'collided' with the second square, but afterwards the two would move together as one. In this case observers would have the impression that the first square carried the second along with it. Michotte called this the **Entraining Effect**. Michotte's conclusions, based upon running countless variations of these experiments under a wide variety of conditions, were that the perception of causality is dependent on a small set of spatial and temporal features stemming from the nature of how space and time are organized cognitively.

Michotte distilled his results into what he called the 'ampliation of the movement'. He broke this down into four basic points:

1) The 'ampliation' involves the appearance of a new event (e.g. the second square starting to move). This event is characterized by change—either spatial displacement or a change in the shape of the object.
2) The new event appears to be a continuation of the existing event.
3) The new event does not involve the disappearance of the previous event. In the experiments, for example, the original movement always persists in some form.

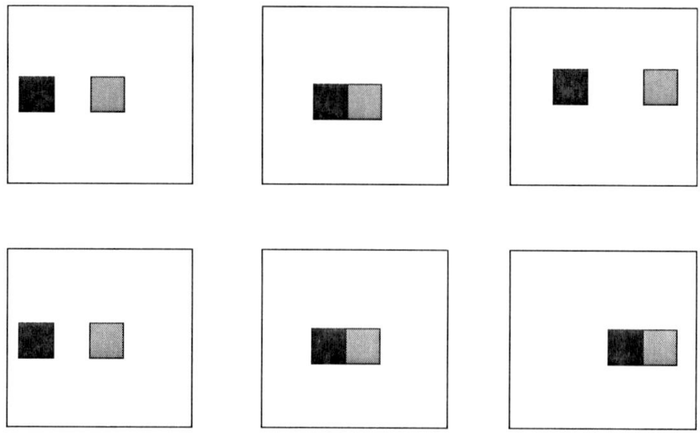

FIGURE 6.7 The top row shows the three key moments of Michotte's Launching experiments. The bottom row shows the three key moments of his Entraining experiments

4) At the moment when the ampliation occurs, there are essentially two events occurring simultaneously.

As Michotte put it 'all of these points can be summed up in one short sentence: the original process develops, and, without ceasing to be what it was before, "becomes also" something else, distinct from itself' (1963: 222).

Michotte's conclusions are virtually in total agreement with the Spatial Prototype theory presented in this chapter. The key factor driving the creation of these EVENTs is change. Each of Michotte's experiments has a few key moments with regard to the change heuristic: (1) the moment when the first square starts moving, (2) the moment right after the two squares meet when the second square starts moving (Michotte's **ampliation**), and (3) the moment when the movement stops. Figure 6.7 shows those moments for each experiment.

There are only a few possible causes for moving from SCENE to SCENE within an EVENT structure: (1) an independent agent (in Michotte's terms, an animal) began moving, (2) an agent affected another object, and (3) the continued movement of an object was great enough to entail the addition of another scene to the structure. Michotte points out that the movement of animals is easily differentiated from other sorts of movement, thus making the first cause simple to detect. Our ancestors did not have machines to contend with, so mechanical movement would not have been a factor. Michotte does point out that there are a few sorts of movements in nature that are initiated by 'invisible' factors such as the wind or gravity. In any case EVENTs will generally be initiated by the new movement of an object. Meanwhile, changes in the object's movements will cause new SCENEs to be created within the EVENT (as well as extended examples of continuous movement). Michotte's concern, of course, is when objects affect each other since that is the case that involves causality. The causality comes in from the contiguity of the originally moving object and the newly moving object.

In a sense, the change heuristic is a way of breaking the world up at its causal joints in much the same way that Gateways break large-scale space up into individual regions. The 'regions' in this case are discrete EVENTs usually initiated by the movement of a person or animal. In some cases this movement causes change in another object and this is where issues of causality arise.

The Michotte experiments can be run with robots as well. Typical results are shown in Figures 6.8 and 6.9. Note that even though the experiments are not nearly as clean in terms of the behaviours of the balls, the real world being far messier than theory, the results are strikingly similar to what is predicted.

One variation that Michotte introduces is to include an interval of a fifth of a second between when the first square hits the second and the subsequent movement. Observers in such a case then don't make the causal link between the first square's movement and the subsequent movement after it 'strikes' the second square. The subsequent movement is seen as independent by observers. From the point of view of the change heuristic, this would be the case as well. We would still have the initial SCENE, and a SCENE of the two squares coming together. At that point, however, the change in the frame would go to zero and the EVENT would be complete. With the subsequent movement a new EVENT would be created. The result is two separate EVENTs, which is exactly how Michotte's subjects saw it. As Michotte points out, this is striking because, even though the events are virtually identical, they are perceived as

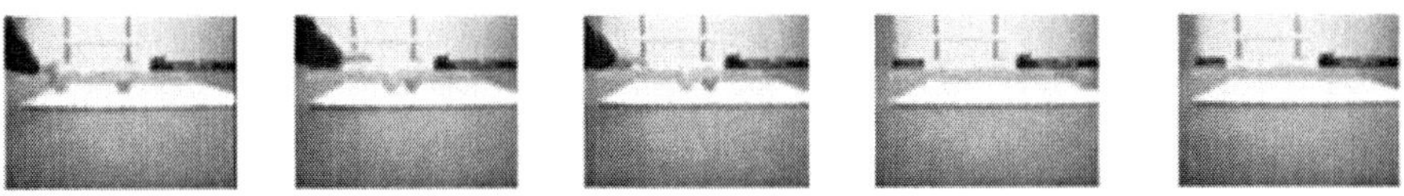

FIGURE 6.8 These five frames were extracted by a robot to serve as its representation when it viewed one of Michotte's Entraining experiments. Note that the second and third frames are fairly similar and both take place near the time of the collision. In practice collisions in the real world are not as clean as theoretical or simulated collisions

FIGURE 6.9 These four frames are the key frames selected by a robot when it viewed one of Michotte's Launching experiments. In this case the orange ball did not have enough momentum to roll completely out of the frame

being radically different. This too is exactly in line with what the Spatial Prototype theory would predict—that event space is broken up according to a measure of change and that when change goes to zero it marks the end of an event.

6.4.2 *Networks of SCENEs*

Virtually every model of human navigation includes at least one type of network structure (Chown, 1999). The network structure is crucial to wayfinding because it provides a mechanism for extracting plans. Traversing a sequence of nodes in a network roughly corresponds to taking a journey along a particular path whether the nodes correspond to landmarks or to Gateways. And, since paths often cross over to other paths, the whole thing occurs in a network. The same holds true at the level of spatial events. As was shown earlier, the individual SCENEs are generic and can stand in for a variety of spatial configurations. This is the crucial property necessary for them to occur in networks.

Having the SCENEs structured into sequences allows them to be used predictively. When the initial SCENEs of a Spatial Prototype are activated, the rest of the Spatial Prototype can be anticipated. The network structure adds an additional level of functionality. Not only can learned sequences be anticipated, the sequences that form out of combinations of learned sequences can be extracted as well. In wayfinding this would be akin to determining the existence of alternate routes to a destination without ever having explicitly taken them.

A simple example of a network structure can be found in Michotte's two base experiments. They start out exactly the same, but bifurcate based on the result of the collision. A network version of the two experiments is shown in Figure 6.10.

Once the EVENTs and SCENEs have been abstracted to the point that they can be networked as Spatial Prototypes they are available for use in general cognition, such as in language.

6.4.3 *Language*

Spatial Prototype theory makes several strong claims that are relevant to work on language. First, according to this theory the schematization of space is not a product of language but comes from several sources. First, the visual system provides an initial type of schematization as it separates visual information into two separate components. This strips detail away from initial spatial representations. Second, the visual stream is broken into discrete pieces according to the change principle. This discretization occurs at both the level of individual SCENEs and the separation of groups of SCENEs into individual EVENTs. Finally these SCENEs and EVENTs are abstracted across many experiences. Such abstractions will lose extraneous details of the individual EVENTs forming prototypical cores of common elements giving rise to the Spatial Prototype name.

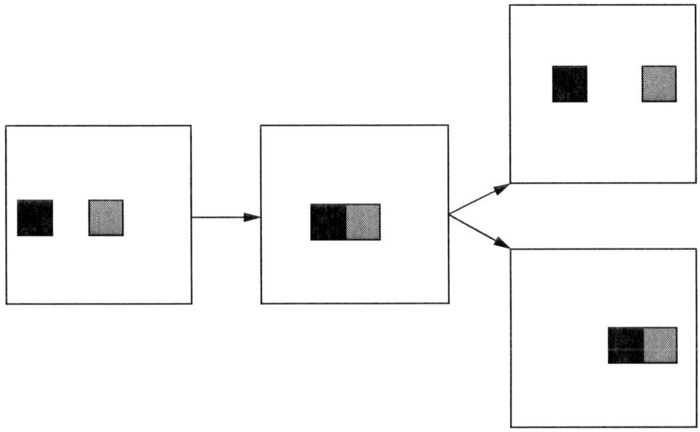

FIGURE 6.10 A networked representation arising from viewing both Michotte's Launching experiment and his Entraining experiment

It follows, therefore, that the relationship of space and language is more about intentionality and grounding than about schematization. These problems are fundamental to language, as language requires representations that are meaningfully connected to things in the world. At its most basic, the theory presented in this chapter is about how that connection is made. Spatial concepts such as 'up' need not be innate as they will arise naturally from perception. This is where the abstraction component of Spatial Prototype theory is so important. Even though every individual has a completely different set of experiences, the abstraction inherent in their representational system will naturally mean they will all have a similar set of basic spatial representations. These common structures can be used to help form the basis of meaningful communication.

It is worth noting that the work in this chapter owes a clear debt to the work on spatial metaphors and image schemata done by Johnson (1987) and Lakoff (1987). The representations proposed here have a great deal in common with the image schemata proposed by Johnson (1987). However, there are crucial differences. The most obvious is that Johnson has no concrete representational underpinning, nor a theory of how image schemata are grounded in perception. Further, Johnson proposes fully dynamic structures whereas the Spatial Prototypes proposed here capture dynamics, but are not truly dynamic. This is not to say that the structures cannot behave as though they were dynamic, as the motions between individual SCENEs can easily be extrapolated.

There is another important relationship between language and space—language is another tool that can be used to acquire spatial information. This is an important issue addressed at length by others in this volume (Avraamides et al.; Barclay and Galton; Denis and Fernandez; Kordjamshidi et al.; Taylor and Brunyé). The chapter by Kordjamshidi and colleagues is particularly relevant because it deals explicitly with mapping natural language on to qualitative spatial representations (QSRs). It is here that the generic nature of spatial cognitive structure shows its value. If people are equipped with a variety of general purpose spatial structures abstracted from many real experiences, it should be possible to use these structures as a framework to quickly build a new structure on the fly. Research on the use of such generic structures goes back to the early days of artificial intelligence, notably including Minsky's Frames (1975), and Schank and Abelson's Scripts (1977) which are discussed in the next subsection. Spatial Prototypes share some key characteristics with such theories, but differ greatly in their relationship to perception.

In this volume Avraamides et al. discuss some of the problems and characteristics of spatial representations constructed from language. They note, for example, 'converging evidence of functional equivalence' of memories encoded by vision and by language. If indeed language makes use of spatial structures already in place then this is exactly what one would expect. Meanwhile, Kordjamshidi et al. (this volume) point out that mapping language to spatial representations rather than, for example, having special purpose spatial linguistic models, 'may also lead to more efficient and effective reasoning strategies'.

6.4.4 General reasoning

Minsky (1975) created Frames as a means of coping with stereotypical situations. Frames contained default object specifications for use in general reasoning, but could be instantiated with other objects, which is why they are called 'slot and filler' representations. In addition, Frames were linked to other Frames, though usually only in a hierarchical fashion. In describing the motivation for developing Frames, Minsky wrote: 'It seems to me that the ingredients of most theories both in artificial intelligence and in psychology have been on the whole too minute, local and unstructured to account—either practically or phenomenologically—for the effectiveness of common sense thought.'

While Frames, Scripts, and other similar representations such as Semantic Nets gained great popularity in artificial intelligence (AI), examining the most popular current AI textbook (Russell and Norvig, 2003) shows that work in this area focuses almost purely on logic and statistics. AI researchers still recognize the need for a 'general-purpose ontology' capable of handling any domain (Russell and Norvig, 2003), but their efforts remain focused on symbol manipulation methods. Meanwhile there is a growing recognition in cognitive science that the fact that the human mind is embodied has a huge influence on how cognition works and how it evolved (Lakoff and Johnson, 1999).

The difference in approaches can consist of examining the differences between Frames and Spatial Prototypes. Among the important differences is that there is no underlying theory of how Frames are created and learned, nor is there any spatial component as Frames are firmly object-based. These differences also apply to Scripts, another slot and filler representation that focuses mainly on actions. The differences are reflective of the longtime disinterest in perception, and for that matter spatial processing, common in AI. It is likely that it was this very disconnect from perception that ultimately limited the usefulness of these data structures.

Nevertheless, the generic qualities that Minsky, Schank, and others were seeking to build into their representations are what gives the human spatial system such power. Humans use their spatial system in virtually all domains, including many that do not naturally seem to have a spatial component (Chown, 1999). Spatial Prototypes have the generic qualities necessary to make this happen. They are grounded in perception, as opposed to AI representations such as Frames which suffer from what Harnad (1990) has termed 'the symbol grounding problem'. Meanwhile, Spatial Prototypes can link to objects and thus can serve as general structures into which objects may be placed. And they are pattern recognizers. Whereas AI research has focused on computation, it has been noted that the speed limitations of the brain suggest that 'instead of computing answers to problems the neocortex uses stored memories to solve problems and produce behavior' (Hawkins and Blakeslee, 2004).

As a simple example of how Spatial Prototypes can be used in a general-purpose way, consider the general memory technique often recommended in learning lists. Subjects are instructed to imagine a familiar place and populate it with the items in the list. This can be done in the Spatial Prototype framework because of this quality of being able to link to any objects. A person can call up the Spatial Prototype corresponding to the specific place. The Spatial Prototype has the details stripped away, but has slots available for new objects. The hippocampus serves both to bring in the correct Spatial Prototype and to link it to the appropriate new objects.

6.4.5 Summary and future work

The Spatial Prototype implementation on a mobile robot shows the strength of the basic principle of using change as an automatic method of parsing event space. It requires no specialized knowledge or reasoning and can serve as a useful heuristic for causal reasoning. Further, the representations created when observing an event serve as a kind of episodic memory that can be abstracted to serve more general functions.

Ultimately this chapter is about the simple idea that tracking the amount of change can serve as the key mechanism for parsing environments and creating generic spatial structures. This principle applies to any environment, including abstract ones, and need not be limited to visual domains. The heuristic is similar to the EST theory developed by Zacks and colleagues (2007). The contribution here is that Spatial Prototypes address what Zacks et al. (2007) called a 'difficult and understudied' problem—schema acquisition. It is important to stress that what has been tested so far is the basic principle and that there are still many questions left to explore. Among these are critical issues such as how to measure change. For example, with a robot change could be measured on a pixel-by-pixel basis from frame to frame. Or it could be measured as an object's relative displacement over time. Experimental evidence suggests the latter is a much better measure than the former as it is less subject to noise. Meanwhile there is the question of how much change is sufficient to (a) drive the creation of a new SCENE and (b) mark the start of a new episode. In Section 6.3.1 it was noted that essentially the change heuristic relies on the second derivative of the visual field, or the change in change. This suggests that the answer to these questions relies on the ability to recognize local minima and maxima in that function. The experiments presented in this paper are part of a long-term research project to explore these ideas. The goal is to start with simple controlled experiments where the observer is stationary as in the Michotte experiments. Over time, as these experiments are better understood, the idea will be to add normal motion as in the ball capture examples in Section 6.3.2. The difficulty in such experiments will be in separating out the change in the visual scene due to the motion of the observer from the change due to the motion of other objects. This is something the human spatial system can do seemingly effortlessly and the mechanisms are the subject of ongoing research (Wolbers et al., 2008). On mobile robots it is typically done with some type of Kalman Filter (Wang et al., 2007) where the robot has a model of its own motion and uses that model to differentiate the relative motion of other objects, or by using techniques relating to *optical flow* (Beauchemin and Barron, 1995).

The model takes strong positions on a number of fundamental questions in spatial cognition including:

1) How is space schematized? A description can be found in Section 6.4.3.
2) Are spatial representations like 'pictures in the head'? In this case the representations share many qualities with pictorial representations, including how they are processed, but they are too schematic to be truly pictorial. Nevertheless, they are clearly not propositional.
3) Are internal spatial representations organized egocentrically? The answer in this case is 'yes'.
4) How are spatial representations perceptually grounded? In this case the change heuristic drives the creation of spatial 'snapshots' that capture the state of the dorsal system at key moments in time.

The proposed theory is not meant to be a comprehensive description of how all spatial structures are formed. Rather this theory represents something of a proof of concept and a starting point for further exploration. Instead of viewing perception as a separate problem from representation, the crucial idea here is that cognitive structure surely is inextricably linked to the structures and information in perception. Indeed rather than trying to figure out how spatial structure is represented starting with a blank piece of paper, it is a much simpler problem to start with a set of constraints.

From this point of view there is a lot of information that can be brought to bear. This starts with an increasing understanding of the kind of information processed in the dorsal system. It continues with functional arguments—as we abstract further and further away from raw perception we should continue to incrementally increase the functionality of the spatial system. Meanwhile these abstractions should be available for use throughout the cognitive system. Finally, a great deal is known about spatial structures and reasoning at fairly high levels of cognition. This knowledge comes from work on image schemata as well as cognitive mapping theory. The emphasis on this chapter is on the formation and abstraction of these structures, it does not contain a theory of how the cognitive system can manipulate them with language. Such a theory is an important future step and will surely build on much of the work (Barclay and Galton; Denis and Fernandez; Taylor and Brunyé) in this volume.

The proposed theory is based upon very simple principles, but has great potential explanatory power in helping to understand how people use space. Among the uses of the representations proposed here are: memory, prediction, as a grounded basis for language, and as a tool for understanding causality. What they are not, is complete. The change heuristic is only one possible way of creating spatial representations. Others may exist as well. Further, the details of what SCENEs actually contain and how they are stored still needs to be worked out. Finally, while much of the abstraction in Spatial Prototypes falls naturally out of the representation, details remain—namely how the process of creating, for example, a single Spatial Prototype across many experiences works; of course this is essentially the same problem still faced in the object recognition world after decades of research. This chapter is more about the power of simple principles, such as the change heuristics, and what sort of representations it might lead to. Despite this, the model is detailed enough that significant pieces have been successfully implemented on mobile robots.

6.5 Concluding remarks

The spatial system is one of the things that helps make human cognition so versatile. While the development of the spatial system may have stemmed from purely spatial needs such as grasping, tracking, and navigation, the structures involved have proven to be so versatile that they are used nearly ubiquitously in cognition. Indeed one of the difficulties of writing a chapter like this is that spatial processing is so pervasive that any treatment of the topic is going to necessarily leave out more than it includes. Nearly any domain, from computer software to pieces of music, can be treated as an environment to be navigated. Many questions remain about how this process works and how the cognitive structures necessary for such reasoning are generated.

The Gateway construct has proven useful in understanding some of the spatial structures used in navigation and this chapter further explores its benefits for other cognitive tasks such as understanding causality and providing structures for use in language that are fully grounded in perception. Spatial Prototypes, and the SCENE and EVENT representations underlying them, can be learned automatically and have a generic quality that makes them ideal for general cognition. The world is not random; there are patterns that occur over and over in a variety of environments and the human spatial system has proven itself more than capable of exploiting this fact. The theory presented here represents one step toward a better understanding of how this works.

7

Learning to interpret spatial natural language in terms of qualitative spatial relations*

PARISA KORDJAMSHIDI, JOANA HOIS,
MARTIJN VAN OTTERLO,
AND MARIE-FRANCINE MOENS

7.1 Introduction

The interpretation of natural language is essential in areas such as artificial intelligence (Roy and Reiter, 2005), computational linguistics (Kruijff et al., 2007), human–computer interaction (Nguyen and Wachsmuth, this volume), the semantic web (Fonseca and Rodriguez, 2007), and geographic information systems (Mani et al., 2008). More recently, the interplay between language and vision is investigated in depth, as, for instance, the contributions by Avraamides et al., Barclay and Galton, Bhatt et al., Chown, and Nguyen and Wachsmuth (this volume) show. In all these fields, different formalisms have been developed that tackle the problem of interpreting natural language expressions in terms of their situation-based or contextualized meaning.

In this chapter, we address a specific aspect of spatial language interpretation as we focus on interpreting (English) spatial language by mapping its situational meaning to qualitative spatial representations (QSR). Our approach is motivated by the potential cognitive adequacy of its components as well as by technical benefits of the component distinctions. The diversity and variety of spatial language usage (for more detailed examples see Denis and Fernandez; Taylor and Brunyé; Seifert and Tenbrink, this volume) requires a flexible mechanism that allows a situation-based interpretation (Bateman, 2010a) resulting in a combination of different spatial components. Here, we concentrate on one component for representing spatial language in an adequate way and another component for representing qualitative spatial information, and we map the first to the second (Kordjamshidi et al., 2011b).

* This research was funded by the DBOF/08/043 grant from KU Leuven and the TERENCE project (EU FP7- 257410).

In our approach, we investigate the use of machine-learning techniques for mapping spatial language to specific spatial qualitative representations. An intuitive argument for applying learning techniques is that it is a difficult task to formulate accurate models and rules that capture the entire knowledge necessary to extract spatial information from language. Humans, however, are able to easily understand and react to spatial language, and consequently, a suitable solution is the generation of many example sentences together with their spatial meanings by humans (i.e. annotated data), and to use these data for learning an automatic mapping from linguistic sentences to specific spatial semantics automatically. This learning procedure can exploit any set of linguistic features that can be extracted from the sentences (e.g. dependency structure or parse tree) or from external resources, and—based on appropriately selected features—generalize to unseen sentences. The novel combination of techniques as put forward in this chapter aims at unifying the advantages from two sides, namely spatial information available from qualitative spatial representations as well as corpus-based information available from linguistic data.

Although the mapping from spatial language to QSR covers only parts of the entire spatial language interpretation process and spatial language is more flexible in its descriptions and expressions than QSR, this mapping is one central task in spatial language interpretation as spatial language primarily describes spatial information in qualitative terms. QSR may also lead to more efficient and effective reasoning strategies, that is, the mapping from spatial language to a qualitative formal model can allow spatial reasoning with this model, whereas spatial reasoning with spatial language itself is hardly feasible (Bateman et al., 2010). Furthermore, QSR is often used in diverse domains and applications for spatial information (e.g. Bhatt et al., Chown, Seifert and Tenbrink, this volume), and our approach aims to provide the link between spatial natural language and the qualitative representations used within those tools and systems.

In previous work (Kordjamshidi et al., 2010a), we advocate a mapping from natural language to multiple spatial calculi models. In their recent publication, Mani and Pustejovsky (2012) consider the interpretation of motion verbs in terms of QSR, based on extensive linguistic analysis of motion verbs and their semantics. Our approach complements their approach from a machine-learning perspective. By recognizing spatial semantics automatically, a number of more fine-grained semantic modelling complications can be ignored to begin with, with a view to further specification in later steps.

In contrast to our mapping approach, some spatial models exist that define spatial relations on the basis of linguistic terms (e.g. Kurata and Shi, 2008). Such models, however, typically reflect not spatial language semantics but specific meanings that are determined by the model's axioms. The diverse use of language requires yet more complex and flexible logical approaches (Bateman, 2010a). These approaches have to take into account contributing aspects for spatial language interpretation, such as context or world knowledge, in order to achieve a mapping from spatial language to qualitative representations (Hois and Kutz, 2008b; Kordjamshidi et al., 2010b).

Our spatial language interpretation approach automatizes the mapping from spatial language to QSR by using machine-learning techniques and a holistic approach for describing spatial language. For this purpose, we distinguish two main levels (Bateman, 2010b):

1) a *linguistic level*, in which natural language is analysed and parts of a sentence, i.e., linguistic features, are syntactically and semantically categorized as having different spatial roles that convey certain spatial information, and
2) a *formal level*, in which the linguistic features are mapped to specific qualitative spatial representations, i.e., formal spatial calculi.

For example, in the sentence 'The book is on the table', the first step is the identification of spatial linguistic features, for example, a spatial relation 'on' that holds between the 'book' and the 'table', and the second step is the mapping to an adequate spatial calculus, for example, a connection calculus that formalizes an 'on'-relation. Both levels use symbolic spatial information for representing the linguistic features and qualitative models, and both levels use machine learning for extracting the linguistic features and mapping them to qualitative models. This allows us to formally structure and present spatial information based on available linguistic characteristics, and to deal with aspects of ambiguity and uncertainty caused by linguistic underspecification, granularity differences, context dependencies, metaphors, etc. This is particularly important as we apply our approach to arbitrary linguistic sources, which is in contrast to using only a limited amount of spatial language or phrases of which it is known that they contain spatial information (Li et al., 2007; Kelleher and Costello, 2009; Kollar et al., 2010).

In essence, the novelty of the presented work lies in the use of machine learning as a methodology for interpreting spatial language in terms of qualitative representations. We analyse three main aspects of this process: first, the performance of automatically mapping linguistic data to linguistic spatial features and to spatial semantics as well as to modalities of a linguistically motivated spatial ontology (GUM-Space); second, the feasibility of learning the mapping of linguistic spatial semantics to a formal spatial calculus, using the Region-Connection Calculus (RCC) as an example for topological information; and third, the interaction between different linguistic features and their contributing meaning during the overall process.

In the next section, we present specific aspects of spatial language that are necessary for understanding spatial language in terms of qualitative spatial calculi. Subsequently, we theoretically introduce our spatial language interpretation approach by presenting the different components of the mapping process from spatial language to spatial calculi. Here, we primarily employ a region-based calculus for topological information, though other types of spatial relations are integrated into our approach as well. Thereafter, we present experimental results and machine-learning outcomes for mapping spatial language to RCC. Here, we also analyse those linguistic features with highest impact, address corpora acquisition and issues, and discuss domain-specific experimental effects. Finally, we conclude with discussions and future work.

7.2 Learning to understand spatial language

Natural language can be used to communicate spatial information, for example spatial relations between objects and their relative location or change in space. For instance, the sentence 'Give me the grey book on the big table' contains information about a spatial configuration of two objects '(book, table)' and an anticipated movement of the book. Different grammatical and semantic linguistic features play important roles in forming the meaning of spatial configurations.

A core task in spatial language understanding is to identify words or phrases with labels that express their *spatial role* in a sentence. For example, in the sentence 'The grey book is on the big table', the preposition *on* indicates a *spatial relation* between *the grey book* and *the big table*. However, not every preposition has a spatial meaning or only one spatial meaning, as lexical entries can generally have multiple meanings and multiple (spatial) roles. For instance, the preposition 'on' in the example above has a spatial meaning but the preposition 'on' in 'I can count on him' has not. The following example (from Bateman et al., 2010) exemplifies such subtle notions.

(1) *He left the institute an hour ago.*

(2) *He left the institute a year ago.*

The first sentence describes a spatial relocation, in which a person leaves a building and goes somewhere else, that is, the person is no longer in the building. The second sentence describes a situation of fundamental change beyond an exclusively spatial meaning, in which the person has apparently quit his job. Another set of examples is:

(3) *The computer is on the table and the mouse is to the left of it.*

(4) *The party could be considered at the far left of the political spectrum.*

Sentence (3) expresses two explicit physical relations about objects on a table. Sentence (4) uses a similar relation 'at the far left of' although with a more conceptual meaning. It is possible to put the party on the left side of it only when explicitly drawing this 'political spectrum' on a piece of paper. Thus, similar lexical items can give rise to very different spatial meanings, whereas two different descriptions can have similar or even identical semantics:

(5) *Looking over his right shoulder he saw his dog sitting quietly.*

(6) *The dog sat quietly on the floor to his right.*

Another important aspect is that spatial information in sentences (1) and (2) is mainly conveyed by verbs, and in (3)–(6) by prepositions. Furthermore, some types of information are not explicitly given by the words but can be inferred, for example, that the dog is back-right of the person in sentence (5). Other linguistic efforts may be needed for such (spatial) inferences, for example the resolution of anaphora.

Despite the many variations in spatial information of natural language expressions, a useful abstraction is that a sentence is able to express spatial relations among objects. For example, sentence (6) expresses a relation toTheRightOf(*dog,he*), in which we can distinguish different spatial roles: toTheRightOf expresses a relation, and *dog* and *he* are arguments of the relation with specific roles. Note, however, that this does not indicate the perspective, from which the relation toTheRightOf (*dog,he*) holds; this perspective information provides yet another spatial role. In this chapter, we use such spatial roles of linguistic elements in order to learn and extract them automatically from any given sentence. In order to apply machine learning effectively, a clear task definition as well as annotated data are required. Semantic hand-labelling of natural language, however, is an ambiguous, complex, and expensive task (Barclay and Galton, this volume). And in addition, to map the extracted linguistic features to qualitative representations, visual data referring to the same linguistic data are required to acquire corresponding qualitative spatial relations.

Automated spatial role extraction from language has been studied in only a few applications such as Li et al. (2007) for Chinese languages. This work focuses on fable stories, however, it only supports limited automatic recognition of spatial elements. The system presented in Kollar et al. (2010) operates in a human–robot interaction context and follows natural language route directions. It is more flexible in extracting spatial information, although it expects only spatial language as input. Work by Zhang et al. (2009) extracts (Chinese) geographical information from language, where spatial relations are used for mapping to general categories of directional, topological, and distal relations using a rule-based system. Finally, in psycho-linguistic studies, the important role of topological relations for expressing static spatial descriptions and even dynamic spatial configurations (Klippel and Li, 2009) and the frequent use of prepositions for topological relations (Kelleher et al., 2009) are highlighted.

7.3 From natural language to formal spatial relations

Our approach for mapping spatial language to qualitative spatial formalizations consists of two levels:

1) relevant linguistic features are selected and detected in natural language expressions, named spatial role labelling (SpRL), and
2) these linguistic features are mapped to their ‘interpretation’ in a spatial calculus, named spatial qualitative labelling (SpQL).

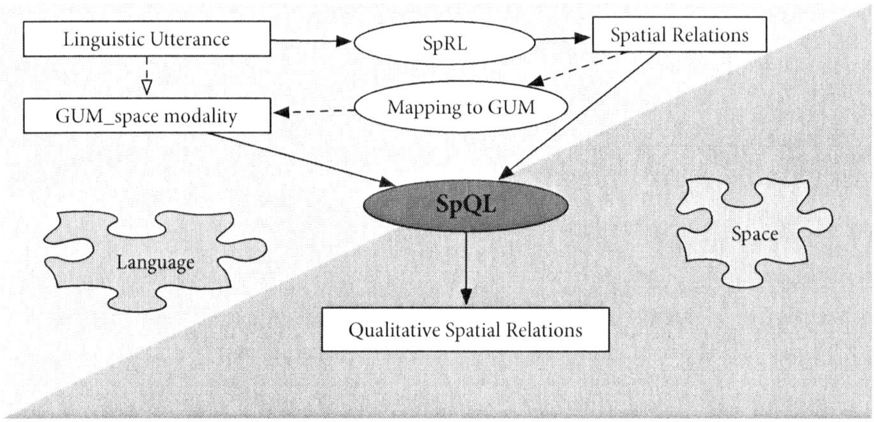

FIGURE 7.1 Mappings from language to spatial calculi

Hence, we regard the feature analysis of spatial language as a separate level from a formal spatial representation (Bateman, 2010b), and for both levels we use machine-learning techniques.

Figure 7.1 illustrates the two-level approach and it shows SpRL and SpQL anchored within the overall structure. The starting point is an utterance in natural language. In general, we aim at 'unrestricted' language, that is, utterances contain arbitrary spatial and non-spatial information. One intermediate level consists of *spatial relations* extracted from the utterance. Such relations are automatically extracted from language (see Section 7.3.1.2). In addition, these extracted linguistic features can be classified according to modalities defined by the linguistically motivated ontology GUM-Space (Bateman et al., 2010). The ontological structure provides abstract linguistic constructs, of which the *spatial modalities* reflect specific types of spatial relations (see Section 7.3.1.3). On the basis of these linguistic analyses, a formalization process maps the spatial language to one or more qualitative spatial relations specified by qualitative spatial representations (QSR). These representations formalize types of spatial relations in a particular formal spatial system together with composition tables (see Section 7.3.2).

The distinction between the two levels in our approach 'allows for the specification of a semantic representation that captures what the linguistic content of spatial expressions is committing to' (Bateman, 2010a); it separates the linguistic content from the formal spatial representation; and it provides a flexible way to map language to other spatial formalisms (Hois and Kutz, 2008b). Moreover, further components can be integrated, for example to support a situated interpretation or to support different types of uncertainties (Hois and Kutz, 2008a). An example of the overall process from the linguistic to the formal component is illustrated in Figure 7.2(a).

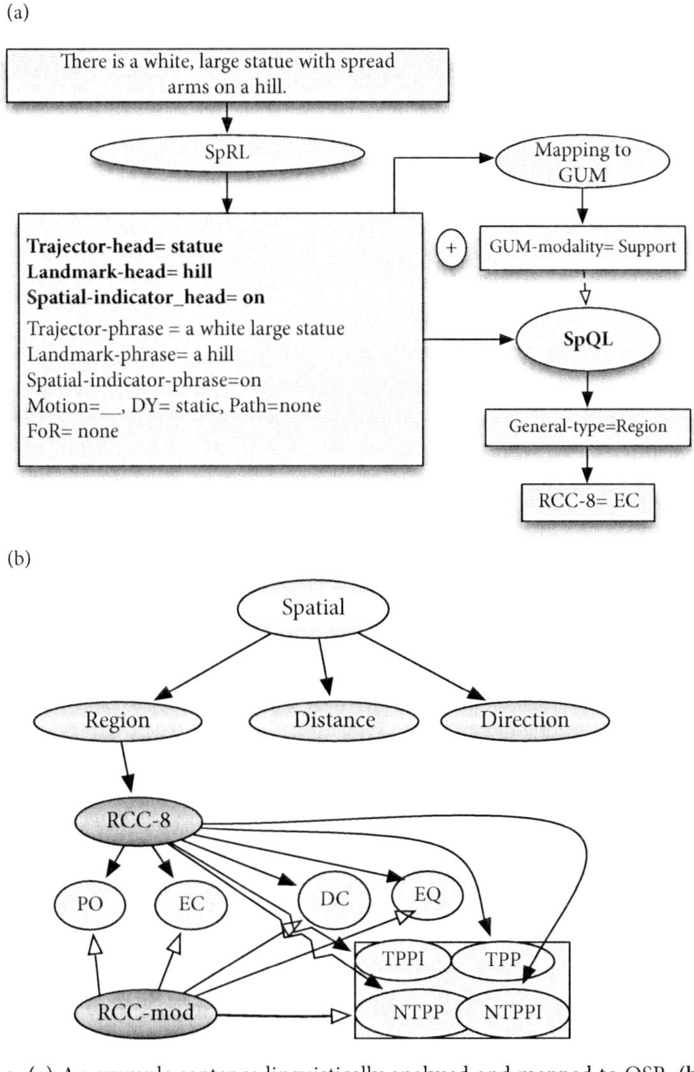

FIGURE 7.2 (**a**) An example sentence linguistically analysed and mapped to QSR, (**b**) General and specific types of QSR

In the following, we discuss our computational approaches for both levels. We start with the spatial role labelling, that is, the extraction of abstract spatial relations from language. This has been presented thoroughly in Kordjamshidi et al. (2011c), and here we (1) describe the task within our overall approach, and (2) show experimental results that demonstrate the feasibility of learning this mapping. Hereafter, we discuss the spatial qualitative labelling, that is, the mapping from abstract spatial relations and GUM-Space modalities to qualitative spatial relations.

This second level, which is the main contribution of this chapter, is demonstrated experimentally by using one specific qualitative spatial calculus, the RCC.

7.3.1 (I) Linguistic level: extracting spatial relations from language

Spatial language has a broad but structured range of ways for relating spatial linguistic entities with each other, both semantically and syntactically (Bateman et al., 2010). This structure provides a classification of linguistic features, and it relates certain spatial roles to entities of a linguistic expression. These *spatial roles* (or features) can be used to relate linguistic expressions to abstract spatial formalizations, for example qualitative spatial calculi. To identify the spatial roles, we build on results from empirical and linguistic research about the way spatial language describes spatial information (Talmy, 2006; Zlatev, 2007), and we automatically detect and classify this spatial information in a sentence by using machine learning. Note, that our goal is to extract abstract spatial relations from language. For example, an abstract relation between *book* and *table* expressed by the preposition *on* can be detected by extracting the spatial roles of the individual words first, in order to then classify the relation *on(book, table)*. The situational meaning of this abstract relation in terms of spatial semantics (e.g. in a topological sense) is detected by the second level in our approach, described in Section 7.3.2.

7.3.1.1 Spatial role labelling We define *spatial role labelling* (SpRL) as the automatic labelling of words or phrases in sentences with a set of spatial roles. The roles take part in one or more spatial relations that are expressed by the sentence. The sentence-based spatial analysis of text deals with characterizing spatial descriptions, for example determining spatial properties of objects or their location. The *spatial indicator* identifies a spatial relation, and other constituents indicate participants in this relation. This is illustrated in the following example:

Give me $[\textit{the grey book}]_{trajector}$ $[\textit{on}]_{spatial\_indicator}$ $[\textit{the big table}]_{landmark}$.

The spatial role set of this sentence contains the spatial roles *trajector* for 'the gray book', *landmark* for 'the big table', and *spatial indicator* for 'on' (see Zlatev, 2007; Kordjamshidi et al., 2010b, for a discussion of these spatial roles). These and more spatial roles are distinguished and defined in the following way:

- **Trajector.** The entity whose location or position is described. It can be static or dynamic; persons, objects, or events. Alternative common terms include local/ figure object, locatum, referent, or target.
- **Landmark.** The reference entity in relation to which the location or the motion of the trajector is specified. Alternate terms are reference object, ground, or relatum.
- **Spatial indicator.** The element that defines constraints on spatial properties like the location of the trajector with respect to the landmark. A spatial indicator

expresses the type of spatial relation and is often expressed by a preposition, but can also be a verb, noun, adjective, or adverb.

- **Motion indicator.** The element that provides specific spatial motion information. It is often expressed by verbs and defines the movement of a spatial entity, and renders a spatial expression *dynamic* instead of *static*.
- **Path.** Start, intermediate, and end points can be expressed to provide path information for the movement, the re-orientation, or the location of an entity.
- **Frame of reference.** In general, a frame of reference defines one or more *reference points*, and possibly a coordinate system based on axes and angles. Three reference types can typically be grammaticalized or lexicalized in English: intrinsic, relative, and absolute (Levinson, 2003). Recently, more detailed distinctions were presented in Tenbrink and Kuhn (2011), where spatial reference frames are represented and systematically specified by the spatial roles locatum, relatum, and (optional) vantage together with a directional system.

These six roles are the most influencing concepts for spatial semantics (Zlatev, 2007). SpRL can be considered as a special type of *semantic role labelling*, which is the task of automatically finding the semantic roles (e.g. agent, patient) for each predicate in a sentence. More specifically, this task determines the constituents in a sentence that are semantic arguments for a given predicate in order to determine appropriate roles for the arguments. Current approaches for semantic role labelling are based on supervised machine learning and require sufficient training and testing material (Màrquez et al., 2008). Similar to semantic role labelling, SpRL identifies and disambiguates the spatial indicator and its semantic role arguments, that is, the trajector and landmark.

SpRL becomes more challenging the more complex a linguistic expression is, for example if an expression consists of several spatial relations. Even in the short example 'The vase is on the ground on your left', grammatical dependencies can cause errors with respect to spatial dependencies. The dependency parser relates the first appearance of 'on' to the words 'vase' and 'ground', which produces a valid spatial relation. However, the parser relates the second appearance of 'on' to the words 'ground' and 'your left', that is, it produces an irrelevant spatial relation, namely 'ground on your left'. For more complex relations and nested noun phrases, the detection of 'spatially valid' relations is not straightforward and depends on the lexical meaning of words. Thus, recognizing the correct prepositional phrase (PP) attachment during syntactic parsing improves the identification of spatial features.

7.3.1.2 Machine learning for SpRL In SpRL we consider the spatial semantics that can be assigned to some words in the sentence as 'role-players'. We assume that the sentence is a priori partitioned into a number of segments (e.g. words, phrases). More formally, let S be a sentence with n segments $s = <w_1, w_2, \ldots, w_n>$ and let the set of roles be ROLES = {TRAJECTOR, LANDMARK, SPATIAL_INDICATOR, MOTION_INDICATOR, PATH, FRAME_OF_REFERENCE, NONE}, then each $w_i \in s$ needs to be labelled with one or more of these roles.

In Kordjamshidi et al. (2011c), this task is formulated as a sequence tagging task, and the triples TRAJECTOR, LANDMARK, and SPATIAL INDICATOR are extracted from the labelled sequences. A conditional random field model (CRF) (Lafferty et al., 2001) is used to tag single words in a sentence with the roles TRAJECTOR, LANDMARK, SPATIAL_INDICATOR, NONE. CRFs are instances of a large class of state-of-the-art machine-learning models called probabilistic graphical models (Koller and Friedman, 2009), among which Bayesian networks are the most well known. A CRF represents input features and outputs (e.g. the role labels) as random variables, and represents a probability distribution over the joint occurrences of these variables using factors representing which features interact (i.e. correlate) and how. In several linguistic tasks, CRFs are used to model sequential relations, that is, relations in which subsequent variables (e.g. words) correlate. In principle, however, any correlation between any set of features can be modelled and learned with CRFs.

Although the whole phrase in a sentence has a spatial role, we label only the *head word* of a phrase in SpRL with its spatial role. A head word is the word that determines the syntactic type of a phrase of which it is a member. The other elements of the phrase modify the head. For example, in 'the huge blue book', the word 'book' is the head word, and 'huge' and 'blue' are modifiers. Our annotated data assign roles to both phrase and head word to maintain the flexibility for future experiments. For learning, we employ sequences containing exactly one (labelled) spatial indicator together with all related trajectors and landmarks. As a sentence can express multiple spatial relations, the sequence tagging extracts all of these relations:

Let k be the number of prepositions in the sentence s, then s gives rise to k examples $e_1 \dots e_k$ where each e_i $(i = 1 \dots k)$ is a sequence $<(w_1, l_1), \dots, (w_n, l_n)>$ in which each word $w_i(i = 1 \dots n)$ is tagged as follows: (i) at most one w_j gets a label l_j = SPATIAL_INDICATOR, (ii) some words get a label TRAJECTOR or LANDMARK if they are trajector or landmark of the spatial indicator w_j, (iii) the yet unlabelled words get the label NONE. In non-spatial sentences all words are tagged with NONE.

Here, we concisely list the linguistically motivated input features that are used for SpRL. These features are inspired by SRL, however, with the spatial indicator as the central feature. As mentioned above, spatial features are defined for head words only. For more details on extracting and labelling spatial features, see Kordjamshidi et al. (2011c).

- Features of a word w are defined by $f_1(w)$, which contains:
  - the *word (form)* of w
  - the *part-of-speech tag*

- the *dependency to the syntactic head in the dependency tree*
- the *semantic role*
- the *subcategorization* of the word

- Features of the spatial indicator s are defined by $f_2(s)$, which contains:
  - the *spatial indicator* word (form)
  - the *subcategorization* of s (sister-nodes of its parent node in the tree)
- Relational features of w wrt s are defined by $f_3(w, s)$, which contains:
  - the *path* in the parse tree from w to s.
  - the binary *linear position* of w with respect to s (e.g. before or not).
  - the number of nodes on the path between s and w normalized by dividing by the number of all the nodes in the parse tree:

$$\text{distance} = \frac{\#\text{Nodes on the path between } s \text{ and } w}{\#\text{Nodes in the parse tree}}$$

Table 7.1 illustrates two label sets for the spatial indicators 'around' and 'in'.

TABLE 7.1. Example of two label sets for the sentence 'A girl and a boy are walking around the desk in the classroom'

Sentence	Label set I	Label set II
A	none	none
girl	trajector	none
and	none	none
a	none	none
boy	trajector	none
are	none	none
walking	none	none
around	spatial indicator	none
the	none	none
desk	landmark	trajector
in	none	spatial indicator
the	none	none
classroom	none	landmark

If an appropriate number of sentences is labelled with information about the spatial indicator, trajector, landmark, motion indicator, path, and frame of reference, an automated spatial relation extraction method can be learned that can be employed successfully on new data. Kordjamshidi et al. (2011c) showed that machine-learning techniques can be used to provide automatic SpRL. Moreover, the SpRL problem is solved by a kernel-based approach and by exploiting a relational language with flexible relational representation of influencing contextual features in Kordjamshidi et al. (2011a). Although the reader is referred to those works for the overall results and more details, Table 7.2 shows part of the results that we have obtained using CRFs. The evaluation is based on a ten-fold cross-validation (see Section 7.4.6.1 for more explanation). These results indicate adequacy of the learning based on two different corpora, namely selections from the HCRC Map Task corpus (Anderson et al., 1991) for spatial routes, and selections from the ImageCLEF IAPR corpus (Grubinger et al., 2006). The learning shows not only high performance for single spatial roles, such as trajector and landmark, but also good performance for the whole relation learning process. The whole relation means having the triples contain the correct trajector, landmark, and spatial indicators simultaneously.

Hence, the linguistic spatial roles can be learned with high enough reliability (precision and recall), and they can thus be used for the mapping to qualitative spatial relations. Also, information from lexical entries has a notable effect on the learning (cf. the discussion on lexical impact in Section 7.4), and automatic (semantic) generalizations and abstractions over the entities involved in the spatial relationships and their contextual implications may improve the learning; this is a direction for further research in this area. Additional analyses demonstrate the adequacy of the employed features: an error analysis demonstrates that (1) half of the errors are caused by misclassification of linguistic terms as none, (2) a quarter of the errors are caused by the complexity of nested spatial information, and (3) one-fifth of the errors are caused by irregularities in the grammar and focus shift in spatial descriptions. In the second step of SpQL, ground truth spatial role labels have been used as input for the mapping to spatial calculi.

7.3.1.3 Linguistic ontology In addition to the spatial features that are directly available from the linguistic surface, a formal semantic analysis can provide further support for the mapping from language to qualitative spatial representations (Eschenbach, 1999; Zwarts, 2005; Pustejovsky et al., 2011). This way, the lexical dependency can be reduced, and semantic categorizations can be used as an additional spatial feature in the mapping process. Here, we apply certain aspects taken from the spatial language categorization of the GUM-Space ontology, the Generalized Upper Model spatial extension (Bateman et al., 2010).

TABLE 7.2 **SpRL results on the extraction of triple relations (trajector, landmark, spatial indicator)**

Corpus data	Precision	Recall	F-measure
HCRC Map Task	0.888	0.904	0.896
ImageCLEF	0.704	0.737	0.720

The GUM-Space ontology is a linguistically motivated ontology that draws on findings from empirical cognitive and psycho-linguistic research as well as on results from theoretical language science. It has been developed specifically for spatial language, in particular, for English and German. GUM-Space categorizes parts of natural language sentences that contain spatial information into groups that share the same semantics. For instance, every starting point of a motion (e.g. 'someone moving away from the door', 'leaving the house', or 'exiting the roundabout') is logically specified in the same way. In particular, GUM-Space is formalized as a description logic, which specifies spatial units that provide particular kinds of spatial information.

GUM-Space's spatial categorization is not based on lexical data, but on the way language characterizes spatial relationships either grammatically or inherently. It specifies information about trajector (LOCATUM or ACTOR), landmark (RELATUM), spatial indicator (MODALITY), motion indicator (MOTION), path (ROUTE), and the frame of reference (PERSPECTIVE), as defined for the SpRL above. Although routes and paths are further distinguished into subcategories in GUM-Space, we focus on the subcategorization of the spatial indicator (SPATIAL MODALITY), as we have only a few dynamic examples in our experiments below.

Figure 7.3 shows the 73 spatial modalities (i.e. spatial indicators) that are distinguished in the GUM-Space ontology based on their hierarchical dependencies (Bateman et al., 2010). Sixty of these modalities can be expressed in natural language directly, the other modalities are generalized supertypes that provide more structure to the categorization. The modalities are not constrained to specific lexical entries, and they provide a distinction of spatial modes of locatings according to their lexico-grammatical expression as well as their semantic implications. Thus, these modalities add further semantic information about the spatial indicator, that is, the spatial relationship between the trajector and landmark. For example, the specific spatial modality expressed by 'on' in the sentence 'The vase is on the ground' is SUPPORT, which defines a physical supportive location of another entity; that is, if the landmark moves, the trajector moves as well (Tyler and Evans, 2003).

Spatial modalities can be computationally specified and accessed using categorical grammar techniques (Bateman et al., 2010). In the following, however, we focus on two aspects with regard to the GUM-Space modalities: first, we investigate if the fine-grained classification of the spatial modalities can also be learned by using machine-learning techniques based on the linguistic features extracted from the SpRL; second, we reuse the resulting spatial modality to map linguistic expressions to qualitative spatial relations. As illustrated in Figures 7.1 and 7.2(a), GUM-Space is used as an intermediate semantic representation between the linguistic features and the qualitative spatial representations. This is motivated by the fine-grained distinction of spatial modalities that reflect the type of spatial relation between trajector and landmark, which is one of the the strongest indicators for the qualitative relation. For this purpose, the corpus data used to learn the mapping to spatial qualitative

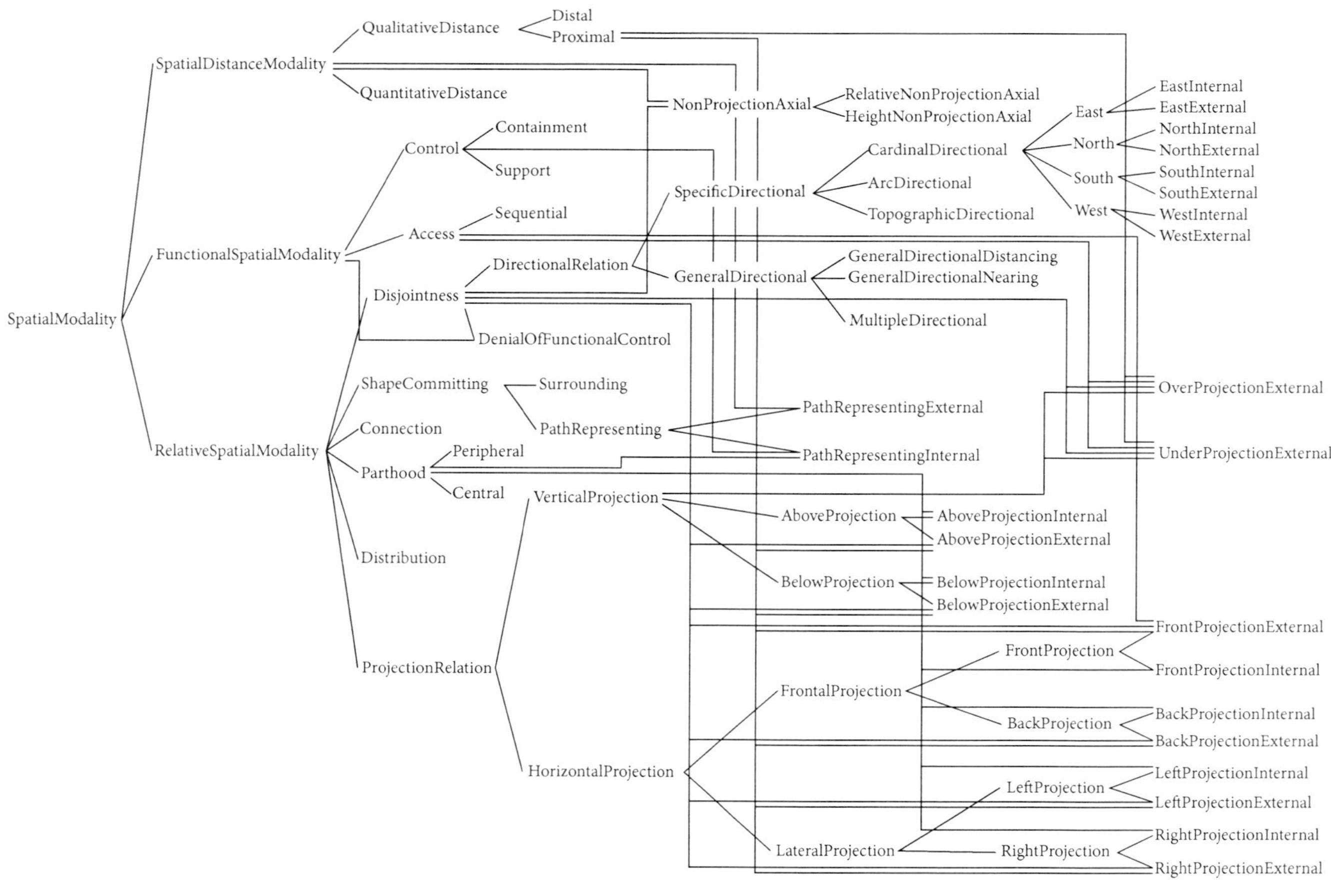

FIGURE 7.3 GUM-Space's spatial modalities

relations (see Section 7.4.2) contains annotations for spatial modalities. These are used together with the other linguistic features during the learning process for the mapping to qualitative representations. The results show an improvement in the mapping performance when using spatial modalities as a learning feature to map linguistic data to a selected spatial calculus (see Section 7.4.6).

7.3.2 (II) Spatial level: mapping linguistic features to qualitative spatial relations

We denote the mapping process from linguistic features to qualitative spatial representations as *spatial qualitative labelling* (SpQL), and define it as the automatic mapping of the output of SpRL and GUM-Space to qualitative spatial representations (spatial calculi). Direct mapping between the two sides is made difficult not only by ambiguity and underspecification of spatial information in natural language but also by over-specification of qualitative spatial calculi (Bateman, 2010b). Most of the qualitative spatial calculi focus on a single aspect, for example topology, direction, distance, or shape (Renz and Nebel, 2007). Spatial language, however, often conveys multiple meanings within one expression (Carlson and Van Deman, 2004). Hence, the SpQL has to cover multiple aspects with a practically acceptable level of generality. Nevertheless, obtaining an annotated corpus is the main challenge to using machine learning for the SpQL, and we built such a corpus in this work. It is essential that such corpora contain not only a sufficiently large number of spatial language texts for learning but also additional (non-linguistic) spatial information such as images or maps, from which a qualitative spatial model can be built that can be related to the textual information. These constraints hold for the selected corpora that we use in our experiments.

To build a corpus that supports supervised machine-learning models to cope with these challenges, Kordjamshidi et al. (2010b) propose an annotation framework inspired by SpatialML (Mani et al., 2008) and a related scheme presented in Shen et al. (2009). Using the training data which is annotated based on that scheme, we map the linguistic features to a *general spatial type*. The general spatial type is primarily expressed by the spatial indicator on the linguistic surface. In our current version, the available general spatial types are Region, Direction, and Distance as these are primarily distinguished by spatial information (Renz and Nebel, 2007). For future work further types will be relevant, such as Shape, Graduation, or Size. However, the three general spatial types cover the coarse-grained aspects of space, for which qualitative spatial calculi are available. In general, every spatial indicator can be associated with one or more of these general types. After mapping the linguistic features to the general spatial type, we map the linguistic features to a *specific spatial type*, which is the qualitative relation available via the spatial calculus used in the mapping process. Figure 7.2(b) illustrates the distinction between general and specific types. For example, if the general spatial type is detected as Region, a specific spatial type of region-based qualitative spatial calculi is detected next.

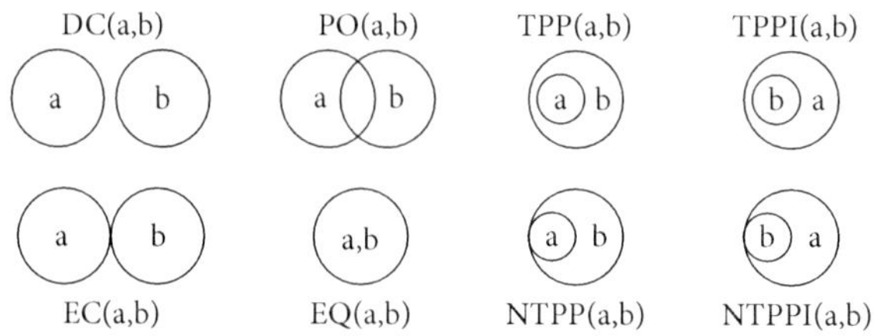

FIGURE 7.4 The RCC-8 relations

In the SpQL experiments presented in Section 7.4.6, we map linguistic corpus data to the Region-Connection Calculus RCC-8 (Randell et al., 1992), which reflects region-based topological relations as introduced in this section (Figure 7.4). The use of RCC-8 in our learning experiments is motivated by several reasons: the linguistic descriptions in the corpus data primarily express static spatial configurations of objects located with regard to a topological relation to another object based on two-dimensional images; topology is 'probably the single most important spatial concept fundamental to cognition in general' (Klippel, 2012, p. 1311); topological or region-based spatial information has been researched in depth in the area of qualitative spatial representation and reasoning; although RCC-8 only formalizes static qualitative relations between regions, its neighbourhood graph provides a method to model movement patterns and spatial change over time (Freksa, 1991), that is, to model dynamic spatial configurations in terms of RCC-8; we assume that the trajectors and landmarks can often be interpreted as spatial regions and, as a consequence, their relation can be annotated with a specific RCC-8 relation.

The last argument is also important, because it allows us to map the linguistic features directly to qualitative spatial relations. For RCC-8, we can thus define which linguistic terms are interpreted as qualitative entities, namely the trajector is interpreted as a region, the landmark is interpreted as a region, and the spatial indicator is interpreted as a qualitative relation. Although our approach is technically able to map linguistic data to other qualitative representations or more elaborate types of spatial formalizations, we expect that more input knowledge will be required to achieve an acceptable learning performance. This additional input knowledge may be based on external resources or ontological information, however, such experimental analyses and the construction of related annotated corpus data are left for future work.

RCC-8 provides eight basic relations (see Figure 7.4): disconnected DC(a, b), externally connected EC(a, b), partial overlap PO(a, b), equal EQ(a, b), tangential proper-part TPP(a, b), non-tangential proper-part NTPP(a, b), tangential proper-part inverse TPPI(a, b), and non-tangential proper-part inverse NTPPI(a, b), which describe mutually exclusive and exhaustive overlap and touching relationships between two well-behaved regions in space.

As our machine learning for the SpQL maps the entire set of spatial features from a sentence to a specific spatial model, we gain a more flexible and powerful mapping, in contrast to mapping only the single spatial indicator to an RCC-8 relation. Furthermore, the SpQL can map linguistic features to more than one calculus (e.g. to RCC-8 and an orientation calculus), which supports the required level of expressivity for a spatial linguistic expression. In the long run, such annotations will enable the use of expressive (probabilistic) logical models that relate linguistic spatial features of a sentence to complex configurations of spatial relations expressed in several spatial calculi simultanously. Learning this multiple mapping can also be used to visualize the spatial semantic contents of the sentence, or—when viewed as probabilistic constraints of these contents—to reason about the most likely inferences that can be drawn. For example, from a sentence 'The mouse is next to the keyboard on the table', one might infer the distance between these objects to be around 20 centimetres based on external knowledge about table sizes and typical configurations of the objects and one might also infer the angulation and positions of the objects with regard to the table. Some of our current investigations concern expressive logical models in a learning setup along the lines of the approach we discuss here; this is beyond the scope of this chapter and is discussed as a future work direction in Section 7.5. In summary, we investigate the following aspects in this work:

Given adequate corpus annotations, is it feasible to map linguistic spatial features to spatial calculi models, are the basic relations of the calculi clearly distinguishable by using the linguistic features, are similar sets of features mapped to similar basic relations, which qualitative relations are available in the corpus, and which linguistic features do we need for an acceptable classification?

As mentioned above, RCC-8 is far less expressive than natural language, for example when comparing its eight basic relations with the 73 spatial modalities semantically distinguished in GUM-Space. As a result, the SpQL leads to loss of information or to a simplification by the qualitative model. However, our experiments show that the mapping process is non-trivial and most of the basic relations can be distinguished clearly with high learning performance when spatial role labels are known.

7.4 Machine learning: from spatial language to RCC

In this section, we present the machine-learning task for mapping linguistic spatial features extracted from natural language to spatial calculi (see again Figure 7.1). In particular, we map spatial language to the region-based calculus RCC-8. For this purpose, we use a general machine-learning approach, in which the linguistic features from the linguistic level are described through feature vectors that form the *input* to the system. The *output* consists of a label derived from RCC-8 for every qualitative spatial relation. This means mapping each spatial relation to one of eight labels from RCC-8={EC, DC, TPP, TPPI, NTPP, NTPPI, PO, EQ, NONE} (see also Figure 7.4). We also describe additional experiments in which we map each spatial relation to five labels from a modified RCC-8 and in which we use the (60 linguistically expressible) spatial modalities from GUM-Space. Note, that the main purpose of our experiments is to show that the mapping from language to space can indeed be learned. For a general analysis, larger amounts of corpus data annotated with several spatial calculi would be necessary. As discussed in Section 7.3.2, we limit our experimental setup to map the linguistic data to RCC-8 relations for practical and theoretical reasons. However, the experimental results still provide enough information to examine the feasibility of learning qualitative spatial relations from a corpus.

7.4.1 Preliminaries

The main hypothesis to test is whether a mapping can be learned from extracted spatial roles and relations (SpRL) to specific models of formal QSR (SpQL). This amounts to learning the real semantics of these abstract roles and relations from annotated data, that is, whether it is interpreted as qualitative spatial types. The spatial roles with a high influence on this task are TRAJECTOR, LANDMARK, and SPATIAL_INDICATOR. Our central experimental investigation is whether detected trajectors, landmarks, and spatial indicators are adequate for determining the specific type of qualitative spatial relations.

In Kordjamshidi et al. (2011c) (and Section 7.3.1 on spatial role labelling), it was shown that these elements can be effectively extracted from natural language by using machine learning. Here, we investigate the learnability question for several fine- or coarse-grained versions of region-based QSR, namely RCC-8 and a corpus-inspired modification of it. We also investigate if the spatial roles MOTION, PATH, and FRAME OF REFERENCE have an effect on this learning process, although motion and path are more applicable for dynamic spatial relations and frame of reference is relevant in projective relations. As most of our data set contains static topological relations, we expect these roles to have almost no influence on the learning for the SpQL. Furthermore, we investigate the influence of GUM-Space spatial modalities in the mapping process. A general problem in many linguistic tasks (using machine learning) is the lack of lexical information in the training data, that is, the sensitivity to specific words. Naturally, we encounter this problem in our approach, in both SpRL and SpQL. Hence, we conducted additional inter-corpora evaluations particularly for SpRL in Kordjamshidi et al. (2011c) to investigate induced errors and to gain insights concerning possible abstractions required for future work.

7.4.2 Data sets

As supervised machine learning requires annotated data, our approach relies on spatial linguistic and qualitative corpus data. As, to our knowledge, these data are not publicly available so far, we have built a corpus based on the annotation scheme in Kordjamshidi et al. (2010b) and spatial annotations of GUM-Space. Our corpus set consists of textual descriptions of 613 images taken from the IAPR TC-12 Image data set (Grubinger et al., 2006), referred to as CLEF. It induces 1213 English sentences and 1716 corresponding spatial relations. CLEF contains images taken by tourists with descriptions in several languages, describing depicted objects and their absolute and relative positions. This makes it a rich resource for spatial information. However, the descriptions contain non-spatial descriptions of images, which makes it a useful corpus for general, less domain-specific approaches. We proposed the shared task of spatial role labelling on the basis of these data for the first time at the SemEval-2012 workshop (Kordjamshidi et al., 2012). In addition to CLEF, we annotated and used 64 selected sentences from the Map Task corpus (which has been used to validate the expressivity of spatial relations in GUM-Space; Bateman et al., 2010) and 124 sentences from the Room Description corpus (Shi and Tenbrink, 2009). The more detailed results are reported for CLEF, as it is a reasonably large data set.

The textual descriptions have been indexed and annotated with spatial roles of trajector, landmark, and their corresponding spatial indicator. Separate roles are assigned to phrases and the head words of the phrases. The verbs and verb phrases are indexed and annotated, particularly when they participate in forming the spatial configurations, and this is mostly the case for dynamic spatial relations. Moreover there are annotations for path in terms of beginning, middle, or end. Each sentence with a spatial relation is annotated as DYNAMIC or STATIC, and each spatial relation is annotated with a GUM-Space modality. Moreover, every sentence with a spatial relation is annotated with general and specific spatial types (see Section 7.3.2). We have measured the inter-annotator agreement for 325 sentences with respect to the spatial roles trajector, landmark, and spatial indicator. With two annotators, the Kappa value resulted in 0.896 with a 95 per cent confidence interval (0.882–0.910).

Our CLEF data set contains 1040 annotated topological, 662 directional, and 91 distance relations. In the current corpus only 50 examples are annotated with more than one general spatial type. For example, 'next to' is annotated as a topological relation DC in terms of RCC-8 and as a distance relation CLOSE in terms of a relative distance calculus:

(7) *Two people are sitting next to her.*
trajector: people,
landmark: her,
spatial indicator: next to,
general type: region/distance,
specific type: RCC-8 / relative-distance,
spatial value: DC / close,
DY: static,
path: none,

frame of reference: none,
spatial modality: Proximal

7.4.3 Two-dimensional versus three-dimensional annotations

Although the textual data used are accompanied by images (CLEF) or maps (Map Task and Room Description), the qualitative spatial annotations for CLEF were based on the text itself. This was done to focus on information that can actually be extracted from the language itself. Nevertheless, human imagination about a described scene can interfere with the textual description, which has resulted in some variations. As an example, take the following sentence and its annotation:

(8) *Bushes and small trees (are) on the hill.*
trajector: bushes,
landmark: the hill,
spatial indicator: on,
general type: region,
specific type: RCC-8,
spatial value: EC,
DY: static,
path: none,
frame of reference: none,
spatial modality: Support

This three-dimensional projection of the description of a two-dimensional image is annotated as externally connected. In the two-dimensional image, however, a partial overlap may also be adequate. In contrast, a two-dimensional map (with an allocentric perspective) of the described scene would lead to a non-tangential proper part annotation. This example illustrates what we have said before; that RCC-8 alone is—quite naturally—not enough to capture adequately all the necessary spatial information, and that in a more general approach more (and combinations of) qualitative spatial calculi have to be used.

7.4.4 Dynamic versus static annotations

In the CLEF data set 25 of the relations are annotated as DYNAMIC, the others as STATIC. If a dynamic situation is annotated with a (static) RCC-8 relations, the qualitative relation can be regarded as a snapshot of the situation. This is shown in the following example:

(9) *People are crossing the street.*
trajector: people,
landmark: road,
spatial indicator: crossing,

general type: region / direction,
specific type: RCC-8 / undefined,
spatial value: EC / undefined,
DY: dynamic,
path: middle,
frame of reference: none,
spatial modality: PathRepresentingInternal

Hence, the annotations refer to time slices for the (linguistic) explanation of the (static) image. This allows a mapping from dynamic descriptions to (static) RCC-8 relations mainly by including the path feature and the relative situation of the trajector with respect to an imaginary path related to the landmark. Allowing RCC-8 annotations for dynamic descriptions is also supported by the conceptual neighbourhood graphs (Freksa, 1991): Every topological change—movements of regions with respect to each other and their changing relations—can be split into a sequence of adjacent RCC-8 relations according to the neighbourhood graph (Klippel and Li, 2009). The annotated RCC-8 relation thus reflects one relation out of this sequence, that is, one moment in time of the topological change. However, we may not predict if the annotations refer to a time slice that reflects the start, intermediate, or end point of the path or the motion process. For instance, Regier and Zheng (2007) show that linguistic expressions seem to focus primarily on the end point of the motion.

7.4.5 Features and experimental setup

In the first set of experiments we apply machine-learning techniques to learn an automatic mapping from SpRL to the following RCC relations.

RCC-8. RCC-8={EC, DC, TPP, TPPI, NTPP, NTPPI, PO, EQ, NONE}

RCC-mod. RCC-MOD={EC, DC, PP, PO, EQ, NONE}, i.e., subsuming {TPP, NTPP, TPPI, NTPPI} under {PP}

The features that are used as the input of the SpQL are the head words that are annotated as TRAJECTOR (tr), LANDMARK (lm), SPATIAL INDICATOR (sp), and MOTION INDICATOR (mo); features that include not only head words but all words in the phrase are TRAJECTOR PHRASE (trPh), LANDMARK PHRASE (lmPh), and SPATIAL INDICATOR PHRASE (spPh); the PATH FEATURE (pa) has one of the values {BEGIN-

TABLE 7.3 **Feature count for CLEF**

tr	lm	sp	trPh	lmPh	spPh	mo	pa	for	dy	gum	Total
354	183	43	258	142	56	12	4	2	2	39	1095

NING, MIDDLE, END}; the frame of reference (for) has one of the values {INTRINSIC, RELATIVE, ABSOLUTE}; the dynamicity feature (DY) has one of the values {STATIC, DYNAMIC}; and the GUM-Space modality (gum) has one of its 60 possible values (see Figure 7.3). An example sentence with its linguistic features is shown in Figure 7.2(a). The occurrences of these features in CLEF are shown in Table 7.3. These features are gradually used to model the learning process, and the performance of the learning is measured by gradually adding linguistic features to evaluate their effect on the learning process. The prior distribution of the RCC-8 relations in CLEF (the main data set) and the other two corpora is shown in Table 7.4. We applied various machine-learning approaches, however, the best results were obtained with support vector machines (SVM), as described in the following. An SVM is a state-of-the-art machine-learning technique typically used for classification and regression problems. It represents examples as points in space and learns to find a boundary in this space that optimally separates positive from negative examples of a concept. New examples are then mapped to the same space and are labelled as the side of the boundary to which they belong (cf. Alpaydin, 2004). The second set of experiments examines the mapping from linguistic features to GUM-Space spatial modalities and to general qualitative spatial types. This is technically implemented in the same way, and the learning model is evaluated by gradually adding input features.

7.4.6 Results and discussion

In this section, the results of the spatial qualitative labelling (SpQL) to RCC-8 and RCC-mod are discussed, and the influence of the linguistic features for the SpQL task is analysed. Also the feature impact on the mapping to general qualitative spatial types and the effect of using GUM-Space modalities is presented.

TABLE 7.4 **RCC-8 annotations of CLEF, Map Task, and the Room Description corpus**

RCC-8 Relations	#Instances(CLEF)	#Instances(Map Task)	#Instances(Room Des.)
DC	147	32	82
TPP	369	3	2
NTPP	12	3	0
EC	458	2	1
EQ	5	0	5
NTPPI	1	0	0
TPPI	25	0	0
PO	15	0	0
NONE	684	29	34
Total	1716	69	124

7.4.6.1 RCC-8 First, we investigate the feasibility of automatically mapping the linguistic features to RCC-8 relations. For this purpose, several classifiers (e.g. decision trees, naive Bayes, and SVM) were used. The results in Table 7.5 present the accuracy in terms of precision, recall, and F-measure over individual classes and the weighted average overall for the best performing classifier (SVM) using ten-fold-cross-validation. This means that the data set is divided into ten parts. Each training time leaves out one of the ten parts as the evaluation set and the model is trained on the remaining nine parts. The overall performance measures are the averages over all ten parts used for evaluation. Precision, recall, and F-measure are computed using:

$$\text{recall} = \frac{TP}{TP + FN} \quad ; \quad \text{precision} = \frac{TP}{TP + FP} \quad ;$$

$$\text{F-measure} = \frac{2*\text{precision}*\text{recall}}{\text{precision} + \text{recall}}$$

with TP = # correctly classified samples wrt the target RCC relation
FN = # samples incorrectly not assigned to the target RCC relation
FP = # samples incorrectly assigned to the target RCC relation

F-measure is the harmonic average of precision and recall (van Rijsbergen, 1979). Table 7.5 shows that the weighted average, in which the performance of each class is multiplied by the number of its instances (i.e. micro averaging), is reasonably high. For classes with more examples in the training data, the performance is higher. The performance over none-RCC-8 relations is rather strong. The performance for DC, EC, TPP, and PO is reasonably high, however for the other relations it is less acceptable.

7.4.6.2 RCC-mod When humans use language to describe objects, it is mostly the salient objects that are chosen as landmarks in the descriptions. The inverse proper

TABLE 7.5 **RCC-8 classification results for CLEF**

RCC-8 Relation	#Instances	Precision	Recall	F-measure
DC	147	0.848	0.871	0.859
TPP	369	0.801	0.897	0.847
NTPP	12	0.667	0.50	0.571
EC	458	0.859	0.841	0.850
EQ	5	0.8	0.8	0.8
NTPPI	1	0	0	0
TPPI	25	0.778	0.28	0.412
PO	15	0.778	0.467	0.583
NONE	684	0.970	0.953	0.962
Total(Weighted Avg.)	1716	0.886	0.886	0.883

part relations thus occur less frequently (cf. Klippel and Li, 2009). This motivates the next experiment in which we combine all variations of proper part including {TPP, NTPP, TPPI, NTPPI} into one class {PP}, resulting in RCC-mod (for RCC-mod, however, no spatial reasoning is directly available). Table 7.6 shows the results over the five classes of RCC-mod in addition to the 'none' class. Although, the overall performance for both RCC-8 and RCC-mod classification is high, there are still two critical relations, namely EQ and PO, with only a few examples in the data set. The confusion matrix indicates that most errors are caused by EC in both experiments, in particular, by two-dimensional object-related distinctions, for example:

(10) 'A ship on the sea ...', which is annotated as PO

(11) 'A bag on the table ...', which is annotated as EC

The classifier could not distinguish between these two examples and classifies both of them as EC. This happens because in the corpus most of these cases are annotated with EC and there are not enough examples to distinguish PO.

The relation EQ, too, is a challenging relation and cannot easily be inferred and described by analysing corresponding two-dimensional images. The equality would mean that two areas have a full overlap, that is, one completely covers the other and vice versa. This will often not be recognizable from two-dimensional images and consequently will not be explicitly expressed in the linguistic utterance. However, EQ occurs for five relations in our corpus, and by looking at these examples it can be observed that there is a high potential of disagreement across the annotations. An example is the sentence '... with his hands in his pockets', which has been annotated as EQ although it could just as well be annotated as PP in RCC-mod.

7.4.6.3 Feature analysis The idea of exploiting the above-mentioned features is inspired by cognitive linguistics and generally derived from the main elements of spatial semantics in spatial language. Here, we present an empirical study to investigate the influence of each of these features on the performance of the machine-learning models for CLEF. We performed two sets of experiments using the RCC-8 classification setting and we looked at the weighted average of all classes with the same SVM classifier.

In the first experiment (Figure 7.5), the spatial indicator is used as the only input feature for classification of the RCC-8 relations. Spatial indicators are mostly prepositions in CLEF, and using only these for the mapping yields an F-measure of 0.474. Adding the head word of the landmark phrase in the sentence increases the F-measure by 0.27. Gradually adding the head word of the trajector adds another 0.08. Adding additional features again increases the performance by 0.044, and accuracy becomes stable. Further feature additions may sometimes even reduce performance due to the combination of noise and lack of data.

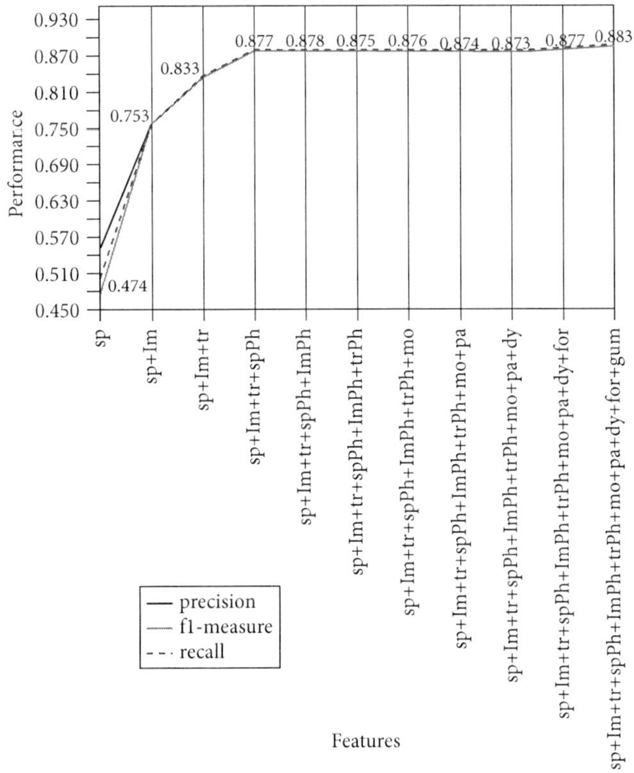

FIGURE 7.5 Performance of the RCC-8 classifier with spatial roles as input

In the second experiment (Figure 7.6), we use GUM-Space modality information as the first step, motivated by the high expressivity of GUM-Space and its in-depth exploration with regard to the spatial relations expressed in spatial language. Using only GUM-Space's spatial modalities yields an F-measure of 0.719. Given the spatial modality, the spatial indicator (mostly the preposition) does not add more significant information and the performance is increased by 0.02. However, the context has an impact on the performance: adding the landmark information increases the performance by 0.07; adding the trajector information gives a 0.07 increase. The remarkable outcome of this experiment is that given the modality feature even trivial fluctuations in the accuracy of the classifiers are no longer observed, and the model seems to behave more robustly against extra and noisy features.

Although trajector and landmark information may not carry explicit spatial indications in terms of QSR relations, using these features increases the overall performance in both experimental setups. This may be caused by the statistical significance for choosing certain reference objects over others, which is shown in Barclay and Galton (this volume). For instance, the absolute sizes of trajector and

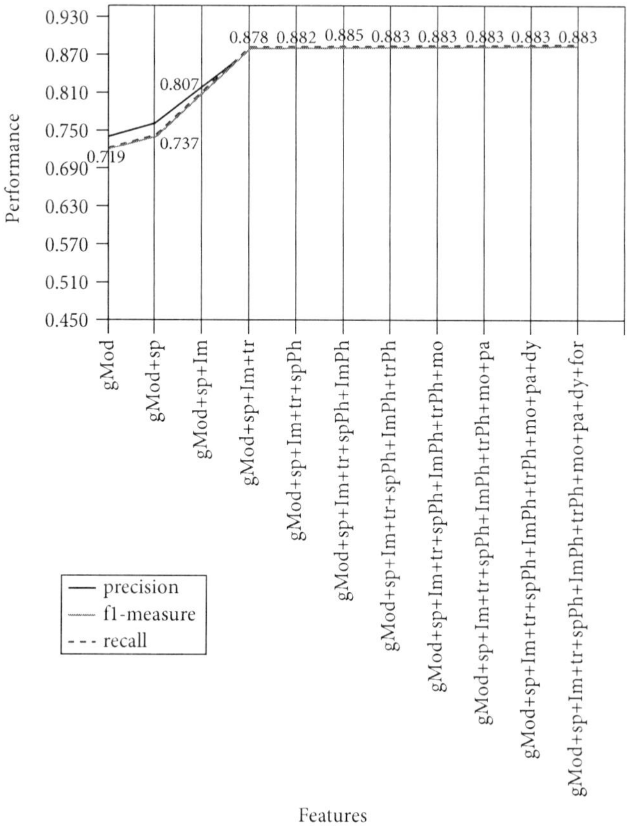

FIGURE 7.6 Performance of the RCC-8 classifier with GUM-Space modalities and spatial roles as input

landmark are important for selecting the reference objects in the initial CLEF annotations of the photographs.

As we have used GUM-Space modalities with only manual annotations so far, the third experiment (Figure 7.7) investigates whether the modalities can also be learned by using the automatically identified spatial roles. Using only the spatial indicator, the performance is 0.55 and thus slightly above chance. However, also using landmark information increases the performance to 0.82, which goes up to 0.96 when using the whole phrase of the spatial indicator. This feature influence seems reasonable as the modalities reflect spatial relations with regard to their landmarks. The experiment shows that indeed spatial modalities can be learned for this data set, however, only 21 from the 60 modality types were used for the annotations.

Adding the phrasal information of the trajector has no effect or even some minor negative effects on the overall performance. In Table 7.6, the five EQ instances are classified with 0.8 accuracy when using all features; when these features are reduced,

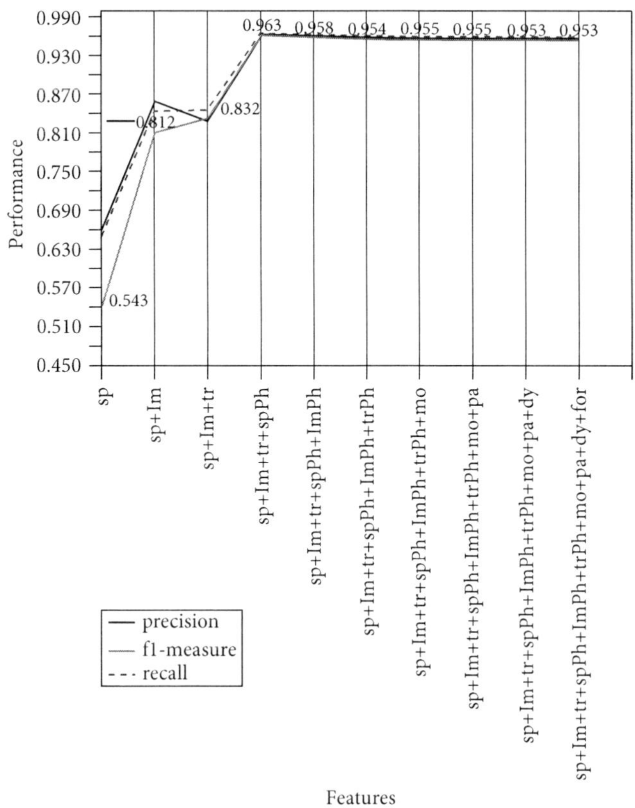

FIGURE 7.7 GUM-Space modality classification

Table 7.6 RCC-mod classification results for CLEF

RCC-mod Relation	#Instances	Precision	Recall	F-measure
DC	147	0.863	0.857	0.86
EC	458	0.858	0.856	0.857
EQ	5	0.75	0.6	0.667
PP	407	0.838	0.887	0.862
PO	15	0.778	0.467	0.583
NONE	684	0.975	0.953	0.964
Total(Weighted Avg.)	1716	0.899	0.898	0.898

the F-measure decreases by 0.4. Reducing the features, however, has a reverse effect on the PO relation and an increase of 0.11 is obtained, caused by the lack of sufficient data for EQ and PO.

Besides the phrasal information, the other sets of features have no significant impact on the performance. According to the reported statistics over the applied data set, this result is plausible because CLEF contains mostly static relations, and only a few motions are annotated in the corpus. Hence, dynamicity and path features do not occur often in the data and are not discriminative features for the type of RCC-8 relation. The frame of reference is often annotated as NONE, as it is not available for topological relations. Consequently, this feature did not have an effect on performance. However, even for directional relations that are based on a frame of reference, this feature might often be ambiguous when given implicitly (see Avraamides et al., this volume).

7.4.6.4 Mapping to general categories of spatial relations In addition to region information, direction and distance relations are relevant for mapping spatial language to QSR (cf. Figure 7.2). This can be considered as a first step towards the SpQL of various types of calculi. As we pointed out before, a spatial relation at the linguistic level can be mapped to various types of calculi since it can be assigned to more than one of these general categories. As discussed above, the expression '*next to*' implies distance and topology information.

Figures 7.8 and 7.9 present this accuracy in terms of input features, using either the spatial modality or the spatial indicator as starting point. As the number of distance relations in the data set is only 32, a lower performance than for topological relations is expected. Similar to the previous experiments, the most influential features are trajector, landmark, spatial indicator, and the phrase of the spatial indicator. Due to the low number of examples, the positive effect of the features is more distinct. Using information of the spatial modality again improves the final performance: knowing the spatial modality for the relation, the general qualitative spatial type is almost certain and the remaining features do not affect this result (cf. Figure 7.8). This result is most likely caused by the specific category types of GUM-Space, which directly imply the general type. However, the confusion matrix shows that distance relations are less well recognized, again due to lack of examples in the data set.

7.4.6.5 Inter-corpora evaluation Even though we use SpRL as an abstract linguistic representation, lexical information has a notable effect on the SpQL. Lexical information specifically from trajectors and landmarks is used for learning the spatial roles in the previous experiments, which did not use abstraction mechanisms, such as word classes. In this section, we examine the influence of lexical information for the two other corpora Map Task and Room Description.

As shown in Table 7.4, the two corpora lack several types of topological relationships because they contain mostly directional relations for route instructions on a

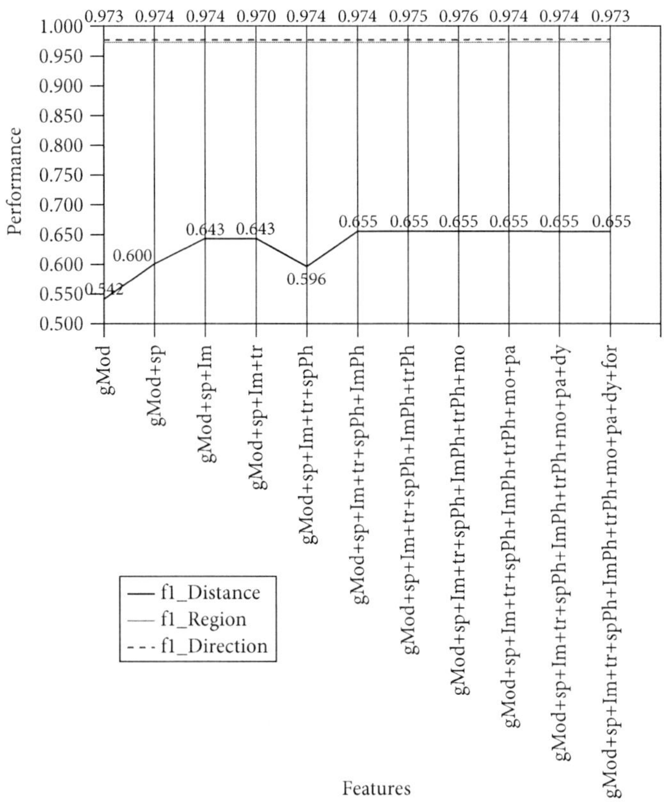

FIGURE 7.8 Performance of the general spatial type classifier starting with spatial modality as input feature

map or in a room, and their relations are annotated mostly as DC in terms of RCC-8 or as NONE. In several tests, we did the same ten-fold cross validation internally on this data set. Here, the F-measure on both was higher than 70 per cent, which is acceptable with respect to the small data sets.

In another experiment, we used the trained machine-learning model for CLEF and used it on those two corpora, which resulted in low performance. The main reason for this outcome is that the lexical entries for the objects described in these two corpora are entirely different from CLEF's lexical entries.

7.5 Conclusions and future work

The aim of this work was to investigate whether a mapping from complex spatial utterances to formal spatial relations can be learned. We have introduced a two-level machine-learning architecture that consists of SpRL (spatial role labelling) to extract

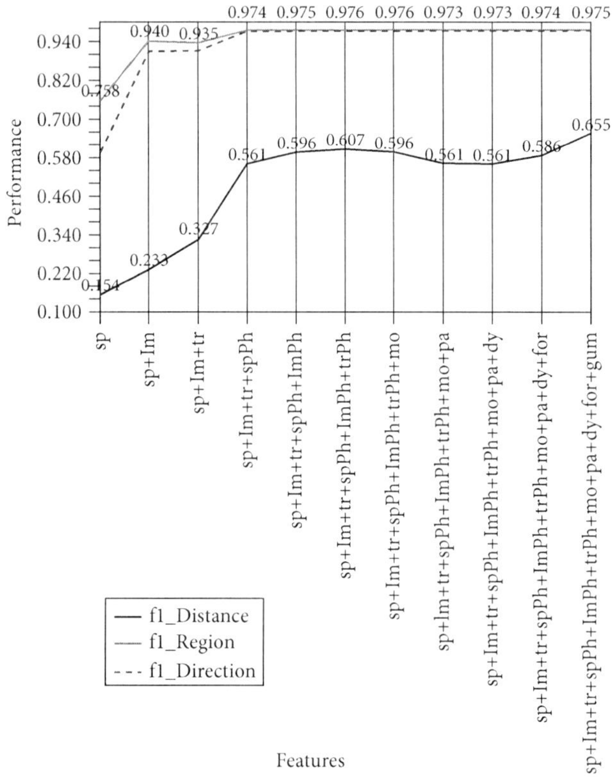

FIGURE 7.9 Performance of the general spatial type classifier starting with spatial indicator as input feature

and identify spatial linguistic features based on a holistic approach and SpQL (spatial qualitative labelling) to map these features to qualitative spatial relations such as RCC.

Our experimental results show, most importantly, that indeed both the extraction of spatial features and the mapping to qualitative relations can be learned, and that the approach is computationally feasible. Feature analysis indicates that the most influential features are trajector, landmark, and spatial indicator. Although phrase information seems to have no effect on learning, it might become more informative when the phrase contains additional functional information. As we mapped primarily static and region-based spatial descriptions in our experiments, the spatial features path, motion, and frame of reference had a negligible effect on the results. An outcome with regard to RCC-8 mappings is that some qualitative relations rarely occurred in the (random) corpus data set (e.g. inverse and equality relations).

The use of formal semantic categories from GUM-Space helped to improve the performance of SpQL. As GUM-Space aims to classify the linguistic types of spatial semantics in a structured way, it appears to be adequately expressive for SpQL, in particular, when mapping to general spatial relation types. Using the GUM-Space modalities, learning becomes more robust when lexical information is fluctuating across the data set. As a consequence, we expect the modalities to have a positive effect when mapping to qualitative relations other than RCC-8. Our experiments showed that it is promising to use the linguistic features for learning the GUM-Space modalities, as these are highly compatible with the linguistic spatial patterns in SpRL. However, mapping to the whole of GUM's ontological information and learning the ontological categories remains for future work.

Another future direction is the mapping to more than one qualitative spatial representation, including calculi for regions, orientations, distances, shapes, sizes, motion, etc. As spatial language is able to express all of these types and combinations thereof, the SpQL task should be able to interpret a linguistic expression more comprehensively when having multiple spatial calculi at hand. Given SpQL, the mapping to multiple calculi is expected to be easy to accomplish with relational machine learning. So far, we have employed so-called propositional representations in our learning approaches, such as SVM and CRF. More expressive relational (also called first-order logical) representations can also be used in the learning process, and these are a more natural way to represent spatial information as it is linguistically expressed in a relational way. Statistical relational learning approaches (De Raedt et al., 2008) combine such representations with state-of-the-art algorithms (e.g. based on the same principles as probabilistic graphical models or SVMs) to exploit logical representations while dealing with uncertainty and ambiguity in natural language. In current and future work we investigate these models for spatial language understanding.

Our inter-corpora evaluations identified challenges with respect to a lack of lexical information when the learned model was used in other domains. This rather unsurprising aspect could be compensated for in several ways. The data set can easily be extended by integrating diverse sources, which is, however, a time-consuming and costly approach. The use of abstractions of lexical information seems to be more applicable, for example abstract features that indicate functional characteristics of trajectors and landmarks may improve the mapping to qualitative spatial relations. This will add another layer of abstract labels for the linguistic data. The annotation types of these abstract labels could be taken from WordNet or word taxonomies, or latent word models could be used. The latent word language model proposes words that can replace the trajector and landmark words so that the set of training examples can be automatically expanded by these replacements (Deschacht and Moens, 2009).

In conclusion, this chapter has shown a machine-learning approach and an empirical study for interpretation of spatial natural language in terms of QSR. Given adequate corpus data, we expect our method to perform adequately when learning the relevant qualitative relations that occur in the data. The kind of spatial language interpretation should be particularly applicable in several domains, whenever a mapping from language to abstract spatial representations is required. Examples are dialogue systems (Kruijff et al., 2007), spatial assistance systems for design or media production (Bhatt et al., this volume), text-to-scene conversion systems (Rouhizadeh et al., 2011), virtual human cooperation (Nguyen and Wachsmuth, this volume), and robotic navigation instructions (Kollar et al., 2010).

8

A computational model of cooperative spatial behaviour for virtual humans*

NHUNG NGUYEN AND IPKE WACHSMUTH

8.1 Introduction

Acting smoothly in space and avoiding collisions with others in their workspace (the space where movements are carried out) is a crucial aspect of natural spatial behaviour of any articulated agent. Just think of two interactants building a toy tower together, where they are able to coordinate parallel actions in shared space without getting in the way of each other. Humans accomplish such tasks without explicit negotiation. For instance, they place needed component parts to areas where the other can reach them or see them and at the same time avoid obstructing each other's workspace or view. Such spatial behaviour in shared space interaction can fulfil two functions: first, the function of coordinating the interactants' actions to successfully accomplish a joint task; second, the function of conveying social signals.

In this chapter we shall demonstrate how a virtual human can cooperate with a partner in building a toy tower together, as one aspect of computationally modelling shared space and spatial behaviour for action coordination. Virtual humans are autonomous agents with human-like appearance and usually human-like multi-modal behaviour like speech, gaze, gestures, and facial expressions. In three-dimensional virtual reality environments, virtual humans can interact with other virtual humans or with real humans. For example, virtual humans like Max (Kopp et al., 2003) can act as co-situated guides in a construction task, or like Steve (Rickel and Johnson, 2000), can act as tutors demonstrating physical tasks to students.

* This work has been kindly supported by the Collaborative Research Center 673 Alignment in Communication, funded by the German Research Foundation (DFG). We thank Kim-Anh Tran for assisting in the research project and this publication.

In the mentioned scenarios, overlapping workspaces were usually avoided by maintaining enough distance between the partners to avoid interferences between their movements. We believe that in natural interaction such interferences have to be dealt with to accomplish cooperative interaction tasks. Thus, we present our work on modelling a virtual human's spatial behaviour in shared near space interactions in order to facilitate the accomplishment of and the partner's engagement in the cooperative task.

Spatial interaction in tasks carried out in distances near to the agent's body usually pose a great challenge to virtual humans. In contrast, humans seem to manage these tasks without much effort. As Tenbrink, Wiener, and Claramunt state in the Introduction to this volume, computational models rely strongly on empirical findings of spatial cognition such as those presented in Part 1 'Empirical Insights'. Such findings serve as helpful evidence for building human-like behaviour for virtual humans. Our approach specifically takes inspiration from findings on how humans process spatial interaction tasks on the level of spatial perception and representation, which again has consequences on the design of virtual humans' spatial perception and representation. Since the findings originate from various disciplines, such as neuropsychology, cognitive neuroscience, and cognitive science, one typical issue for computer scientists is to translate the concepts into technical models. Two main aspects related to human spatial cognition are technically modelled in the present work. The first aspect is concerned with how the reach-space (**peripersonal space**) is represented and structured in humans (see Section 8.4.3.1). The second aspect deals with how humans represent the peripersonal space for a proximate partner when interacting with her. This matter is addressed under the concept of **spatial perspective taking** in Section 8.5.1.

Although our computational model integrates theoretical concepts from non-technical disciplines, one main goal for technical systems eventually is the application in interaction with a human partner. One common view, especially in the field of virtual human development, is to build agents that are able to assist and guide humans. This view is also reflected in Part 3 'Human-centred Assistance'. Following the definition of Bhatt, Schultz, and Freksa (this volume) our virtual human application could therefore be conceived of as a Spatial Assistance System, within the domain of *Ambient Assisted Living* (AAL). According to Bhatt and colleagues, AAL applications usually involve software systems or robots which are supposed to assist and empower humans within private spaces in everyday life. We adopt this definition and in addition believe that in AAL applications especially with embodied agents, the engagement of humans increases since humanoid robots and virtual humans imply more immediate spatial interaction. In our application, Max builds a hypothesis from the partner's spatial perspective and matches it with the next actions of the joint task. If a needed object is not reachable for the partner, Max places the object to locations that are reachable for her. We believe that embodied agents that are supposed to share the same space with humans will need to deal with the spatial perspective of the partner as well as with social aspects. In some cases, especially when it comes to the private spaces mentioned by Bhatt et al., these factors might turn an assistive agent into a more cooperative and thus more acceptable partner.

Improving virtual humans in movements carried out in their individual workspace is a classic problem (Huang et al., 2003; Zhao et al., 2005). We put the focus of this chapter on two issues of this challenge. One issue is to improve the virtual human's sensorimotor and perceptual abilities, which are crucial for body action/motion planning and control. The virtual human's workspace is the space where sensor modalities have to focus and where possible objects have to be observed or manipulated by reaching, grasping, or avoiding them. Sharing parts of this space with other agents makes interaction only more challenging, which leads to the following second issue. As mentioned before, the partners' actions and movements have to be coordinated, thus, the virtual human needs a representation of the shared near space in order to perform smooth, effective, and also cooperative interaction. We focus on delimited near space arrangements (e.g. a table), involving mainly the virtual human's upper part of the body.

In our work we connect the two issues of modelling the space surrounding the body with regard to an individual virtual human and modelling the same space with regard to interpersonal interaction. Accordingly, our goals in developing a virtual human are first, to enable the virtual human to learn and adapt to its reaching space, that is, the virtual human knows from its sensory modalities whether objects are in the near space immediately surrounding its body or whether they are beyond this space, second, to facilitate its perception and action within its reaching space, that is, to control the virtual human's perceptual and sensory attention with taking account of possible objects and hindrances, and third, to facilitate cooperation in shared space, for example by sustaining actions toward locations reachable by all interaction partners.

In the presented approach we approve the recent work outlined by Lloyd (2009) claiming that the principles underlying the individual representation of the space surrounding the human body also mediate the space between interacting human partners. This idea is also valuable to provide virtual humans with the abilities we aim to model. We present how our work on the reaching space of an individual articulated agent's body—the **peripersonal space**—is used to model the shared reaching space of cooperative interaction partners, which we define as **interaction space**.

Motivated by research from biology, neuro-, and cognitive science our model of peripersonal space is in particular derived from the cognitive model of egocentric space by Previc (1998) and takes input from the virtual human's sensor modalities to learn its reaching and lean-forward distances. One key feature of our computational model is the partition of peripersonal space into different subspaces. Although peripersonal space concepts have been studied in different research fields, they are widely unexplored in situations of face-to-face interaction. With regard to this aspect our work on interaction space is developed as a supplement to Kendon's F-formation system, a concept describing and analysing spatial arrangements in human interaction (Kendon, 1990). The F-formation system describes how humans arrange their body orientation and position to each other when cooperating in physical space. In our approach, we use potential field functions to facilitate the virtual human's behaviour strategies in peripersonal and interaction space. Depending on its own interaction goals, layout and position of the interaction space, the virtual human can plan its actions, for example, relocating object positions toward or away from locations reachable for all partners. This application demonstrates how the virtual human facilitates cooperative interaction in a joint task with other partners.

The remaining chapter is organized as follows. In the next section we briefly explain the terms and concepts from other research disciplines on which we base our presented work and we describe related work in modelling artificial humanoids. In Section 8.3 we propose an interpretation of the concepts, suitable for a technical framework. In Section 8.4 the work on how a virtual human learns its peripersonal space is presented. Based on the learned reaching distances, we show how information from multiple sensor modalities is organized in spatial maps to help maintain the virtual human's attentional focus and perception in peripersonal space. In Section 8.5 we present our novel approach on a computational model of interaction space by supplementing Kendon's F-Formation system using potential fields. Finally, in Section 8.6 we summarize the major aspects of our approach.

8.2 From body space to interpersonal space

In this section we highlight relevant definitions and valuable findings from technical as well as non-technical research areas on the space immediately surrounding a body. In the following we use the term *body space* when generally referring to this space, to avoid misunderstandings. It can be observed that individual body space is often analysed in terms of sensorimotor and perceptual characteristics, and commonly termed as peripersonal space, for example in engineering, cognitive neurosciences, or biology. In contrast, when body space co-occurs in interaction with others, it is usually analysed as a social phenomenon and treated in terms of social relationships depending on body distances and orientations. Of particular interest are recent works that highlight the so far unattended influence of shared body spaces on social cognitive processes.

8.2.1 Body schema and peripersonal space

Holmes and Spence (2004) presented evidence for a neural multi-sensory representation of peripersonal space that codes objects in body-centred reference frames and defines humans' actions in near space: 'Objects within peripersonal space can be grasped and manipulated; objects located beyond this space (in what is often termed 'extrapersonal space') cannot normally be reached without moving toward them' (Holmes and Spence, 2004: 94). As stated by Biocca et al. (2005) findings from research in human spatial cognition suggest that objects are represented using different types of spatial reference systems. For example, according to Longo and Lourenco (2007) peripersonal space is represented differently from extrapersonal spaces, which also supports prior work by Previc (1998). Vallar and Maravita (2009) propose that the spatial environment proximate to humans is represented in egocentric (i.e. body-centred) frames of reference, whereas distances where spatial navigation is carried out are represented in allocentric (i.e. independent from observer) frames of reference.

Previc introduced a comprehensive theoretical model of humans' three-dimensional spatial interactions containing four different realms. His model is a synthesis of existing models and findings from cognitive neuroscience and neuroscience. In addition to peripersonal (PrP) space he distinguishes three extrapersonal spaces differing in function and extent (i.e. size). Of particular interest is that he defines PrP's lateral extent as being 60° central in front of the body, corresponding to the extent of human stereoscopic vision. PrP, together with one of the extrapersonal spaces, also includes movements of the upper torso, for example leaning forward to reach for objects, which Holmes and Spence assign to extrapersonal space.

One foundation of Previc's interaction-focused model comprises findings about sensor-modality specific space representations. More recent findings in this vein are presented by Avraamides, Mello, and Greenauer (this volume). In particular the reviewed literature on the functional equivalence between different sensor modalities in building a unified spatial representation, are consistent with the multi-sensory representation of peripersonal space. One aspect discussed by Avraamides et al. is the functional equivalence between sensory and linguistic space representations which we also address in this chapter.

Work on utilizing the concept of peripersonal space as a way towards naturally structuring visual object recognition tasks in artificial systems has been conducted by Goerick et al. (2005). We use peripersonal space to structure the space covered by multiple sensor modalities.

Although research on the different representations and functions of peripersonal and extrapersonal spaces allow deeper insights, the boundaries between them is still a matter of discussion. For example, work from Caggiano et al. (2009) suggests that peripersonal space's boundary is limited to the arms' reach. In contrast, Longo and Lourenco (2006) did not discover a categorical limit but instead found that peripersonal space is scaled as a ratio of arm's length and gradually transitions to extrapersonal spaces. Their work raises a less rigid definition of peripersonal space which changes with movements of the torso and tool use. In humans, the representation of peripersonal space is intimately connected to the representation of the body structure, namely the body schema. A comprehensive discussion on body schema, as a neural representation, which integrates sensor modalities such as touch, vision, and proprioception, was provided by Gallagher (2005). This integration or mapping across the different modalities is adaptive to changes of the body, that is, if the structure of the body changes, the representation also changes. Findings from Maravita and Iriki (2004) revealed the adaptivity of body schema in tool use. They showed that the body schema representation extends and incorporates a tool wielded by monkeys. This is in line with findings by Longo and Lourenco (2007) stating that when wielding a tool with the hand, the tool is integrated into the body schema which can be interpreted as a manipulation of the arm length and therefore extends the size of peripersonal space. A lot of research in robotics was inspired by the concept of an adaptive body schema which offers a mechanism to learn tool use and to save engineers from laborious work on predefining an artificial articulated agent's—possibly changing body structure (Nabeshima et al., 2006; Fuke et al., 2007). More recently, work with different approaches on connecting body schema learning with interpretations of peripersonal space for articulated agents have also been presented (Hersch et al., 2008; Fuke et al., 2009). Our current research addresses this aspect (Nguyen and Wachsmuth, 2009), which is reconsidered in more detail in this chapter.

8.2.2 *Interpersonal space*

In this section we introduce how body space is defined when occurring in interpersonal interaction. A prominent model on interpersonal space is Hall's model of proxemics (Hall, 1966), which describes interpersonal distances starting from what he calls an intimate distance of a few inches to large-scale distances of 25 feet (7.6 m) and more. The range of peripersonal space falls roughly into the scope of intimate and personal distance. Hall's theory is a taxonomy which maps interpersonal distances to human social relationships. Therefore, it does not aim at analysing the cognitive relevance of the spaces. An example of robots changing their locomotion in the presence of humans, depending on social spaces, has been presented by Sisbot et al. (2006). We will not focus on locomotion or navigation, but instead we focus on face-to-face interaction in which a virtual human changes its motor actions depending on the space it shares with others.

Aware of the two isolated fields of neural analysis of peripersonal space and research on interpersonal behaviour, Lloyd proposes a framework that aims to investigate and interpret the 'neural mechanisms of "social space"' (Lloyd, 2009: 298). In her hypothesis she argues that the mechanism explaining how interactions with inanimate objects affect body space, can be applied to interactions with human partners. Krueger (2011: 2) addresses the same issue by stating: 'Yet little attention has been paid to what sort of role these shared action-spaces *themselves* might play in driving various social cognitive processes.' This issue is a major aspect in our framework.

Kendon (1990) presented a notably relevant work on observable patterns, called **formations**, when humans orient and group themselves in physical space. He defines an **F-formation** as a pattern, which 'arises whenever two or more people sustain a spatial and orientational relationship in which the space between them is one to which they have equal, direct, and exclusive access' (Kendon, 1990: 209). He describes in particular three typical F-formations, namely vis-à-vis, L-, and side-by-side arrangements, as depicted in Figure 8.1. Kendon also mentions an activity space in front of a single interactant, which he calls a **transactional segment**. This space somehow corresponds to peripersonal space, as previously defined. In arrangements, where several interactants' transactional segments overlap, the intersection is called **o-space** (see the grey regions in Figure 8.1). Kendon mentions, but does not elaborate on the two spaces. We amend these aspects by focusing on the space between F-formations in Sections 8.4.3 and 8.5.

Other work has been presented, using Kendon's F-formation system for proximity control of agents, which navigate in space in the presence of humans or other agents. In the work of Yamaoka et al. (2008) a robot engages the most appropriate position and orientation to present products to customers or exhibits to museum visitors. Virtual humans have also been provided with navigation models which incorporate F-formations. Pedica and Vilhjálmsson (2009) aim at modelling groups of virtual characters that show a certain degree of social presence in virtual reality environments like games. To achieve their goal they presented a behaviour model which controls group formations of virtual characters. In contrast to these works, we do not deal with creating an F-formation, but with extending o-space and sustaining cooperative interaction, once an F-formation is established.

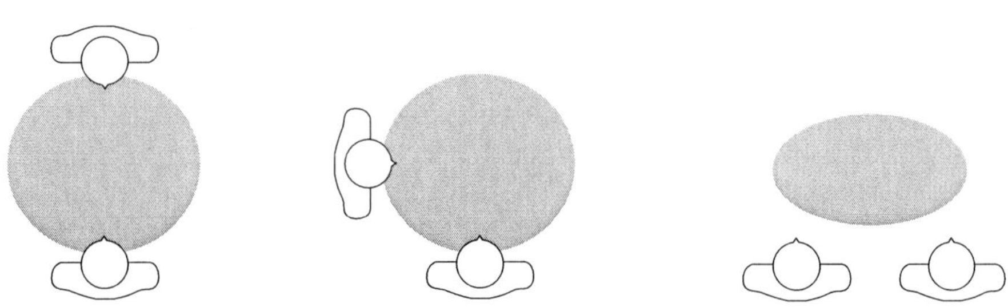

Figure 8.1 Spatial arrangements typical in F-Formations. From left to right: Vis-à-vis, L- and side-by-side arrangements

8.3 Technical framework

We first present an overview of the architecture to realize a technical system which models peripersonal space and interpersonal space at the same time (see Figure 8.2). In the following subsections we will describe the different parts in more detail. The findings from other research fields, presented in the previous section, are incorporated into our framework.

8.3.1 Body schema

The virtual human learns its body structure and the kinematic functions of the limbs by means of a recalibration approach involving tactile and proprioceptive sensor data. Thus, the limb lengths and joint positions of the kinematic skeleton are learned. This part is described in Section 8.4 and corresponds to the findings in humans, stating that body schema is learned from sensorimotor information, coding the body's kinematic structure and is adaptive to bodily changes.

8.3.2 Peripersonal space

In the technical framework one main aspect is the partition of the realm of peripersonal space into different subspaces according to findings from an interdisciplinary background. Extracted from the learned body schema they differ in spatial range, extent, and frames of reference. The core spaces are determined by their predominant sensor modality and comprise of a **touch space**, a **lean-forward space**, and a **visual attention space**. The subspaces are in line with the finding of multiple

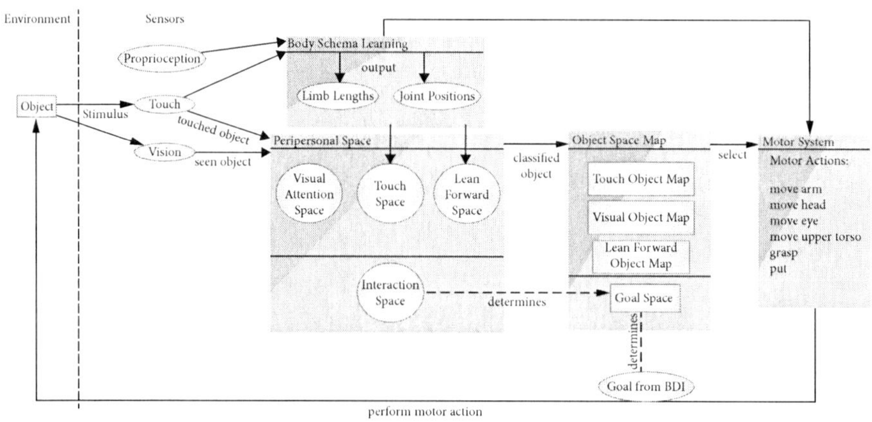

Figure 8.2 Technical Framework Overview. Information from body schema learning is utilized to build peripersonal subspaces. Objects perceived from different sensor modalities are classified into the subspaces and are maintained in object space maps. Objects outside the goal space induce a motor action, leading to a new sensor input

sensor modalities being involved in peripersonal space. In particular the lean-forward space allows for the gradual transition between peripersonal and extrapersonal space as described in Section 8.2.1. For a technical system, where sensor modalities do not necessarily cover the same spatial regions, the subspaces facilitate a more robust representation of peripersonal space. More details are described in Section 8.4.3.

8.3.2 Object space maps

Since an object can be perceived with different sensor modalities, it can be represented in different peripersonal subspaces. Each perceived object is maintained in object space maps, corresponding to the sensor modalities it was perceived from. The advantage is that the virtual human can keep track of whether objects are within its visual or touch space. Thus the virtual human can select its next movement, for example, forward-leaning or reaching for an object. As an additional spatial map we define a **goal space** within the peripersonal space. This space defines a region in peripersonal space, which the virtual human should direct its attention to, for example to objects related to a task on a table in front of the torso. The extent and location of the goal space can be determined through different factors, for instance, a new goal from the virtual human's Belief-Desire-Intention framework. The maintenance of the object space maps will be described in Section 8.4.3.

8.3.3 Motor system

Information about object positions from the object space maps is used to choose an appropriate motor action. For example, if an object has been touched, but not seen so far, the motor system will generate a head or eye movement in the direction of the touched object. By means of this, the visual attention space is shifted to cover the new object. If the object lies outside the goal space, a motor action is generated to grasp the object and put it into the current goal space.

8.3.4 Interaction space

If one or more articulated agents are entering the virtual human's peripersonal space, it assumes that they are also surrounded by a peripersonal space. The peripersonal spaces, in a first simple approach, are simulated as large as the peripersonal space of the virtual human. The overlapping spaces form the space reachable to all participants. In cooperative interaction this space is then marked as a new **goal space**. The virtual human would now centre its attention to the new space and would, for example, place objects into it, supporting the interaction. We describe this issue in Section 8.5.

8.4 A computational model of peripersonal space for a humanoid

In this section we present our computational model of peripersonal space for Max (Kopp et al., 2003), our virtual human. Multisensory abilities are a crucial factor in our framework, thus the demands we make on a virtual human's sensor system are described in Section 8.4.1. On the one hand sensor data are used to learn Max's kinematic structure using data from virtual touch and proprioception sensors, described in 8.4.2. On the other hand, since sensor modalities do not necessarily cover the same space, their combination accounts for establishing a comprehensive perception of Max's peripersonal space, described in Section 8.4.3. In the interaction scenarios we assume that peripersonal space interaction with objects usually involves a plane, lateral in front of a virtual human's body, for example, a table. In order to decrease the complexity of the model, we therefore focus on peripersonal space on a two-dimensional plane lateral, in front of Max's upper torso. The range of the spaces defined in Section 8.4.3 is thus projected on this two-dimensional plane.

8.4.1 Sensory requirements for a virtual human

In previous work (Nguyen and Wachsmuth, 2009), we developed and technically realized touch receptors for Max's whole virtual body. These receptors allow for differentiating between different qualities of tactile stimulation. Biological findings

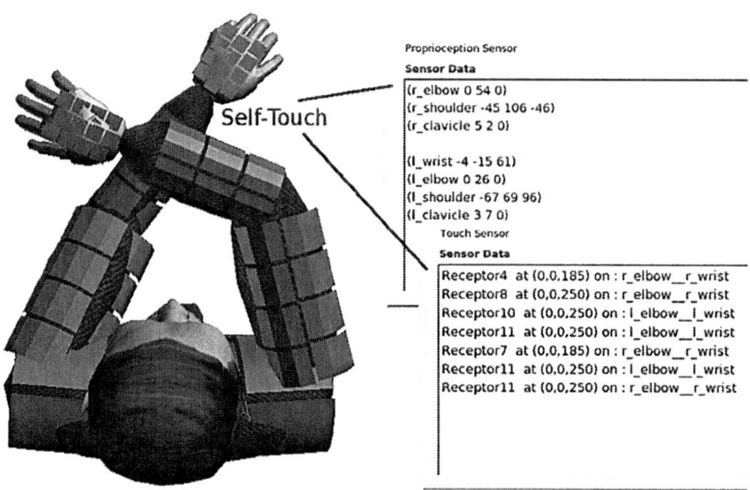

Figure 8.3 Tactile body schema learning: For each random posture, sensor consequences are output by the sensor systems. The touch sensor provides an ID of the receptor, the limb it is attached to, and the position in the frame of reference (FOR) of the corresponding limb. Angle data for the involved joints are output by the motor system, representing the proprioceptive information. (Figure taken from Nguyen and Wachsmuth, 2009)

on the human tactile system were incorporated to build an artificial sense of touch for Max. The virtual skin consists of flat quadrangle geometries varying in size, each representing a single skin receptor. Altogether the virtual skin consists of more than 200 virtual skin receptors. Max's tactile system provides information on which body limb a virtual skin receptor is attached to, together with the position in the limb's frame of reference (FOR), allowing for determining where Max is being touched. In addition to the tactile system, the virtual agent's body has an underlying anthropomorphic kinematic skeleton which consists of 57 joints with 103 Degrees of Freedom altogether. Every time Max executes a movement, the joint angle information of the involved joints is output. Synchronously with the tactile information, the proprioceptive information can be observed.

In this work, Max's virtual visual field of view corresponds to human stereoscopic vision (Previc, 1998), required for effective hand–eye coordination and thus is limited to an angle of 60°, lateral attached to his head. Head and torso movements are translated to the virtual visual field, changing its position. The angle of view is projected onto a two-dimensional plane when he is sitting or standing at a table. Objects perceived in his virtual view are represented in head-centred coordinates.

8.4.2 Tactile body schema learning for a humanoid

The model for learning the body structure takes input data given by touch sensors and joint angle data given by the proprioception sensors. In a first step, Max executes random motor actions resulting in random body postures. For each posture he perceives proprioceptive data from his joints and tactile stimuli when randomly touching positions on his body (see Figure 8.3). As described in our previous work (Nguyen and Wachsmuth, 2009) we consider the body schema as a tree of rigid transformations. In our case this kinematic tree is prescribed by the skeleton of the virtual human Max. In the initial kinematic tree, the number of joints linked in their respective order with the number of limbs are known, but the joint orientation and positions are unknown. In our model the touch receptors are attached to the limbs and their position is represented in the limb's FOR. In the kinematic tree representation, the touch receptors can therefore be represented as located along the edges.

In order to learn the real positions and orientations of the joints which also determine the limb lengths, we make use of the algorithm proposed by Hersch et al. (2008). It is a novel and general approach in online adapting joint orientations and positions in joint manipulator transformations. Our challenge in using this algorithm was to adapt it to a case different from the one to which it was originally applied. In our case we did not use visual and joint angle data, but instead replaced all visual information by tactile information in order to update all the transformations along the chains in the kinematic tree. Each time Max touches his positions on his body, the two involved skin receptors' positions are used as an input for the update algorithm.

Due to the fact that the approach takes knowledge from the body structure in advance and does not learn sensorimotor mapping, this learning method is in the strict sense a recalibration mechanism, which corresponds to the definition of body schema which adapts to changing body limbs. By means of this, the limb lengths of Max's articulated skeleton are learned, and then used to determine Max's reaching distances (Nguyen and Wachsmuth, 2011) and thus, his peripersonal space boundaries can be determined as suggested in Section 8.2.1. This aspect is described in the next section.

8.4.3 Structuring peripersonal space

According to Previc's (1998) cognitive model, each type of realm egocentrically surrounding a human is associated with certain predominant behavioural actions, for instance, visuomotor object-manipulation is predominant in peripersonal space and locomotion in action extrapersonal space. More precisely, in his model he ascribes a set of sensory-perceptual and motor operations and a predominant FOR to each realm. In order to technically realize this idea, and focusing on peripersonal space only, we decomposed his definition of peripersonal space into three major sensor components, namely vision, touch, and proprioception. Each of them spans a realm with a specific extent, FOR and predominant motor actions.

In this section the technical framework outlined in Section 8.3 and in Figure 8.2 is specified in more detail. In Table 8.1 characteristics of the spanned three subspaces of

TABLE 8.1 **Characteristics of sensor subspaces of a virtual human's peripersonal space**

	Visual attention space	Touch space	Lean-forward space
Function	Visual search, visual control	Grasping, placing, manipulation	Grasping, placing
2D location, extent			
Vertical		Lower field, Projection on frontal 2D plane	
Origin	Head	Shoulder, trunk	Shoulder, trunk
Lateral	Central 60°	360°	Frontal 180°
Radial	0–2 m	Length: shoulder joint to hand palm	Length: hip to hand palm
Frames of reference	Head centred	Limb centred	Limb centred
Motor action	Head, eye movements	Arm movements	Upper torso movements

peripersonal space are presented. The results from the learning algorithm described in the previous section determine the boundaries of the subspaces. In the next subsection we explain the content of the table and will describe in Section 8.4.3.2 how the subspaces influence spatial object maps. Finally, we show how the object maps together with motor actions, delineated in Section 8.4.3.4, satisfy a defined goal realm, which is specified in Section 8.4.3.3.

8.4.3.1 Subspaces in peripersonal space The subspaces we define within peripersonal space are deduced from Previc's (1998) work and are adopted to the technical conditions determined by Max's sensor system. The major sensor modalities assumed to be involved in peripersonal space are determining the three subspaces. Vision is mainly utilized in object search and visual manipulation control and determines a **visual attention space**. Touch is mainly utilized in object manipulation and grasping, determining a **touch space**. The function of proprioception is always utilized in peripersonal space, but plays a particular role in the placing and grasping of objects at the boundaries of peripersonal space when efforts have to be made by leaning forward; in our framework it therefore determines an additional **lean-forward space**. As mentioned previously, this subspace constitutes the gradual transition between peripersonal and extrapersonal space.

The characteristics of peripersonal subspaces we define for our virtual human are listed in Table 8.1. Their technical counterparts are shown in Figure 8.2. Each subspace defined here is associated to a main function determining the predominant motor actions carried out in the specific subspace. As mentioned at the beginning of this section, the boundaries of the subspaces are projected on an assumed two-dimensional plane on a table in front of Max. Hence, the vertical extent of each subspace is projected on a lower radial 180° two-dimensional plane. A schematic layout is depicted in Figure 8.4.

The **visual attention space**'s origin lies in the centre of the head. Its lateral extent is projected to the touch and lean-forward spaces. Stimuli perceived in Max's 60° field of view are represented in a head-centred frame of reference.

The **touch space**'s boundary is limited to the lengths of the arm limbs which were learned from the body schema. It radiates from the trunk's centre with the maximal distance covering the range between the shoulder joints and the palms of the hands. Although tactile stimuli may also affect the back of the body, that is, touch space covers a space with a lateral extent of 360° around the trunk's centre, in the following scenarios we only examine 180° of the lateral, frontal space.

The **lean-forward space**'s boundary is limited to the maximal reaching realm of the upper torso, when bending forward. From the learned body schema we compute the maximum range achieved with the arm limbs together with the spine joints which begin above the hip joint. This space thus extends touch space. Objects and stimuli perceived in both subspaces are represented in a limb-centred frame of

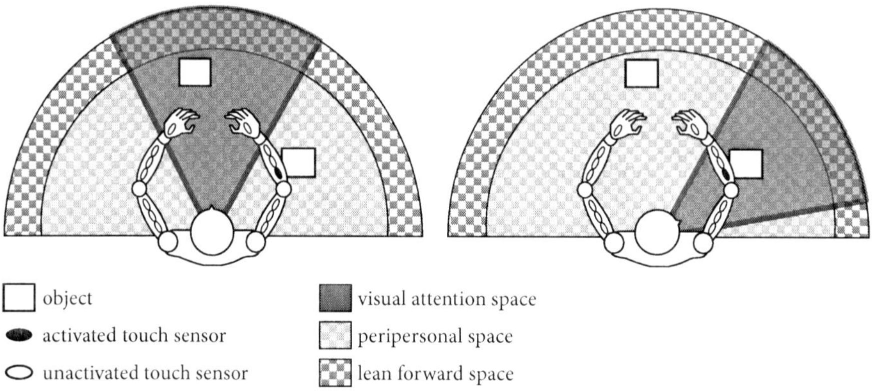

Figure 8.4 The virtual human directs its sensor attention to an object. Left: the virtual human perceives an object with the skin sensors beyond its visual attention space. The object is registered in the touch object map. Right: a motor action is selected and shifts the head and the visual attention space toward the touch-location. The object elicits a visual stimulus and is then registered to the visual object map

reference. Compared to touch space, the function of object manipulation is not predominant in lean-forward space.

In addition to the mentioned spaces, other subspaces which potentially structure Max's peripersonal space can be established in our framework. As soon as other virtual or real human(s) enter Max's proximity, we assume that they are also surrounded by peripersonal spaces. The intersection of their overlapping peripersonal spaces are registered as an **interaction space**. Depending on the sensor modality from which an object was perceived, which subspaces the object is located in is evaluated. The classified object is then registered to the corresponding object space maps (see Figure 8.2).

8.4.3.2 Object space maps Sensor information may differ in its reliability, for instance, the cause of a tactile stimulus might not be an object but a person touching Max's arm to attract his attention. Thus, in order to keep track of the objects in Max's peripersonal space, the sensor modalities have to cover the objects, depending on a predefined sensor hierarchy. An example of objects being located in different peripersonal subspaces is shown in Figure 8.4. Although the literature reviewed by Avraamides et al. suggests the functional equivalence between perceptual-derived spatial representations in humans, the authors also state that the 'representational and computational underpinnings of equivalent performance' (Avraamides et al., this volume: 18) still needs to be clarified. In our framework we assume that the visual modality may provide more immediate information about objects—for example size and shape—than the touch modality, where acquisition of the same information requires more effort by grasping the object, exploring the surface, and building a mental image. Thus, visual search is preferred over tactile manipulation, and tactile manipulation is preferred over leaning forward. In the example shown in Figure 8.4, a virtual human like Max is accidentally touching, but not seeing a virtual object, since its visual attention space at that moment is not covering the object behind its arm. In our framework, the object would be listed in the **touch**-, but not in the **visual**- or **lean-forward object map**. Due to the mentioned hierarchy, a motor action would be triggered to sense the object with the visual modality. In this case a motor action is selected to turn the virtual human's head to the location where it touched the object, which leads the visual attention space to shift to the object location. If the stimulus was caused by an object, it is additionally registered to the visual map. In the current sensor hierarchy we consider factors like immediacy/effort and reliability in the acquisition of object information (e.g. position and size). According to such factors, further object properties such as audio (e.g. spatial sound is perceived from the direction of a newly placed object) or spatial language (e.g. the interaction partner verbally describes a newly placed object location) cues might also be integrated into the sensor hierarchy.

8.4.3.3. Goal space In order to avoid collisions with objects when interacting, the virtual human may reorganize the object positions in its peripersonal space. For this purpose an additional spatial map, a **goal space** is defined, which describes his region of attention. In the example shown in Figure 8.4, we assume that the goal space is set to a default spatial region on the table, with an angle of 60° central in front of the virtual human, so that objects are easy to see, reach, and touch, and the virtual human's motions are less prone to hindrances. All sensory modalities have a preference to cover the goal space as long as no external spatial interferences or constraints are given. Each time an object is perceived, the goal space map is compared to the object space maps. If differences between the maps are found, a motor action is selected to bring the virtual objects into Max's current goal space. In the schematic layout on the left in Figure 8.4 the default goal space is the space where visual attention and touch space overlap. Due to the preferences defined for the sensor modalities, the virtual human would turn its head to the location where the touch stimulus occurred. In a next step, due to the goal space definition, described in detail in Section 8.4.4, another motor action is triggered to grasp and put the object into the goal space.

8.4.3.4 Motor actions As outlined in the previous example, motor actions are selected depending on the subspaces. Another factor in the selection of the appropriate motor action is the superposed potential fields, the topic of the next section. In touch space, arm movements are predominant motor actions for fulfilling the functions of grasping, placing, and manipulation. In lean-forward space, arm movements are combined with upper torso movements, like leaning forward, in order to grasp for or place an object. Object manipulation is not predominant in this space, since objects are more likely to be brought to touch space. Visual attention space relies on motor actions like eye movements to control the gaze and head movements to shift the entire space. Furthermore, the replacement of objects relies on the information of the potential fields defined by the goal spaces. The information from the body schema is used to translate object positions from one frame of reference to another, since the subspaces code objects in different coordinate systems.

8.4.4 Selecting actions in peripersonal space with potential fields

In order to trigger appropriate motor actions with regard to objects at each location in peripersonal space, we used the method of artificial potential fields. This method is very common in obstacle avoidance and motion and path planning for artificial agents (Khatib, 1986). An artificial potential field is an array of vectors, which defines a spatial region in which each location of the field is exposed to a force vector, describing the direction and the strength of the radiating force. For example an object's direction and the velocity of a motion can be controlled depending on the length and the direction of the force vector. Multiple potential fields can be defined for the same spatial region. By adding the fields together, a new field with attenuated or amplified forces is built.

Goal space and Max's peripersonal space are modelled as artificial potential fields. The peripersonal space is described as a repulsive field F_{peri}, defined by Equation 8.1 with tangential directions covering a semicircle, defined by Equation 8.2. The field is visualized in Figure 8.5, left. A vector between the centre of peripersonal space and any location in space is denoted by position vector **p**. We calculate the force vector $\mathbf{v}_{peri}(p)$, that is currently affecting **p**, using Equation 8.3. The parameter ξ denotes a positive scalar which influences the length of the resulting force vector. The force vectors $\mathbf{v}_{peri}(p)$ point to the frontal, sagittal mid-line, described by vector $r_{perimid}$. The field covers all **p**s within an angle of 90° to both sides of this mid-line. The regions beyond the radius r_{peri} of peripersonal space are not affected by the potential field. Therefore any $||\mathbf{p}||$ that is greater than r_{peri} results in a zero force vector. The default goal space is modelled as a selective attractive field F_{goal} defined by Equation 8.3. The field covers the angle Θ_{goal} with an angle bisector denoted by $r_{goalmid}$, and force vectors pointing away from the centre in (see Equation 8.4). The default goal space has an angle of $\Theta_{goal} = 60^\circ$, and is visualized in Figure 8.5, middle. The sum of the two fields are shown in Figure 8.5, right.

Each time Max perceives an object, the current force vector v_{res} impacting on the object is calculated using Equation 8.5. Objects outside the goal space, that have to be relocated, would be affected by force vectors, describing a path which leads in the direction of the inside of the goal space. With decreasing distance to the centre, the

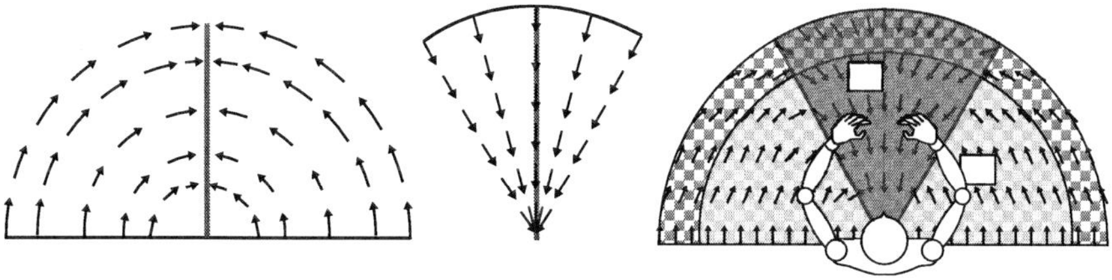

Figure 8.5 Peripersonal space modelled as tangential potential field with $r_{perimid}$ depicted as a grey line. Middle: Default goal space modelled as selective attraction field with an angle Θ_{goal} of 60° and $r_{perimid}$ depicted as a grey line. Right: Addition of the two fields shows the resulting peripersonal space field

strength of the potential field disappears, ending the path. Max is not exactly following the path, but uses the force vectors as a trigger to select a grasping motion. The end position of the path is used as a target position for a placing motion. Objects located within goal space are represented with repulsive potential fields, which prevents new objects being placed at their location. This example shows that potential fields are a suitable method to associate each point in peripersonal space to a specific behaviour, in this case motor actions. By superposing several potential fields, behaviours can be combined, allowing for more sophisticated actions. It is important to note that the potential fields facilitate the selection of an appropriate action with regard to the space it is executed in. The question of whether and when the action is executed is determined by the actual situation and interaction context.

Goal spaces in general can be determined by a new goal, raised by the Belief-Desire-Intention system or by a new established subspace of the peripersonal space. In particular a new established interaction space as described in Section 8.4.3.1 holds interesting potential field combinations and associated motor actions that we describe in Section 8.5.2.

$$\mathbf{F}_{peri}(\mathbf{p}) = \begin{cases} \xi(\frac{1}{\|\mathbf{p}\|} - \frac{1}{r_{peri}})\frac{\mathbf{p}}{\|\mathbf{p}\|^3} & \| \mathbf{p} \| \leq r_{peri}, \\ 0 & \| \mathbf{p} \| > r_{peri} \end{cases} \tag{8.1}$$

$$\mathbf{v}_{peri}(\mathbf{p}) = \begin{cases} -(\frac{\pi}{2}) * \mathbf{F}_{peri}(\mathbf{p}) & \forall \mathbf{p} | \measuredangle(\mathbf{r}_{perimid}, \mathbf{p}) \leq -(\frac{\pi}{2}), \\ (\frac{\pi}{2}) * \mathbf{F}_{peri}(\mathbf{p}) & \forall \mathbf{p} | \measuredangle(\mathbf{r}_{perimid}, \mathbf{p}) \leq (\frac{\pi}{2}), \\ 0 & else \end{cases} \tag{8.2}$$

$$\mathbf{F}_{goal}(\mathbf{p}) = -\xi \frac{\mathbf{p}}{\| \mathbf{p} \|} \tag{8.3}$$

$$\mathbf{v}_{goal}(\mathbf{p}) = \begin{cases} \mathbf{F}_{goal}(\mathbf{p}) & \forall \mathbf{p} | \measuredangle(\mathbf{r}_{goalmid}, \mathbf{p}) \leq (\frac{\Theta_{goal}}{2}), \\ 0 & else \end{cases} \tag{8.4}$$

$$\mathbf{v}_{res}(\mathbf{p}) = \mathbf{v}_{peri}(\mathbf{p}) + \mathbf{v}_{goal}(\mathbf{p}) \tag{8.5}$$

8.5 A computational model of interaction space for a humanoid

Until now we have modelled the individual peripersonal space for a virtual human with potential fields to control its actions. We will now propose how to computationally model the space between a virtual human and its interaction partners. As mentioned previously, we base our work on Kendon's F-formation system.

8.5.1 Extending the F-formation system

With our model we aim to supplement the F-formation system by adding the aspect of a measurable shared space, suitable for computational applications. In Figure 8.6 we show how we modelled the space between interactants. Compared to Figure 8.1, Kendon's o-space is now defined as the intersection of the interactants' overlapping peripersonal spaces (Figure 8.6, striped regions). We define this space as their **interaction space**. Since our definition is the intersection of all interactants' reaching realm, it conforms to Kendon's definition of the interactive space as that which is equally and exclusively reachable to all interactants, and in which they cooperate. In order for a virtual human to sustain an F-formation arrangement, once it is established, we incorporate interaction space into our described framework.

When Max perceives an interactant within an F-formation, he projects his own peripersonal space to the partner, in order to build a representation of the partner's reaching space. This process is similar to a mechanism in human cognition which is usually referred to as **spatial perspective taking**. By means of spatial perspective taking, Max is able to build a hypothesis on how far the partner can reach with her hands or by additionally leaning forward with the upper part of her body. Together with the representation of the objects, Max builds a hypothesis on which objects are or are not reachable for the partner. The fact that Max is simulating the partner's perspective by using his own body structure is commonly known as **embodied simulation** (Gallese, 2005) and is a hypothesis of how humans understand others. Gallese views the mechanism of embodied simulation as one basis for social cognition. Studies by Kessler and Thomson (2010) support the position that spatial

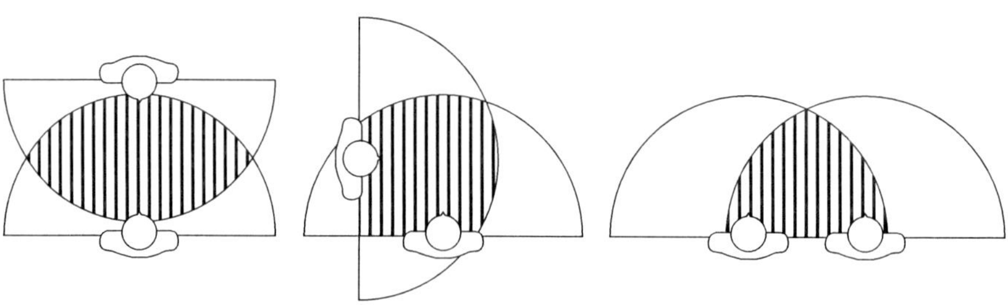

Figure 8.6 Kendon's o-spaces modelled as interaction spaces (striped regions). Interaction spaces are established by the intersection of the interactants' overlapping peripersonal spaces

perspective taking is an embodied cognitive process which might still be rooted in embodied representations, which supports our approach. However, at the current stage, in the framework we project Max's peripersonal boundaries to another partner's body structure manually. Further steps are discussed in Section 8.6.

8.5.2 Selecting and modulating actions in interaction space with potential fields

Figure 8.7 (left) shows a vis-à-vis F-formation between Max and another articulated humanoid in a virtual reality scenario. In this scenario both partners are standing at a table and cooperate in an object manipulation task where they have to cooperatively build a tower with toy blocks. The tower is to be located in the partners' interaction space. In order to build the tower, the partners have to put one tower block upon the other. A newly placed tower block has to be smaller than the previously placed one. The tower blocks are randomly placed at free locations within the partners' peripersonal spaces. In this scenario, Max helps his partner, for example, by placing into the interaction space a tower block which is not within reach for, but is needed by, his partner.

As soon as Max's and his partner's peripersonal subspaces overlap (see Figure 8.7, middle) an interaction space is established, which is defined as the new goal space. Therefore Max directs his sensor attention to this space. Max's and the partner's peripersonal spaces are modelled as selective repulsive potential fields, as shown in Equation 8.1. Their interaction space is modelled as an attractive potential field F_{inter}, as described in Equation 8.3, with its centre being the centre of a circle, which approximates interaction space. The range of the F_{inter} covers all interactants' potential fields. Thus, each force vector within their peripersonal spaces is distracted in the direction of the interaction space, as depicted in Figure 8.7, right. Within interaction space, the field strength disappears so that objects are placed within the

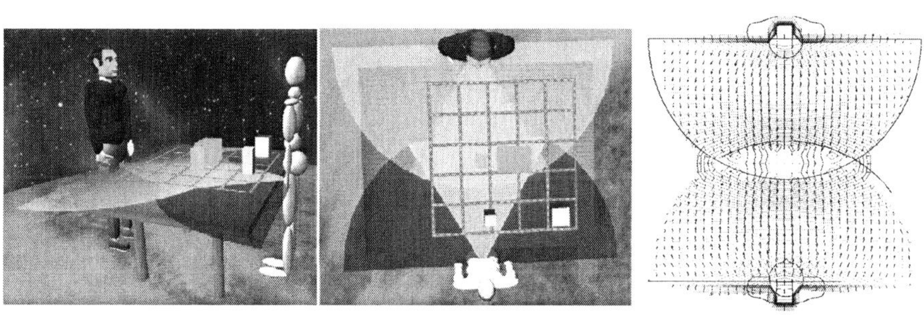

Figure 8.7 Left: Max (left) and an articulated humanoid (right) interacting in a virtual environment with visualized peripersonal subspaces. Middle: Bird-view perspective in the vis-à-vis arrangement with interaction space between the interactants. Right: The resulting potential field as a superposition of interactants' selective repulsive fields and one attractive potential field within interaction space

space. If Max is currently performing a motor action to place a tower block into interaction space, the partner could in parallel lean forward to place another tower block. This would immediately result in a larger interaction space, which leads Max's current placing motion to end before the initially planned one. Thus, interferences between Max's and his partners actions are avoided without Max having to interrupt his entire action.

8.5.2.1 Modelling cooperation and competition in F-formations In the scenario described so far, Max acts in a cooperative way as soon as an F-formation yielding an interaction space is established. The fact that Max's peripersonal space is modelled as a repulsive potential field, can be interpreted as his potential to share tower blocks with his partner, that is, to put tower blocks into interaction space, where they are accessible to the partner. However, Max's cooperative behaviour can be modulated, or even be inverted to competitive behaviour. The following parameters can be changed to control Max's behaviour.

The radius r_{peri} of the potential field in Equation 8.1 is related to the end point of the chosen motor action. By decreasing or increasing the radius of the field, the distance of a chosen motor action can be modulated. This may trigger motor actions associated with more effort, for example bending of the whole torso to lean forward. The more effort Max makes to help his partner—that is to place objects nearer to the partner—the more his action could be interpreted as helpful. Decreasing the parameter ξ in Equation 8.1 leads to a less repulsive field, whereas increasing ξ leads to a more repulsive field which again decreases or increases the lengths of the resulting force vectors. In our approach, force vector lengths control the velocity of a chosen motor action. A chosen motor action affected by a strong repulsive field would be executed in a fast motion, whereas a less repulsive field would induce a slower motion. In the example of Max being helpful by placing objects nearer to the partner, a slow motion could be interpreted as Max being more reluctant, whereas a faster motion could be interpreted as Max being more cooperative. Finally, changing the repulsive field into an attraction field may reveal Max's competitive behaviour by taking all the objects from the interaction space to his peripersonal space, where only he can access them.

8.6 Discussion

With our presented approach we aim to make two contributions. The first contribution is a computational model of peripersonal space to be employed by an artificial agent. The model allows for a multi-modal and comprehensive perception of the space immediately surrounding the agent's body and for naturally structuring the agent's behaviour. The second contribution is a model of interaction space suitable for coordinating an agent's actions performed in shared space in the pursuit of cooperative tasks. While the approach could also serve for a physical robot, our methods were developed for a virtual human.

In a first step we realized the individual body space of a virtual human in terms of a multi-sensor representation, involving touch, vision, and proprioception. This concept, commonly known as peripersonal space, takes its information from the body structure, known as body schema. Changes in body schema also affect peripersonal space, which we realized by a recalibration algorithm. In a second step we divided peripersonal space into subspaces corresponding to each sensor modality. In a third step we modelled the behaviour within peripersonal space and interaction space. The method of potential fields proves to be applicable not only for modelling the peripersonal space of a virtual human, but also for the space it shares with others, that is, the interaction space. This aspect corresponds to work by Krueger (2011) who raises the issue of connecting individual and interpersonal space in order to study social cognition. Finally, we showed how our model of interaction space for virtual humans supports their cooperative behaviour in shared space and also implies a broader range of social behaviour. With this work we demonstrate how virtual partners can show adequate spatial behaviour even in limited near-space interaction to facilitate the accomplishment of cooperative tasks. In the future, Max's spatial behaviour has to be evaluated in cases where the partner's actions are very quick. Max's performance should then also keep up with the course of the interaction. Another issue is to evaluate Max's perception when his peripersonal space is crowded with objects. Max should then still be able to form an adequate representation of the objects within peripersonal space.

Further technical work is conceivable in order to enable Max to infer the spatial perspectives of the partners and to anticipate their actions. Work on perceptual spatial perspective taking, together with action recognition between two robots, has been presented by Johnson and Demiris (2005). Work on action anticipation was presented by Gray et al. (2005) in which the robot Leonardo anticipates a human's task plans in a cooperative interaction task. However, their robots are not humanoid and the two works need to be modified and combined to be applicable to spatial interaction scenarios as described in this chapter. Another aspect to be improved was mentioned previously: in the interaction scenario between Max and another humanoid partner, Max's peripersonal space boundaries are manually projected to another partner's body structure. In a next step Max should automatically recognize the partner's bodily structure and then simulate the partner's peripersonal space. Max could also adapt to the individual bodily differences by inferring and ascribing, for instance, a smaller reaching space to people with shorter arms. By simulating the peripersonal spaces of the partners, Max may, for instance, infer whether objects are obstructed or not reachable from the partners' perspectives and hence may better help them.

One final aspect to address is the multi-modal representation of peripersonal space. In Section 8.4.2.3 we mentioned the possibility of incorporating into the framework further modalities, such as auditory and lingustic input, in order to build a more comprehensive spatial representation of Max's surroundings. However, linguistically derived spatial representations in technical systems have to be carefully modelled in terms of computational performance and its contribution beyond the perceptual derived spatial cues. Such factors need to be analysed to assure consistent sensor attention and spatial behaviour of embodied agents like Max. More discussion about the relation between perceptual and linguistic space representation is presented by Avraamides et al. (this volume). We agree with Avraamides and colleagues that a more comprehensive understanding of how different modalities establish space representations in humans improves the spatial behaviour of technical systems. We consider this viewpoint as important in enabling humanoids like Max to support humans in spatial interaction tasks.

Part 3

Human-centred assistance

9

The '*space*' in spatial assistance systems: conception, formalization, and computation*

MEHUL BHATT, CARL SCHULTZ,
AND CHRISTIAN FREKSA

9.1 The spatial premises: introduction and overview

Spatial thinking, conceptualization, and the linguistic communication of commonsense as well as expert knowledge about the world is one of the most important aspects of everyday human life. Philosophers, cognitive scientists, linguists, psycholinguists, ontologists, information theorists, computer scientists, and mathematicians have each investigated space through the perspective of the lenses afforded by their respective field of study. Interdisciplinary studies on spatial cognition, for example 'language and space', 'spatial memory', 'spatial conceptualization', 'spatial representations', 'spatial formalizations', 'spatial reasoning' are extensive and enormous to say the least. Within this book itself, other chapters present an elaborate review of the state-of-the-art for some of these fields of study.[1]

We address 'space' from a formal modelling and computational viewpoint, that is, space, as it is interpreted within the computer science disciplines concerned with the investigation of artificial intelligence and knowledge representation in general, and formal methods in spatial representation and reasoning in particular (Freksa, 1991; Aiello et al., 2007; Cohn and Renz, 2007; Bhatt et al., 2011a). Indeed, the connections

* We gratefully acknowledge the funding and support of the German Research Foundation (DFG). The work described in this paper has been conducted as part of the DFG-funded SFB/TR 8 Spatial Cognition project [DesignSpace], www.sfbtr8.spatial-cognition.de/designspace.html.

[1] In particular, chapters most directly related to our work include: (1) the chapter by Gallay et al. on navigation assistance systems for blind pedestrians; (2) the chapter by Barclay and Galton on reference object selection for locative expressions; and (3) the chapter by Taylor and Brunyé on the cognitive underpinnings of survey and route description comprehension. Further comments on the relationship with these works are included as they accrue in the rest of the chapter.

between formal models of space, and space as it occurs in language, are themselves a matter of intense research activity, for example within initiatives concerned with generalized linguistic ontologies of space—this is an aspect that we do not directly address in this chapter. However, the chapter does not lose sight of the inherently 'linguistic' aspects and presents several examples of spatio-linguistic discourse material within a broad range of case studies coming from domains with a real application impact. Against the backdrop of this material, we build practically grounded case studies and strive to concretely present the relationship between the conception, formalization, and the computational aspects of '*space*', as it occurs within systems of human assistance, assurance, and empowerment.

9.1.1 Assistance systems

The core of our perspective in this chapter is rooted in our interpretation of a general class of systems concerned with assistance, assurance, and empowerment of humans in their everyday professional and personal lives. Confining the discussion to the spatial domain, these are systems of human–computer interaction involving the representation of space from several different perspectives—the psycho-linguistic and cognitive perspective of humans, the abstract knowledge-centric (symbolic) perspective of an intelligent software construct, and the all too complex and inherently numerical or quantitative perspective of the computer. In our terminology, spatial assistance systems are cognitive agents that 'know' the properties of physical space and are skilled to deal with them in such a way that they can support humans. A special requirement for spatial assistance systems is that they are able to empathize with their human partners to a certain extent; that is, they should adapt to the needs of people rather than require people to adapt to their needs.

9.1.2 Multi-perspective semantics

In order for (our select class of) assistance systems to achieve their functional objectives of human assistance, assurance, and empowerment, the perspective of each stakeholder within the human–computer interaction process has to be adequately accounted for. This must be achieved in a manner of representation and ontological conceptualization that is consistent with the respective *spatial* interpretation of a specific stakeholder. We refer to this notion as *multi-perspective representational semantics*, or simply, multi-perspective semantics for short.

Take the case of an *architecture design assistance system*: here, a designer conceptualizes and abstracts the *structural form* of an environment with respect to an abstract and innate understanding of the shape, layout, and the connectivity of a *spatial structure*. For instance, the *designer* represents one perspective consisting of concepts and relationships from the design domain. By a complex mysterious process we term 'creativity', the designer's abstract notion of structural form is then translated into a real design. She or he uses a tool for spatial design in the form of a floor plan. This plan contains a geometric feature hierarchy consisting of points, line segments, and polygons. These geometric elements are a part of what constitutes another perspective with the system, namely, the perspective of the *design tool.* Now imagine a symbolic design reasoning module that is aimed at deriving the independent as well as inter-related inferences about the perspectives of the designer and the design tool. For example, the reasoning system may be entrusted with the responsibility that a particular design requirement such as 'spacious' or 'private' in the context of a particular 'Room' within the design are indeed satisfied. Here, the conceptual constraints expressed by the designer need to be validated with respect to the realizations within the design tool, with respect to the perspective of the *design reasoner.* In the case of this example, the design reasoner utilizes its own perspective, minimally consisting of qualitative abstractions about the topological, relative orientational, and other spatial aspects of the quantitatively modelled design.

What multi-perspective semantics suggests is that the ontological viewpoints of each of the stakeholders—*designer, design reasoner, design tool*—involved within the assistance system are important, and have to be accounted for in their own right. Taken together, these perspectives constitute the essential nature and character of 'space' within a spatial assistance system. Multi-perspective semantics is further elaborated in Section 9.4.

9.1.3 Language as a representation of structural form, and behavioural function

Continuing with the architecture design assistance scenario, consider an architect or an interior designer confronted with the objective to design an office environment in response to a pre-specified set of client requirements. The expert designer, either individually or within a group interaction, conceptualizes the design task bearing in mind the structural form of the environment with respect to the corresponding functional expectation, which the conceptualization is expected to produce. That is, a structure is envisioned with respect to an anticipated behaviour, and the behaviour or a set of behaviours satisfy a desired function.[2]

9.1.3.1 Interior space description As an example, consider either an architect, an interior designer, or more ambitiously, a system concerned with interpreting, annotating, or understanding the spatial semantics of an interior space description:

> As you enter the office, there are two tall bookcases directly to your left. Sitting on the top of the bookcase <u>farthest from you</u> is a potted plant that hangs down almost to the floor....In front of you from the door is a small <u>round</u> table with two chairs <u>on either side</u> of it....

[2] This view of the *functional* aspects of a design bears close relationships to the ontological framework of the Function-Behaviour-Structure (FBS) (Gero, 1990; Gero et al., 1991) model of the design process.

Further into the room is a small leather couch. A desk is directly behind the couch. There is also a floor lamp between the couch and the whiteboard. The four bookcases to the right of the couch extend along that wall so that the last one is parallel to the desk. Across from that bookcase, on the other side of the desk, is one last tall bookcase....There is an open path between the couch and the bookcases to get to the far end of the office. Across from the large table on the same wall as the whiteboard is a second desk that is arranged perpendicularly to the first desk.... The back wall of the office is mostly composed of a large window that extends almost to the floor. A radiator with a flat top is under the window. Several more potted plants are sitting on the ledge of the radiator as well as some coffee mugs and a coffee maker.

(Volen 258 Center Description, ISO-Space 2010 Workshop Documents, Pustejovsky, 2010.)

We have derived a design of the interior description on the basis of the Volen 258 text (the text was provided as a handout at Airlie ISO-Space workshop, see Pustejovsky, 2010), illustrated in Figure 9.1: Figure 9.1(a) is a two-dimensional floor plan corresponding to the spatial description, whereas Figure 9.1(b) is its corresponding three-dimensional interpretation. Finally, Figure 9.2 illustrates the spatial descriptors—features and relationships—that essentially determine the overall constitution of the interior space. It is easy to see the crucial role of paths and patterns such as 'circularity', 'extends_along', 'further_into', 'open_path', 'composed_of'.[3] From the viewpoint of computing spatial relations for design, the spatio-linguistic, semantic, qualitative, and quantitative interpretation of descriptors such as these acquires a significant role within systems, that is, automated reasoning processes. In the domain of design assistance, these processes are concerned primarily with two key aspects:

- *design engineering*: this refers to the creation of a structural form that conforms to a behavioural and functional specification;
- *design reverse engineering*: given a structural specification, this corresponds to inferring the extent to which a precise structure fulfils a set of anticipated functional requirements. In this form of analysis, often the easier approach is to look for malfunction, as opposed to an extensive enumeration of the functional aspects.

The above described notion of structural forms and their corresponding behavioural and functional entailments is not confined to the domain of architectural design. In this chapter, we will illustrate the utility of this line of thought to a completely different design domain, namely the creative design of media. Furthermore, outside of the design domain, we also illustrate their applicability for the domain of real-time emergency assistance.

[3] The design exercise illustrated in Figure 9.1 has been made taking into account the 'narrative-like' structure of the text as a whole, and a consolidation of all the 'perspectives' that are offered therein.

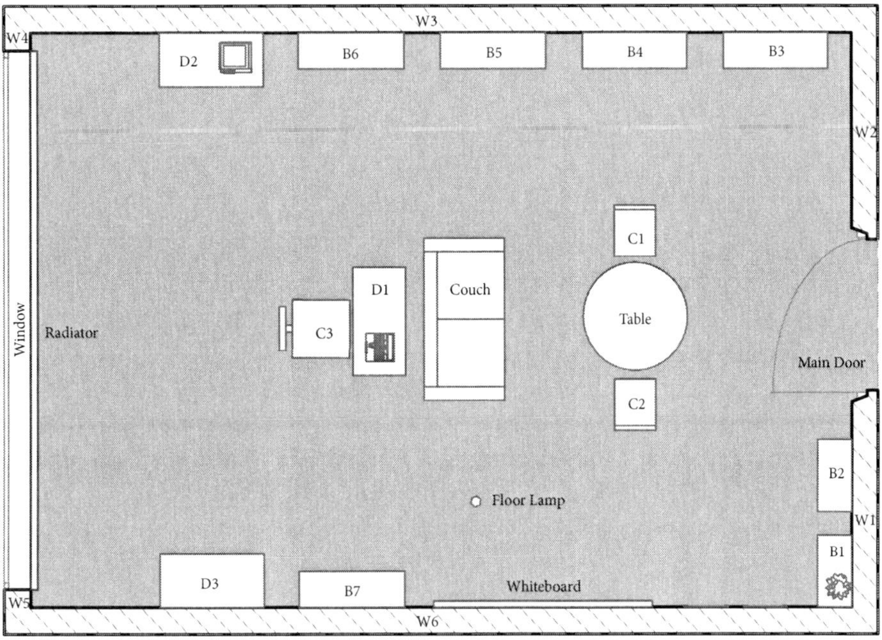

(a) 2D floor plan

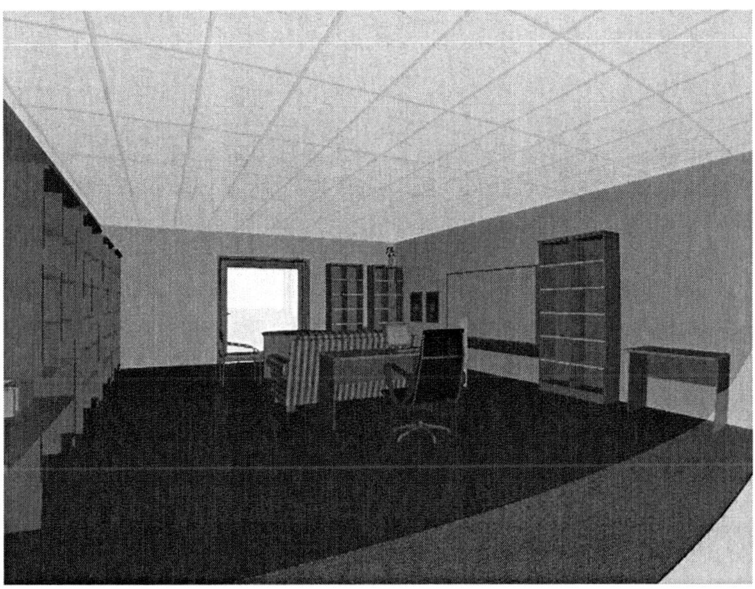

(b) 3D conceptualization (view from the window)

FIGURE 9.1 *Our interpretation of the Volen 258 Center Design*: Role of paths, patterns and commonsense knowledge in design

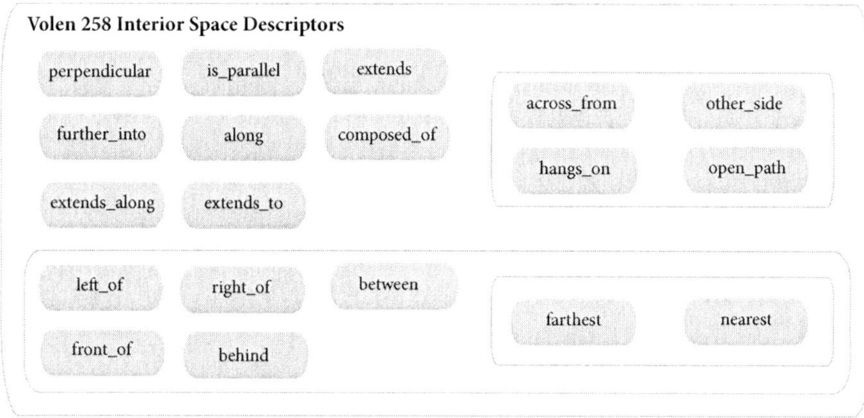

FIGURE 9.2 Key spatial descriptors

9.1.4 Aim and organization

The aim of this chapter is to: (a) pragmatically illustrate the manner in which spatio-linguistic conceptions occur within a range of spatial assistance systems involving human assistance, assurance, and empowerment; (b) provide an abstraction and formalization mechanism that may be used as an interfacing mechanism between the spatio-linguistic conceptualization of humans, and their precise interpretations within a practical system; (c) with respect to our perspective on spatial information handling for a stated purpose, illustrate the nature, role, and significance of computing spatial relations within practical spatial assistance systems.

The chapter is organized as follows: Section 9.2 lays out our interpretation of a spatial assistance system. Section 9.3 focuses on select application areas from the range introduced in our general discussion of spatial assistance in Section 9.2. Section 9.4 develops the formal framework that is required to operationalize the abstractly identifiable notion of the structural form (also developed in this section). The crux of this section lies in the elaborations on multi-perspective semantics and qualitative abstraction mechanisms in the context of structural forms. Section 9.5 demonstrates the spatial computing and underlying reasoning patterns that may be realized with the formal model presented in the chapter. Finally, Section 9.6 includes a brief summary of the chapter, together with a discussion of broad perspectives and aspects not covered in the chapter in detail.

9.2 What is a spatial assistance system?

A spatial assistance system (SAS) is a computational manifestation of the spatial decision-making and other forms of analytical and creative abilities situated along a multi-dimensional and mutually interacting spectrum: on the one hand are those abilities that typically require extensive domain-specific training, knowledge, and expertise (e.g. an architect, a cinematographer), on the other are those abilities that merely require the natural intelligence that humans are equipped with by virtue of their everyday existence (e.g. wayfinding tasks). Regardless of the precise domain of application, the crucial developmental aim of a spatial assistance system is to transfer the cognitive stress involved in a human analytical activity onto a system, by externalizing and operationalizing the decision-making processes involved therein. In essence, spatial assistance systems are instruments of human *assistance*, *assurance*, and *empowerment* and they serve one or more of these functions depending on the precise area of their application.

This chapter is focused on computational systems and frameworks for spatial awareness capabilities comparable to those of humans. Some instances of SASs include decision-support tools that require specialized *spatial reasoning capabilities*, for instance, in the manner as defined for the domain of intelligent spatial design assistance (Bhatt and Freksa, 2010). As a broad range of examples, consider the following application areas where the aforestated notion of spatial assistance may be clearly identified:

- *Spatial design assistance.* The domain of spatial design assistance encompasses all those areas where humans engage in a creative spatial design or planning activity of some form. As a key example, consider the case of architectural design assistance systems where a work-in-progress design of a building within the context of a Computer-Aided Architecture Design (CAAD) tool has to be reasoned about. Other scenarios include urban planning, tasks involving spatial conceptualization and visual thinking about spatial structures, for example preproduction assistance in media design.
- *Real-time emergency assistance.* These are systems that provide intelligent assistance for emergency scenarios such as fire-fighting, rescue, evacuation, and paramedic-support in emergency scenarios. This domain is characterized by the need to provide high-level strategic-planning assistance, for example in order to meet the immediate challenges of navigating and overcoming physical obstacles in an unfamiliar and hostile (e.g. low-visibility) environment.
- *Ambient intelligence (AmI), pervasive, and ubiquitous computing (UbiComp).* These include a broad range of technologically driven systems involving the modelling of, for instance, action, change, interaction, situational context, and the semantics of *space* and *place* within practical deployments in the real world. Some instances include mobile and location-based services, systems of activity modelling, and behaviour interpretation and analysis.
- *Indoor navigation assistance.* These are systems that provide specialized wayfinding and navigation support mechanisms for built-up environments such as

malls, exhibition centres, museums, airports, train stations and other categories of built-up public spaces.

- *Ambient Assisted Living (AAL).* This is a class of practical systems aimed at serving an empowering and assuring function within private spaces such as residences and offices. Typically, these systems involve interactions between humans, robots, and software systems.

Whereas there exist distinct categories of spatial assistance systems differing in the domain of application and the precise manner of intelligent assistance they provide, there are several fundamental similarities underlying the systemic and information-theoretic aspects of these systems. Primarily, the similarities pertain to the ontological, representational, and computational foundations that underlie their practical design and implementation. Specifically, central to these categories of assistance systems is a common foundational basis consisting of representational modalities and computational capabilities:

- from a *representational viewpoint*, modalities for semantic modelling, multi-perpective representations, and qualitative spatial abstractions acquire a central significance;
- from a *computational viewpoint* and closely connected to the representational modalities, computational techniques for conceptual and spatial reasoning define the essential character and nature of the (spatial) analytical and assistive capability that is implemented.

In essence, one may identify several fundamental capabilities with respect to the spatial conceptualization, modelling, and computing capabilities within the context of the range of systems identified herein. For instance, consider the case of spatial design assistance and emergency and navigation assistance systems. The information and computational requirements for *spatial reasoning* in the context of an indoor *Structured Spatial Environment* (SSE) bears close relationships and similarities[4]—several capabilities, for example by way of wayfinding complexity analysis, (real-time) wayfinding assistance, data analysis, artefactual simulation, virtual reality, and natural human–computer interaction that may be operationalized within these systems can be easily identified. Furthermore, the conceptualization and representation of quantitive descriptions of real (i.e. already existing) or hypothetical (i.e. being designed) indoor spatial environments is also based on shared foundations. For the case of indoor or built-up environments, and for spatial assistance scenarios such as those aforementioned, it may be presumed that geometric model(s) of the environment under consideration are available, for example by way of accurate building and

[4] These models of SSEs need to be grounded to industrial data representation standards designed for community-wide tool compliance and interoperability. Relevant remarks concerning this aspect may be found in Section 9.6.

floor plans (CAAD, design assistance), graph-based models (wayfinding assistance), and finite-element models based on computation fluid dynamics (for structural analysis, cost estimation, phenomenal studies to simulate fire spread). These models may pertain to real spatial environments that have been built (e.g. a museum), or they may pertain to an arbitrary environment that is undergoing initial conceptualization, prototyping, and design.

Spatial reasoning (for spatial awareness), however it may be defined from a cognitive, ontological, or computational viewpoint, does not differentiate between real and hypothetical environments. That is, different types of analytical capabilities that may be deemed to be within the purview of a particular interpretation of spatial awareness have to be based on high-level quantitative and qualitative perspectives that are grounded to a geometric model of an environment that may exist either in reality, or merely as a hypothetical construction within a system.

9.3 The spatio-linguistic markers within spatial assistance systems: select case studies

The aim of this section is to present high-level, yet concrete examples of the nature of spatial assistance that is applicable within a select category of spatial assistance systems. The approach here is to illustrate the spatio-linguistic conceptions that accrue within the scope of our selected case studies. These case studies are then further elaborated on in the rest of the chapter.

9.3.1 Architecture design assistance

Spatial design in general, and architectural design in particular, as a problem-solving activity typically consist of the conception—modelling—evaluation—re-modelling cycle. Essentially, a designer in this case is considered to be an *agent of change*, who may be intuitively regarded as traversing a complex configuration space of possibilities, and selecting one course of action (guided by domain knowledge, expertise, cognitive capabilities, specialized requirements, aesthetic preferences, and so forth) that produces a desired product or design. Since architectural design tasks are concerned with a spatial environment, formal representation and reasoning along conceptual and spatial dimensions is essential to ensure that the designed model satisfies key requirements that enable and facilitate its intended function.

9.3.1.1 A design task As a use case, consider the task of initial conception and design of a museum. A museum is an instance of a *structured spatial environment* that not only has a desired form and function, but is also constructed keeping in mind pre-determined aesthetic, cultural, psychological, and other subjective parameters. For example, consider the high-level spatio-linguistic conceptualizations of sobriety, austerity, and comfort with respect to low-level functional features such as connectivity, spatial distribution, and organization:

the larger trees are located in an area that is more elevated than the whole northern rim of the plot...covering creates a gentle artificial elevation that perspectively accentuates and enhances the whole architectural composition. The distribution of the construction volumes fundamentally followed a desire for horizontality, allowing one to read the continuity of the green space beyond the construction and in all directions.
This sober, rational and markedly horizontal structure is distinctive for its laminar exterior, its modular repetition, austere design and the hard quality of the materials which shape it, concrete and glass.
The main matrix-like structure had long units and was spatially austere, linking public and private spaces while at the same time delineating the two inner patios of the Museum.

(Museum Gulbenkian, Lisbon. Architecture and Landscape (Tostoes et al., 2006: 22, 23, 26, 27))

9.3.1.2 Statutory building codes In addition to high-level conceptions of the design, designers are confronted with mandatory regulations that enforce several structural constraints at different levels of complexity:

Steps of a staircase may not be connected directly to a door that opens in the direction of the steps. There has to be a landing between the staircase steps and the door. The length of this landing has to have at least the size of the door width. (Staircase / Treppen (§35(10), pg. 24. Bremen (Germany) Building code (BremLBO, 2003))

9.3.1.3 Design guides In addition to the statutory codes, one may also identify design requirements emanating from expert recommendations. For instance, the following (statutory or optional) requirements may be identified from the US Courts design guide (US GSA, 2007), and Alexander's *pattern language* (Alexander et al., 1977):

Courtroom areas used by the public must be accessible to people with disabilities. Private work areas, including the judge's bench and the courtroom deputy, law clerk, bailiff, and court reporter stations, must be adaptable to accessibility. While all judges' benches and courtroom personnel stations do not need to be immediately accessible, disabled judges and court personnel must be accommodated.

All architectural elements must be proportional and arranged hierarchically to signify orderliness. The materials employed must be consistently applied, be natural and regional in origin, be durable, and invoke a sense of permanence (US Courts Design Guide 2007 (US GSA, 2007))

Place the main part of the kitchen counter on the south and southeast side of the kitchen, with big windows around it, so that sun can flood in and fill the kitchen with yellow light both morning and afternoon. (Alexander et al., 1977)

Environmental feature descriptions such as the ones here mentioned refer to abstract, high-level spatial design patterns that correspond to specific structures at a quantitative level. For instance, it is noticeable from the descriptions of the museum that early design and conceptualization involved high-level feature descriptions of the *structural form* of the environment. Spatial features such as *continuity, spaciousness, symmetry*,[5] *modular repetition, elevation, relative positioning of entities, visibility relationships, (barrier-free)*, and *accessibility* may be easily identified. Section 9.4.2 will further illustrate the manner in which such high-level features may be modelled and reasoned about using the abstraction mechanisms that we propose in this chapter.

Contemporary professional design tools, and the precise quantitative modelling paradigm that they are based on, are incapable of exploiting the correspondence between high-level descriptions of spatial concepts and features. Such tools simply lack the ability to exploit the expertise that a designer is equipped with, but is unable to communicate to the design tool explicitly in a manner that is consistent with its inherent human-centred conceptualization, that is, *semantically* and *qualitatively* (Bhatt and Freksa, 2010). This chapter, in its remaining parts, illustrates the manner of formalization and computation that is needed to develop the assistance capability to solve design problems of this nature.

9.3.2 *Creative assistance in media design*

We interpret *creative assistance in media production* as the capability of computational tools to augment the creative capabilities of experts and artists such as cinematographers, directors, script and screenplay writers, and storyboarding artists at several stages within the media design and creation process. Consider the domain of film and comic book pre-production. Here, one may identify several forms of assistance at the production phase, for example virtual cinematography, storyboarding, and scene visualization from scripts and automatic camera control in the animation domain. Some examples from domains of our active interest follow (Bhatt and Flanagan, 2010).

9.3.2.1 Scenario description in media pre-production As an example, consider a typical creative design process between a script or screenplay writer, and storyboarding artist and cinematographer or director. The illustration in Figure 9.3 is a snapshot from the freely available media pre-production software CelTx, which is designed from the viewpoint of a wide range of design domains. It facilitates creation and organization of media projects like screenplays, films, videos, stageplays, audio plays, documentaries, machinima, comics, games, and podcasts.[6]

The illustration of Figure 9.3 corresponds to parts of a script and screenplay for a comic strip, as reproduced below:

[5] For example, one way to ensure a sense of '*justice*' and '*fairness*' within a court-room is to ensure a symmetric balance and equal access to (elements from) both sides of the courthouse.

[6] Celtx. A media pre-production environment. www.celtx.com/

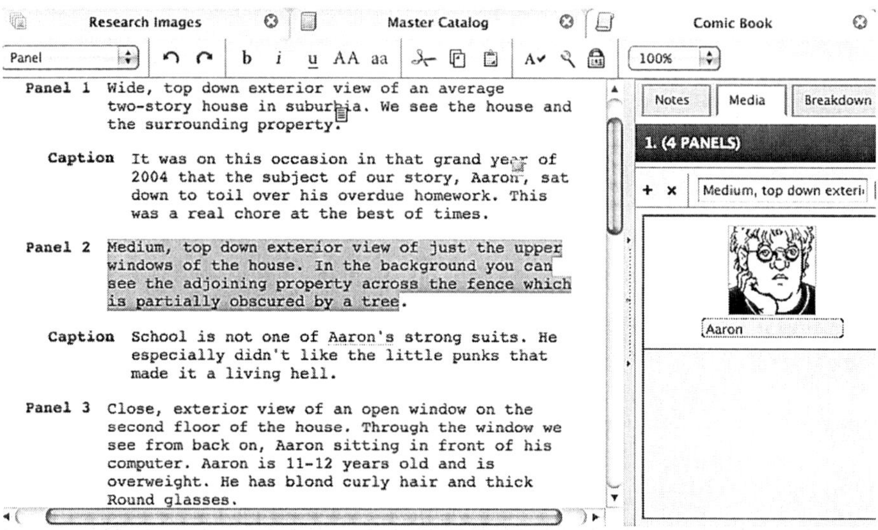

FIGURE 9.3 CelTx: A (free) media pre-production software.

Scene: Wide, top down exterior view of an average two-story house in suburbia. We see the house and the surrounding property.

Scene: Medium, top down exterior view of just the upper windows of the house. In the background you can see the adjoining property across the fence which is partially obscured by a tree.

Scene: Close, exterior view of an open window on the second floor of the house. Through the window we see from back on, Aaron is [. . .]

Scene: A wide panel that uses the lower half of the page. A close-up view of Aaron. He is staring directly ahead [. . .]

(The Mechanical Shakespeare; text source: Comic strip, Celtx pre-production software.)

As a basic level of assistance, the capability to automatically produce scenario visualizations based on the semantics of the spatial content in the discourse material is useful. For instance, it is typical for cinematographers and set designers to use the skills of storyboarding artists to start conceptualizing the precise manner and technical method of actually filming or animating a sequence. Consider the example storyboards included in Figures 9.4(a)[7] and 9.4(b).[8] Here, one may imagine the automatic production of such storyboards on the basis of the *structural form* that is semantically interpreted and derived from the scenario description of the scene.

[7] Art credit: Wikimedia, Wikiversity. www.wikiversity.org/

[8] Art credit: Peter Rubin; Goldman et al., 2006.

(a) A storyboard

FIGURE 9.4 Automatic storyboarding

(b) Action-annotated storyboards

FIGURE 9.4 Continued

The generation of these storyboards is connected to the much broader goal task of scenario and narrative completion, as explained below.

9.3.2.2 Scenario and narrative completion In general, the field of automatic cinematography aims to derive a sequence of camera shots (i.e. the camera's orientation to the actors, camera's focus, angle of view, etc.) from descriptions provided in a script (Drucker and Zeltzer, 1995; Christianson et al., 1996; He et al., 1996; Lu and Zhang, 2002; Bhatt and Flanagan, 2010). Most automatic cinematography involves using a knowledge base of filming heuristics to control the perspective or placement of a camera based on contextual cues of the scene. In this context, a film can be viewed as a hierarchy (He et al., 1996); the top of the film hierarchy is the script, which consists of a sequence of time-ordered narrative descriptions (Bhatt and Flanagan, 2010), referred to as scenes. Each scene, in turn, provides contextual information, in the form of actions and events that can be used to derive a specific camera shot. The objective of each camera shot is to capture the sequence of events in a manner that is cinematically 'pleasing', that is, it achieves a pre-determined aesthetic, dramatic, or emotional effect.[9]

[9] For elaborate studies on the theory, art, and semiotics of narration, and the 'language games' that underlie this domain, the collected works of Edward Branigan (1984, 2006) should be investigated.

(a) Establishing shot

(b) External shot

(c) Close-up / reaction shot

FIGURE 9.5 '*Film Idioms*' are heuristics for determining the relative positioning of actors and the camera within a scene

As an example, consider the simple but common scene in Figure 9.5[10] depicting a group of two actors. Within this scenario, the context of each scene is based on the current state of each actor with regards to their participation in the conversation, that is *talking*, *listening*, or *reacting*. Below is a sample script that involves two actors,

[10] Stills credit: '*Mesrine: L'instinct de mort*' (2008), France. Director: Jean-Francois Richet. www.mesrinemovie.com/

Kendra and Annika, engaged in a conversation. In the example, contextual cues are provided as key words that indicate the current state of each actor: 'Kendra starts to *talk*' and 'Annika *reacts* with astonishment' and so forth, whereas from the screen-play writer's perspective, the manner in which the scene has been conceptualized is based on heuristics guiding the placement of the actors and entities in relation to the location of the camera:

Act: Kendra and Annika
[Establishing-shot] – Kendra and Annika
Kendra starts talking to Annika – ['*dialogue*']
[Cut: mid-shot] – Annika reacts anxiously to Kendra
Kendra continues talking to Annika
[Cut: Close-up] Annika responds to Kendra – ['*astonishment*']
End.

(A sample narrative)

As the scenes progress and the conversation develops, the states of the actors change. From this information, it is the job of the (automatic) cinematographer to decide on an appropriate sequence of camera shots to properly depict the conversation. The result of this process is similar to the storyboard found in Figure 9.5, or the virtualization depicted in Figure 9.6. They show the perspective of the camera throughout the key moments of the scene. Because this scenario is so common in film, cinematic patterns have emerged that define heuristics to capture this particular type of situation, referred to by cinematographers as a *film idiom* (Arijon, 1976). These idioms have been defined for many typical cinematic situations, such as groups of actors in a conversation, or an action sequence. For instance, the 'spatial structure' associated with a film idiom may be formalized using qualitative spatial abstraction in a manner depicted in Figure 9.6. In general, a film idiom can be seen as a set of declarative rules that specify a mapping between the use of camera shots to a situational context. We formally build up on these aspects in the sections to follow, and in Section 9.5, illustrate the nature of spatio-temporal computation that is necessary to perform reasoning (e.g. scenario consistency, scene interpolation, or spatio-temporal abduction) in this particular domain of interest.

9.3.3 Real-time emergency assistance

The domain of emergency rescue work is characterized by urgent, safety-critical decision-making based on very limited and vague information that is dynamic, volatile, and possibly erroneous. These challenges exist at a range of abstraction levels, from the high-level strategic planning of firefighting captains to the immediate challenges of navigating and overcoming physical obstacles in extremely hostile, noisy, and low-visibility environments faced by firefighters. Navigation requires a

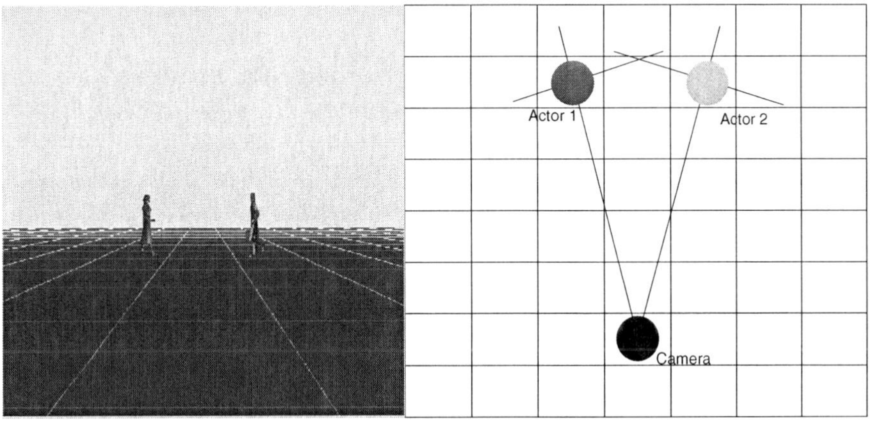

(a) Establishing shot

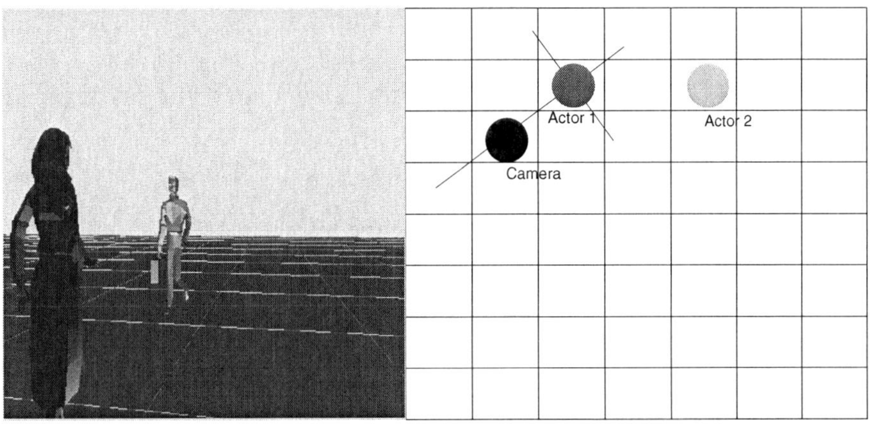

(b) External shot

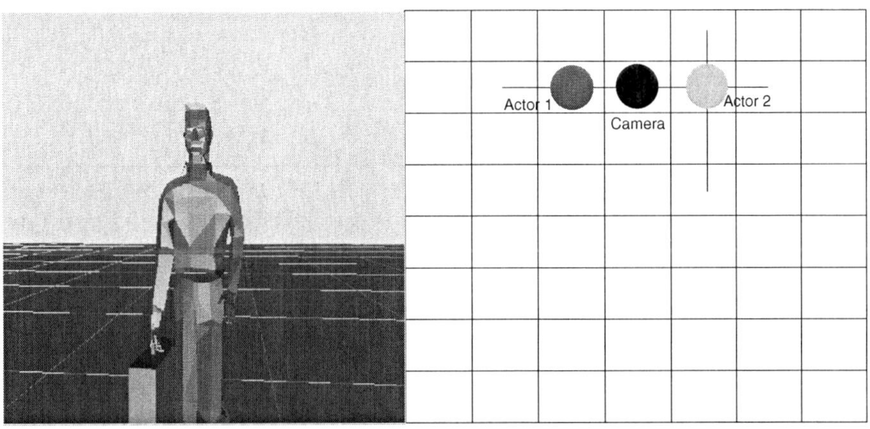

(c) Reaction shot

FIGURE 9.6 Minimal structural form of film idioms modelled using ternary orientation relationships. Virtual cinematography; *2 Avatars and 1 Virtual Camera*; animation shows perspective of the black circle, which is a virtual camera

combination of spatial modalities; in particular, the standard definition of route graphs (Werner et al., 2000) that is based on the connectivity of regions is effective at the geographic level, but is too general for providing the spatial references and directions needed by firefighters as they navigate through a burning building with zero visibility. Instead, a more specialized definition of route graphs is required that is modelled on the spatial language that firefighters use when communicating and navigating through hazardous indoor environments. For example, consider two firefighters navigating through one of the buildings at a power station site in search of victims (Figure 9.7); the following communication takes place:

FFB: [. . .] we're standing by the door where we came in . . so now we can put the right hand on the wall here

⋮

FFA: do you feel a wall on the right side?

FFB: yeah, I've had contact with the wall the whole time there was a shoe shelf on the right side so

⋮

FFB: <[. . .] we've gone into the reference room again, so now we're going into the second door on the left side>

FC: <ah, you went into door two I heard, over>

⋮

FFA: [. . .] ok . . then we have one room left in the first room, there was a door, we haven't been in there

⋮

FFA: <[. . .] we went straight on the first door on the left side and searched it>

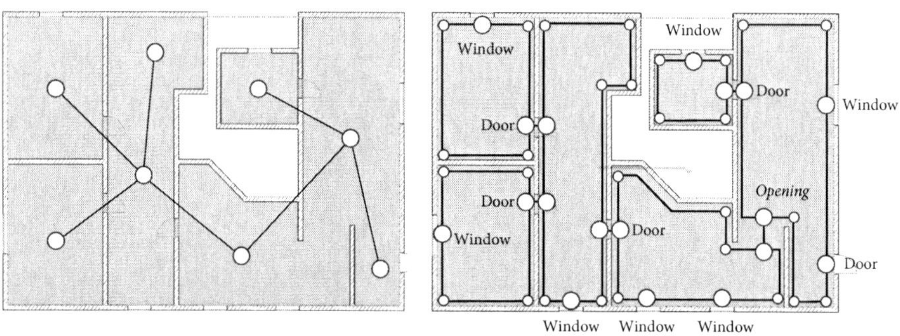

(a) Region-connectivity route graph of a building in the power-station complex

(b) Specialized route graph based on the positioning of features along walls of rooms in a building in the power-station complex

FIGURE 9.7 Specialized route graphs provide a more effective modality that supports firefighters' current approaches to navigation compared to the standard definition of route graphs (floor plan adapted from the ground floor of the Stockholm Fire Department training facility in the Agesta training centre, Kuipers et al., 2003; Lindgren, 2004)

FC: <eh, can you repeat, over>
FFA: <[...] we took the left hand on the wall and [...] went into the first room>
⋮
FFA: [...] I've reached a doorway on the left side and we are going in there and we're holding the right hand on the wall [...] and see if we find the fire-extinguisher, over

(communication recorded during firefighter training exercise by Lindgren, 2004; Lindgren et al., 2007)

As is clear from the above communication, the firefighters' sense of orientation depends heavily on reference features such as doors, walls, corners, and large pieces of furniture. As illustrated in Figure 9.7(a), the standard route graph does not provide the type of information that a firefighter needs when navigating through a building. A more effective, domain-specific route graph is defined by the arrangement of salient features such as doors and windows along room walls, as illustrated in Figure 9.7(b).

The chapter by Gallay et al. (this volume) focuses on devices that provide navigation assistance for blind pedestrians. Within this context, the task we present for first response emergency rescuers can be characterized as requiring global navigation assistance via either visual or non-visual interfaces in a low-to-zero visibility environment; as a contrast to the chapter by Gallay et al., we focus on assistance through a visual interface.[11] For example, consider the environment from a firefighter's perspective as they enter the building, as illustrated in Figure 9.8. A navigation assistance system is mounted on each of the firefighter's helmets with a small transparent display on the mask. The mask display lists the features of the room that are useful for orientation, ordered using the specialized route graph in Figure 9.7(b). The standard route graph can be used when the firefighters can move freely through the room, although when visibility is significantly reduced, the specialized route graph is more effective for navigation, as the firefighters need to rely on walls for orientation.

The role of (assistive) spatial reasoning is to augment the firefighters' sense of the surrounding space and enhance their comprehension of the environment, for example, by informing them about local reference objects and their relative qualitative spatial relationships. Effective information technology tools can assist in strategizing about search routes within the building, taking into account temperature sensor readings, fire spread models based on the known characteristics of the incident floor plans of the building, and so on.

[11] In contrast to pedestrian navigation, the relatively fine-grained localization required for indoor navigation can be achieved using a combination of dead reckoning strategies (Miller, 2006).

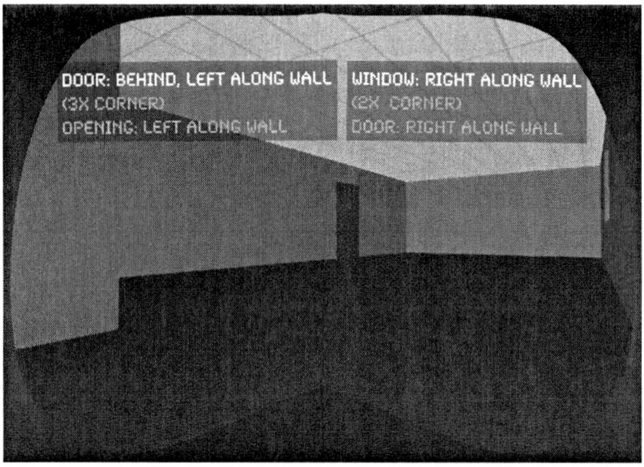

(a) Firefighter perspective with no smoke; the standard route graph is applicable for providing navigation assistance

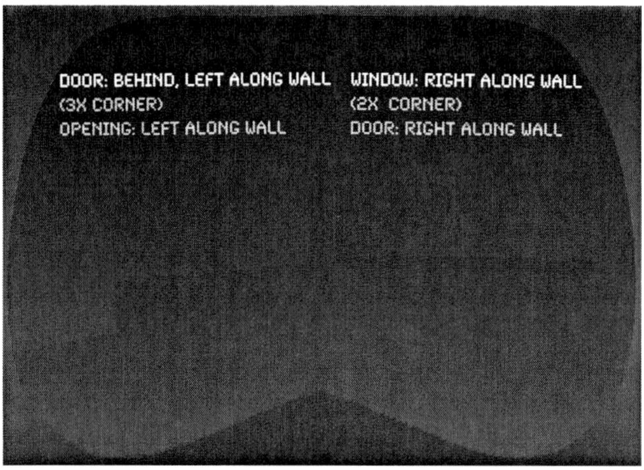

(b) Firefighter perspective in burning building with smoke-filled interiors; the specialized route graph is required for navigation assistance

FIGURE 9.8 Navigation assistance as firefighters enter a burning building

9.4 Structural form: multi-perspective representational semantics and modal abstraction

Having exemplified the spatio-linguistic markers that are identifiable as a point of human and spatial assistance system interaction in Section 9.3, the aim in this section is to elaborate on the internalization mechanisms of the respective linguistic markers within the system. The section situates itself in the context of the range of spatial assistance systems introduced in Section 9.3.

9.4.1. Formalizing spatial structure

Reconsider the range of spatial assistance systems introduced so far in Sections 9.3.1–9.3.3 by way of the architecture design assistance, creative media design assistance, and the real-time emergency assistance domains. In each of these domains, a human stakeholder is involved in the spatio-linguistic conceptualization and its communication with a system using some human–computer interaction modality. In essence, there exists some human-centred conceptualization of the spatial structure of a real or hypothetical environment, scene, or more abstractly, a *structural form*.

The following abstract notion of the *structural form* of an environment is identifiable:

> The *structural form* of an environment is an abstraction generally corresponding to the layout, shape, relative arrangement, composition at the common-sense level, of spatial entities, artefacts, and anything else—abstract or real—that may be geometrically modelled, interpreted, or derived within a design system. The only conceivable premise underlying this notion is that it should be possible to communicate the conception of the structural form using one or more spatio-linguistic modalities—e.g., spatial prepositions, path and pattern descriptions, region and point-based abstractions—that may be wholly or partially grounded to an underlying physical structure either in metric space, or in an abstract qualitative space.

For instance, the structural form may be minimally interpreted as a constraint network that determines the relative qualitative spatial relationships between the real and artefactual entities (Section 9.4.3; Figure 9.10) contained within a design. A *scene description matrix*, that is, a two-dimensional table characterizing the spatial relationship of each entity with every other entity in the model, could be the minimal basis of qualitatively abstracting an indoor spatial model. Indeed, from a formal modelling viewpoint, the qualitative model would be based on a semantics that is cognitively and linguistically grounded, and conforms to the formal relational semantics of a spatial logic (Aiello et al., 2007).

The above-stated notion of structural form based on a complete scene model is indeed minimal, and several possibilities exist for further refinement. For instance, as opposed to a complete scene description that characterizes the spatial relationship of an entity with every other entity, one may resort to a hierarchical model that exploits the natural order of organization in the physical environment. Similarly, hierarchies may be generated not only based on physical structure, but also on the basis of semantic organization. It is useful to be able to characterize *spatial patterns* such as 'circularity', 'extends_along', 'further_into', 'open_path', 'composed_of', etc. that were identified in the context of the Volen 258 scenario illustrated in Figure 9.1 in Section 9.1.

9.4.2 Multi-perspective semantics and representational modularity

Given the interpretation of structural form, an abstraction such as a 'Room' or 'ArchitecturalEntity' may be identified semantically by its placement within an ontological hierarchy and its relationships with other conceptual categories. This is what a designer must deal with during the initial design conceptualization phase. However, when these notions are transferred to a tool (e.g. a CAAD or a media pre-production tool), the same concepts acquire a new perspective. In the case of the CAAD tool, the designer must deal with points, line-segments, polygons, and other geometric primitives available within the feature hierarchy of the design tool, which, albeit necessary, are in conflict with the mental image and qualitative conceptualization of the designer. Likewise, a 'Floor' at the designer's conceptual level is abstracted as a 'Region' at the qualitative level of a reasoner and as a 'ClosedPolygon' within the geometric perspective of the design tool, thereby preserving the geometry at the quantitative level of a CAAD-based feature model (Figure 9.9). Multi-perspective representational semantics enables a knowledge-based system to make inferences about the conceptual design and its geometric interpretation within a CAAD model in a unified manner. On the representational front, the concept of multi-perspective semantics may be readily translated to representational modularity, as identified within the ontology/knowledge engineering, and conceptual modelling disciplines (Stuckenschmidt et al., 2009).

9.4.3 Multi-modal abstractions

This abstract view of spatial structure can be grounded to reality via the medium of modalities, namely, semantic, qualitative, and other forms of (graph-theoretic) abstractions that serve as an interface between the spatio-linguistic conceptualization of structural forms, and their concrete interpretation within a spatial assistance system. The following abstractions may be identified.

9.4.3.1 Spatial artefacts Semantic descriptions of designs and their requirements acquire real significance when the spatial and functional constraints are among strictly

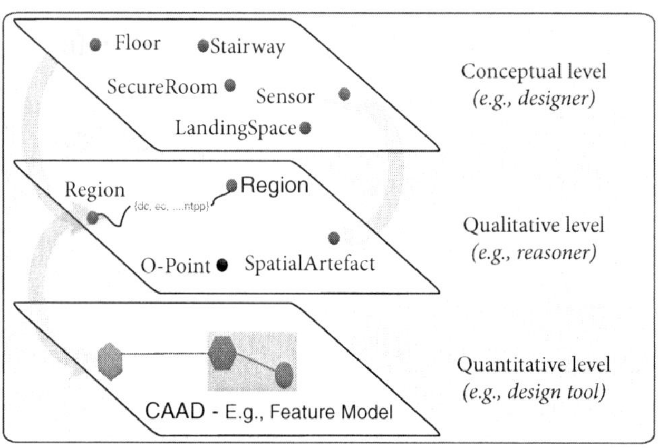

FIGURE 9.9 Multi-perspective representational semantics

spatial entities as well as abstract *spatial artefacts*. For instance, it is possible to model the spatial layout of an environment at a fine-grained level; but it is not possible to model spatial artefacts such as the range space of a sensory device (e.g. camera, motion sensor, viewpoint of an agent) in the same way. Spatial artefacts do not necessarily have a material existence; nevertheless they need to be treated as such. In general, architectural working designs only contain physical entities. Therefore, it becomes impossible for a designer to model constraints involving spatial artefacts at the design level. For instance, consider the following constraint: '*the motion-sensor should be placed such that the door connecting room A and room B is always within the sensor's range space*'. The following spatial artefacts may be identified (Figure 9.10).[12]

A1 the *operational space* denotes the region of space that an object requires to perform its intrinsic function that characterizes its utility or purpose;

A2 the *functional space* of an object denotes the region of space within which an agent must be located to manipulate or physically interact with a given object;

A3 the *range space* denotes the region of space that lies within the scope of a sensory device such as a motion or temperature sensor, or any other entity capable of visual perception. Range space may be further classified into other categories, such as *observational space* (e.g. to model the concept of the *isovist*[13]).

9.4.3.2 QvGraphs We propose Qualitatively Annotated Visibility Graphs (QvGraphs) as an extension to the concept of a *Visibility Graph* (Lozano-Pérez and Wesley, 1979; de Berg et al., 2000). In computational geometry, a visibility graph of a polygonal scene shows the intervisibility relations between a set of points (indicating locations, obstacles, and so on) in a scene, as geometrically constituted within the Euclidean plane. Specifically, visibility graph vertices correspond to point locations and edges represent a visible connection between them. QvGraphs extend visibility graphs by deriving and annotating the *visibility link* with (potentially disjunctive) knowledge about spatial relationships pertaining to one or more spatial domains such as topology, orientation, and distance. Figure 9.11(a) illustrates an example of a visibility graph of a museum lobby. The direction of the edges indicates the direction of the binary qualitative relations; for example, the 'ReceptionDesk' is 'right_of the LobbyEntrance', indicated by the direction of the edge in the QvGraph, although the 'visible' relation in this example is symmetric.

9.4.3.3 Spatial sequence graphs In natural language, it is common to refer to a sequence of objects, where the objects are ordered along some path through the environment. Consider the following expressions:

[12] Formal definitions of spatial artefacts may be found in Bhatt et al. (2009).

[13] An isovist is the set of all points visible from a given vantage point in space and with respect to an environment (Benedikt, 1979).

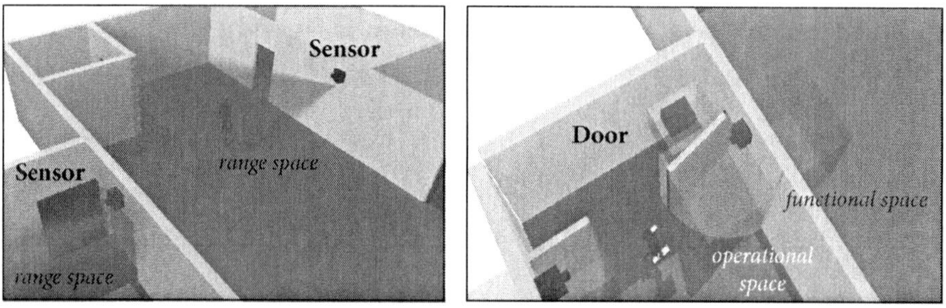

(a) Implicit artefacts within a design

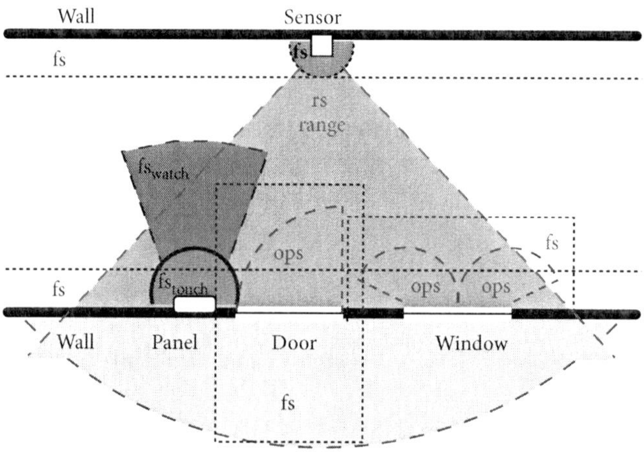

(b) Floor plan perspective of the implicit artefacts within a design

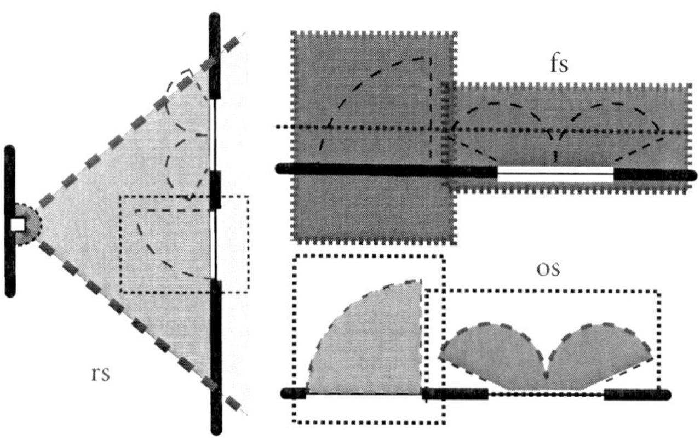

(c) Range space (rs), Functional space (fs), Operational space (os)

FIGURE 9.10 Spatial artefacts are entities which, unlike regular spatial objects, do not have a physical manifestation in reality (or within a design), but need to be treated as if they do for all practical/reasoning purposes

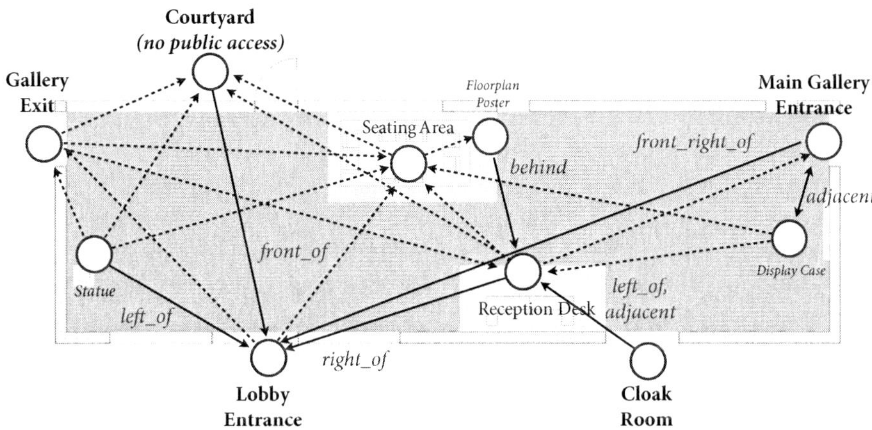

(a) Partially annotated QvGraph of the lobby. The user has specified that orientation and topological relations are relevant for this QvGraph (the qualitative annotations on the dashed edges have been omitted for clarity)

(b) The real environment: lobby area

FIGURE 9.11 Lobby area at the Gulbenkian Museum—QvGraph analyses

- Numerous paintings are mounted along the wall.
- The Far East art section is down the room after the Oriental-Islamic and Armenian rooms.
- Rivets have been placed evenly along the edge of the column.
- Further into the room is a group of partitions.

In each of these examples a virtual *path* has been implicitly defined, and the objects have been partially or totally ordered along this path. The paths typically follow the shape of some reference object such as a wall, beam, table surface edge, and so on. Moreover, the path is *directed* giving meaning to the terms *before* and *after*; one example is by specifying the start of the path to be the object that is nearest to the person referring to the sequence of objects. Note also that paths may be a simple cycle consisting of a loop involving all objects, for example, art pieces positioned along the complete perimeter of a gallery room.

This notion is formalized as a *spatial sequence model* where vertices represent objects and directed edges represent the object ordering. Edges are optionally annotated with any useful additional qualitative spatial relations between the ordered objects. Figure 9.12(a) illustrates an example of two spatial sequence models in one of the Gulbenkian Museum gallery rooms.

9.4.3.4 Route graphs A route graph, as defined in Werner et al. (2000), corresponds to a cognitively and linguistically motivated spatial representation of an environment that focuses on qualitatively capturing different routes an agent can use for navigation. The standard definition of route graphs is based on the connectivity of spaces (rooms for example), such that an agent can move freely from one space to another without necessarily passing through an intermediate space. For example, Figure 9.13 illustrates the route graph (from the perspective of art gallery visitors) of the entire Gulbenkian floor plan, and Figure 9.14 illustrates the route graph of the Gulbenkian lobby.

There is a strong connection between properties of structural layout (such as regularity), floor plan complexity, and properties of navigability that a building design affords (i.e. separate from other means of navigation such as signage) (Peponis et al., 1990; O'Neill 1991a, 1991b; Baskaya et al., 2004; Werner and Schindler, 2004). Route graph analysis can greatly assist in tasks such as building design and navigation when the route graph is derived from the appropriate structural properties according to the particular application domain. That is, domain-specific specializations of connectivity are required to model movement in different applications. Moreover, these heterogeneous route graphs must be integrated in a manner that corresponds to a person's local and more global topological comprehension of an environment (Haq and Zimring, 2003).

9.4.3.5 Flow vectors The topological information represented in route graphs is not rich enough (or at least does not make the necessary region distinctions) to specify certain qualitatively significant movement patterns of people and objects, such as modelling airflow in a relatively confined room connected to the building ventilation system (Kowadlo and Russell, 2006). Such movement patterns cannot be sufficiently expressed using route graphs without first introducing new approaches for partitioning a room into regions that are only relevant for adequately modelling

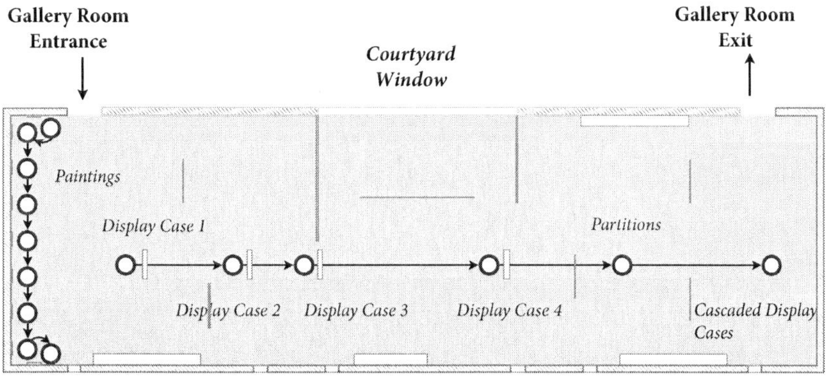

(a) Two spatial sequence models

(b) The real environment

FIGURE 9.12 Oriental-Islamic and Armenian Gallery Rooms at the Gulbenkian Museum—spatial sequence graph analyses

some particular movement phenomenon. Rather than introducing numerous specialized region distinctions, these distinctions can be implicitly embedded in the definition of a new type of model called the *flow vector graph*. Flow vector graphs are derived by directly focusing on the physical movement patterns of agents and objects rather than on the a priori definition of connectedness of the spaces that the agents and objects are moving through. Flow vector graphs are closely tied to the underlying geometry of spaces and the semantics of objects. That is, rules for deriving flow vector graphs can specify different movement patterns depending on

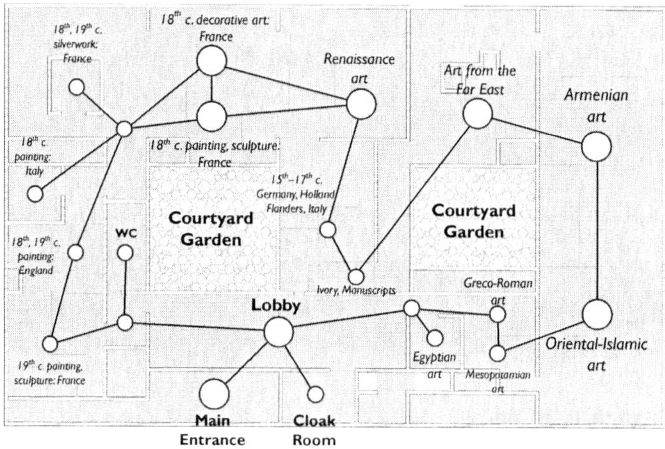

FIGURE 9.13 Route graph of the Gulbenkian floor plan from the perspective of visitors

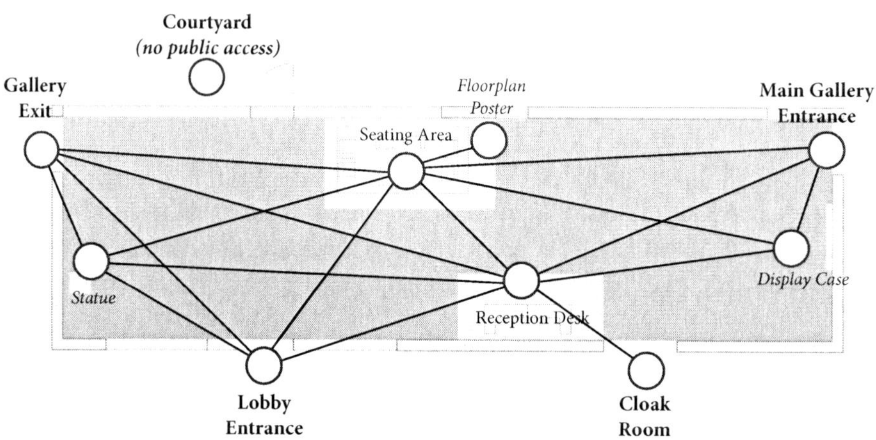

FIGURE 9.14 Route graph of the Gulbenkian lobby from the perspective of visitors

whether an object is a *statue* or a *chair*. As with route graphs, flow vector graphs typically either specify movement between different spaces within a building or specify local movement within a space (such as a room).[14]

9.4.3.6 Hierarchical models The data access framework provides access to a *hierarchical* and *multi-domain* model of space. From the viewpoint of hierarchization, the aim of this work is to develop an organization of qualitative spatial information that

[14] Flow vector graphs are further elaborated on with an example in Section 9.5.3; also see Figure 9.18.

splits the related entities into independent subsets and allows for solving spatial reasoning tasks at an adequate level of granularity. The resulting hierarchical representation should support the same reasoning and design tasks that would be possible with a flat qualitative representation but do so in a more efficient and intuitive way.

Hierarchization of buildings is defined based on spatial containment and spatial aggregation, such that a higher-level feature contains or consists of the lower-level features in the hierarchy. These hierarchical models can be based on either structural or semantic relationships. Structural containment refers to the strictly physical aggregation and containment of the components of a building; for example, a building is composed of floors (or storeys), each floor is composed of spaces (such as rooms and corridors), where each space contains features such as furniture, windows, and so on. Figure 9.15 illustrates the structural and semantic hierarchical models for a section of the Gulbenkian Museum. Semantic containment is based on the logical grouping of building components regardless of the physical relationship between those spaces. For example, two different art gallery rooms may be located

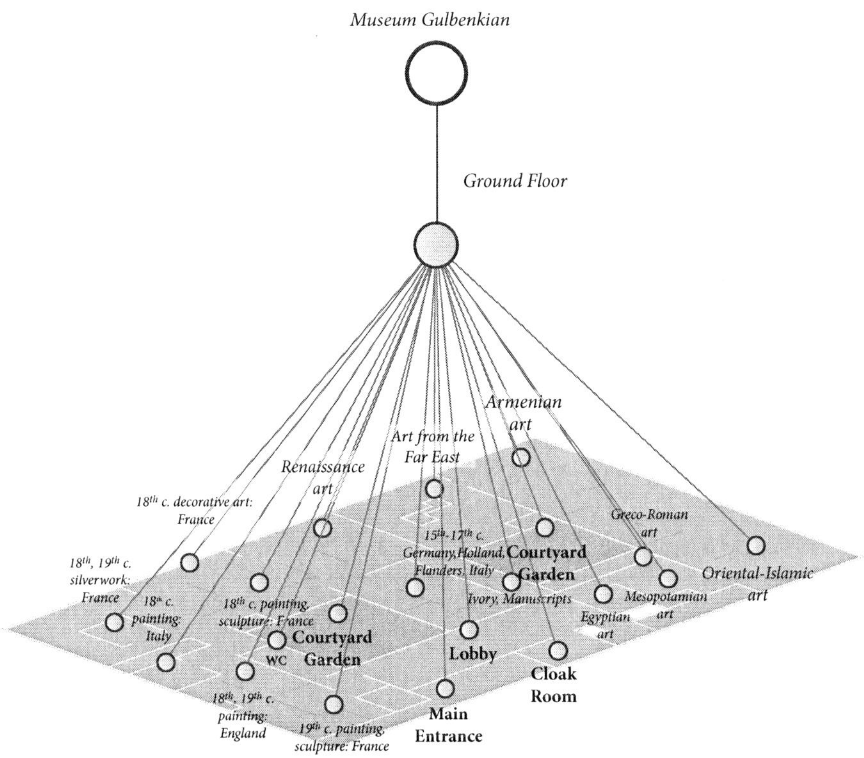

FIGURE 9.15 (a) Structural hierarchy

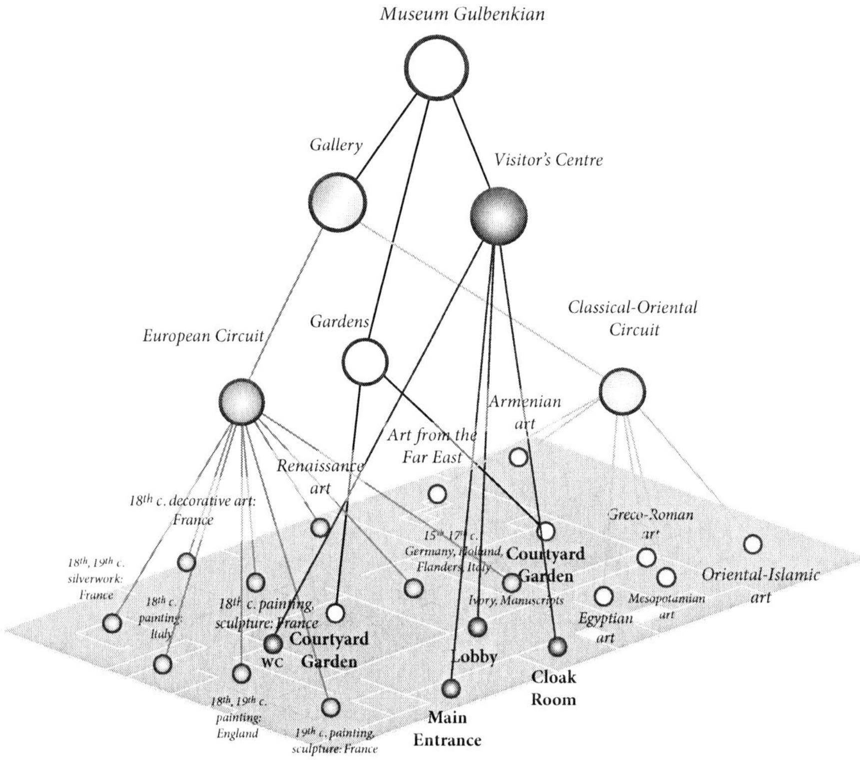

(b) Semantic hierarchy

FIGURE 9.15 Hierarchical models of the Gulbenkian Museum

on separate floors (or in completely different sections of the building), but are logically grouped together as belonging to the same art era. Another example is the set of components that are associated with *utilities* at the highest abstraction level such as water, heating, electricity, or the ventilation system, each of which, in turn, consists of utility-specific components distributed throughout the building such as air vents, ventilation shafts, terminals for controlling airflow, and so on.

9.5 Spatial computing within assistance systems

In Section 9.3, we presented the spatio-linguistic conceptualizations that occur within the range of the select application domains, namely, the architecture and media pre-production design domains, and the real-time emergency assistance scenario. Section 9.4 introduced the key notion of the 'structural form' that accrues as an abstraction mechanism for the formalism of spatial structure. Here, the focus was on the multi-perspective, qualitative, and multi-modal characterizations that constitute one model for our notion of structural form. In this section, we now turn to some instances of the types of spatial reasoning tasks that may be achieved with the abstraction and formalization mechanisms illustrated so far: Section 9.5.1 attempts to take stock of the paradigmatic underpinnings and fundamental problems within the scope of spatial reasoning. Sections 9.5.2–9.5.4 provide concrete examples grounded to the application domains introduced in the chapter.

9.5.1 Spatial computing: guiding principles and fundamental problems

The kinds of fundamental reasoning tasks that may be identified within the purview of spatial reasoning span a wide spectrum, for example including reasoning patterns such as spatial property projection, spatial simulation, spatial planning (e.g. for configuration problems), explanation with spatial information (i.e. causal explanation, hypothetical reasoning) to name a few. Both within and beyond the range of domains identified in this chapter, these are reasoning problems that involve an inherent interaction between space, actions, events, and spatial change in the backdrop of domain-specific knowledge and commonsense knowledge about the world (Bhatt, 2010).

Our notion 'spatial computing' has at least two semantic interpretations: (1) it refers to computing spatial relations, and (2) it refers to using spatial structures to do the computing (Freksa, 2013). When we use the notion 'spatial computing', we actually refer to both meanings simultaneously, that is, we use spatial structures to compute spatial relations. For instance, a simple and well-known example of spatial computing heavily used in architecture is constructive geometry: here we use a flat sheet of paper, an architect's plan, as a spatial structure that is made to correspond in certain aspects to the spatial structure of the floor of a building. In particular, angles on the plan are identical to the corresponding angles of the building. Distances typically intentionally are not chosen to be identical, to make it easier to handle the plan and to obtain a good overview of the depicted structure; however, the ratios of corresponding distances and the ratios of corresponding areas are identical.

The important aspect of the architect's representation of the floor layout of the building is that many correspondences 'automatically' are generated by their plan without requiring any computing time: in the moment the architect has correctly drawn the lines for parts of the layout, the corresponding angles, the corresponding ratios between line segments, the corresponding areas, and all other spatial correspondences are established. In other words, the architect's plan is an extremely powerful 'computer'.[15] The computing power of the architect's plan comes at a

[15] A rather disconnected, but interesting analogue may be identified here in the context of the ISO 216 Standardisation of paper sizes: all ISO paper sizes in the *A*, *B*, and *C* series have the same aspect ratio of $1:\sqrt{2}$. If a sheet with this ratio is divided into two equal halves parallel to its shortest sides, then the halves will again have the same ratio. The practical advantages of this are many, but most importantly, in

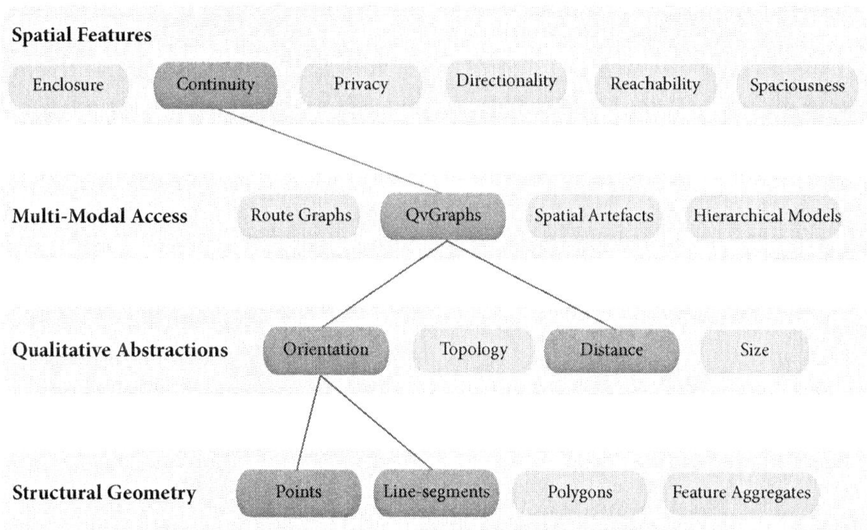

FIGURE 9.16 Spatial qualities within the architecture design assistance domain as viewed through multi-modal lenses

cost, though: you can't automatically do all the things that a regular computer can do; it is specialized at computing spatial relations.

9.5.2. The behaviour and function in spatial structure, as viewed through multi-modal lenses

The illustration in Figure 9.16 is an elaboration of the concept of multi-perspective semantics from Section 9.4.2: the top-most component comprises spatial features that may be identified at an initial design conception stage. For instance, a certain (set of) spatial structure(s) or its components may be desired to fulfil one or more of the qualities exemplified in Figure 9.16.

As one goes lower down the abstraction hierarchy of Figure 9.16, one moves close to the quantitative perspective of the design tool, which is where the precise geometry of a spatial structure resides. The basic idea is short, and simple: high-level spatial features and qualities that exist within the conceptualization of a human (e.g. designer, creative artist) correspond to one or more spatial structures at a lower level of abstraction. Some examples from the domain of architecture follow.

automatic scaling (e.g. *A*3-to-*A*4 on copiers) without compromising layout, paper, or space wastage, and estimation (e.g. postal and publisher estimates, organization and binding ease for libraries), etc. All this is achieved by just one structural constraint, which in some sense, is similar to our semantic characterization of spatial structure being a rather powerful computer in itself.

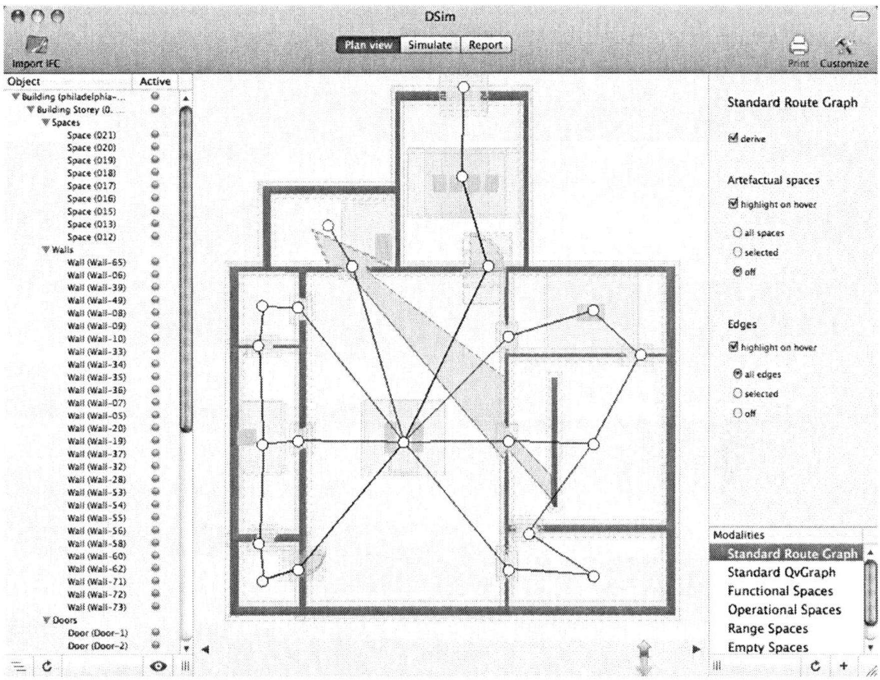

FIGURE 9.17 DSim—a design assistance system (prototype; 01.01.2011)

9.5.2.1 Privacy (Security) A typical design requirement may entail that certain parts of the environment may or may not be visible or readily accessible. For instance, it may be desired that the 'WashRoom' is as isolated as possible from other work areas, and that it may not be within the reach of sensing apparatus such as in-house cameras. As an example, such a constraint may be directly encoded at a higher level of abstraction within a rule-based programming mode (also see Figure 9.17):[16]

secure by (Door, Sensor):-
structural_geometry(Door, SGeom),
operational_space(SGeom, OpSpace),
range_space(Sensor, RgSpace),
topology(OpSpace, RgSpace, inside).

The example constraint utilizes the spatial abstractions that were introduced in Section 9.4.3 by way of the *spatial artefacts*: that is, the requirement of visibility or

[16] Indeed, such a rule-based model would technically rely on the underlying representational framework that has been discussed herein. Precise details are not relevant for this chapter, but it may be noted that we realize such rule specifications within the Constraint Logic Programming framework (Bhatt et al., 2011b). Also refer to the concluding discussions on this topic in Section 9.6.

invisibility is formulated in terms of a topological constraint between two spatial artefacts, namely, the *range space* (of a sensor), and the *operational space* (of the door). Note the manner in which even a simple rule such as this utilizes concepts and data structures from three different perspectives: the quantitative geometry coming from the CAAD model, the qualitative abstractions by way of spatial artefacts and topological relationships (e.g. *inside*), and the conceptual level of the designer, consisting of elements such as 'Doors' and 'Sensors'.

9.5.2.2 Continuity Continuity among a set of entities or locations may be, for instance, identified as mutual visibility among the entities and locations under consideration. In addition to mutual-visibility as an interpretation of continuity, one may expect additional constraints involving relative and absolute positional constraints among the entities. Additionally, one may further refine the notion by the inclusion of distance constraints, for example '*X should not only be visible from Y, it should also not be too far away...*'. At a lower level of abstraction, the notion of continuity therefore translates to a set of visibility, orientational, and distance constraints over a spatial structure, which is precisely the perspective offered by the modality of a QvGraph (see Figure 9.11(a); Section 9.4.3).

9.5.2.3 Spaciousness One may interpret a high-level spatial feature such as *spaciousness* as denoting the sense of volume and openness felt by an occupant, that is, spaciousness too may be interpreted as being related to the notions of visibility and the arrangement of objects within a room (Flynn et al., 1973; Flynn, 1977). For instance a spacious environment can have many objects around the perimeter of the room, but critically must have no large objects, or very few, in the central region of the room. Centrally located objects can occlude mutual visibility of large portions of a room, and objects positioned on walking paths can create a sense of clutter (Key, 2009).

Our objective in the above examples pertaining to spatial features has been to illustrate the usability of our multi-perspective and multi-modal abstractions as an interface between structural form, and the behaviour and function that it entails. The detailed examples for spatial features such as continuity and spaciousness are rather involved from an implementation viewpoint and have therefore been omitted for the ongoing discussion.

9.5.3 Expected navigation and movement patterns through art exhibition spaces

The manner in which a museum space is navigated by a visitor will have a direct impact on their experience of the exhibits (Wineman and Peponis, 2010). Moreover, structural features of the layout often dictate, to a greater or lesser degree depending on the agenda of the visitor, how the space is explored, and in turn, affect the degree of exposure that particular exhibition pieces are afforded. For example, Melton

(1933) presented these ideas in a seminal article after conducting a number of studies on movement patterns of museum visitors.

Consider the situation where an art director of the Gulbenkian Museum is planning to hold a temporary exhibit. The art director needs to make decisions about the placement of the temporary pieces to get the desired effect; for example the art director may want to

- elicit a sense of impact and boldness with the new pieces, or alternatively, introduce them in a subtle way;
- maximize the exposure of temporary exhibits, or alternatively, strike a balance between the prominence of the new exhibits and the permanent works.

The SAS with spatial modalities can be used to predict the expected movement patterns of visitors and thus assist in the art director's task of placing the new exhibits to evoke the desired impression in the visitors. Modelling the principles of movement patterns presented by the architecture research community requires combinations of spatial modalities.

At a global level, as people move between different spaces within the museum they tend to establish a primary set of paths referred to collectively as the *skeleton* (Kuipers et al., 2003) of the environment in their comprehension of the space. Using this skeleton as a reference, visitors explore more localized collections of gallery rooms. Kuipers et al. (2003) propose that the skeleton that emerges as a person explores an environment tends to consist of *major* paths and locations that have a relatively high degree of topological connectivity to other paths and locations (i.e. boundary relations). Wineman and Peponis (2010) link the notion of a skeleton to research on the movement of visitors in museums such as Choi (1999); they emphasize the role of *accessibility* and *visibility* between spaces in characterizing major, integrated paths that form the skeleton (Wineman and Peponis, 2010). By analysing combinations of spatial modalities, in particular, route graphs, QvGraphs, and semantic hierarchical models, an architect can build up a picture of the accessibility and visibility afforded by the museum layout. Thus, the architect can make reasonable predictions about visitor movement patterns, for example, by automatically deriving a skeleton according to connectivity and visibility metrics. Moreover, researchers have investigated structural characteristics that influence the relative duration of occupancy of a space; for example, Choi's (1999) studies lead to the following principle (as expressed succinctly in Wineman and Peponis (2010):

Principle 1: Visitors stop more often in spaces that have greater visual connections to other spaces; they also stop in spaces that are visually connected to [spaces on the skeleton]

Spaces in which visitors are likely to stay for relatively longer periods can be identified by analysing route graphs and derived skeletons in conjunction with

QvGraphs. Thus, the art director can use the spatial modalities to determine the expected global movement patterns of visitors through the museum, and then decide where to place the new exhibits according to their desired aims. For example, by placing new exhibits in areas of high visitor traffic, the art director can ensure high exposure and conjure a sense of impact surrounding the new pieces.

At a local level, people move through a gallery room based on the layout of exhibits and the shape of the gallery room. Spatial assistance systems can employ key spatial modalities to model these movement patterns. We will consider a number of principles that have been proposed in the architecture research community for visitor movement behaviour and model these using a vector flow graph. Bitgood (1995) has compiled a collection of principles that govern visitor movement within a gallery room based on well-known architecture research. For example:

Principle 2: Visitors tend to turn in the direction of the closest visible exhibit, all other factors being equal [...]

This indicates that *distance* plays a role in exhibit selection. This can be used to divide exhibits into those that are either 'near' or 'far' from a given visitor location based on a threshold shortest-path distance through the route graph, where 'near' exhibits have priority over 'far' exhibits. Exhibits that are 'near' are considered to be in the same equivalence class with respect to distance, and thus the following principles can then take effect in determining further selection.

Wineman and Peponis (2010) have proposed a collection of principles that govern visitor movement in very open-plan gallery spaces (i.e. where the visitor is not explicitly guided along a restricted set of paths):

Principle 3: The more accessible an exhibit element is from all other exhibit elements, the more likely it is to be visited. This provides the [...] hierarchy of the likelihood that an exhibit will be perceived in spatially guided movement.

Accessibility refers to how *easy* it is to navigate to an exhibit from other locations in the room. Using the route graph where each node corresponds to an exhibit, accessibility of a given exhibit can be measured by the graph-theoretic centrality value (many possible centrality metrics can be experimented with, for example the degree, betweenness, and closeness). Thus, if a visitor is presented with a selection of exhibits, all other factors being equal, they will tend to choose the most accessible exhibit.

Bitgood (1992) has proposed a precedence ordering on the decisions that museum visitors make, again based on a number of well-known architecture studies (typically taking place in North American or Western European museums).

Principle 4: [...] visitors exit a gallery by the first open door they encounter

If a visitor is near an exit door, then they will take it.

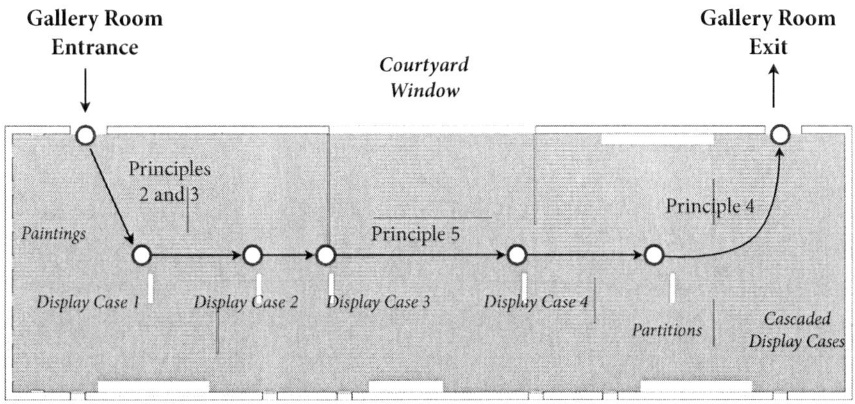

FIGURE 9.18 Expected visitor movement pattern through the Oriental-Islamic and Armenian gallery rooms

Principle 5: [. . .] people tend to walk in the same direction. Thus, if a visitor enters a gallery along the left-hand wall, [they] continue walking along this wall

This introduces a notion of visitor trajectories, such that the direction of the path defined by the previously visited exhibit e_1 and the current exhibit e_2 will influence the decision for the next exhibit e_3.

Figure 9.18 illustrates the application of these principles in generating a flow vector graph of the expected visitor movement pattern. We can observe that the central display cases dominate the paintings on the right-hand side of the entrance as visitors enter the gallery room due to proximity and accessibility (Principles 2 and 3). As the visitor travels down the gallery room, they are expected to exclusively take the left-hand path; this is because a movement trajectory is established (Principle 5) by the regularly placed display cabinets (each providing a natural *next location* due to proximity and accessibility, i.e. Principles 2 and 3). As soon as the exit becomes visible the visitors leave the room (Principle 4), ignoring the cascading display cabinets along the back wall. Using this predicted movement pattern, the art director can decide whether to place new temporary exhibits along the expected path, or perhaps position them in a more subtle position such as the courtyard window, or near the paintings at the entrance.

9.5.4 Scenario and narrative completion by spatio-temporal abduction

Re-consider the illustration in Figure 9.6 for the domain of automatic cinematography that was introduced in Section 9.3.2: the world consists of three point-abstracted entities—2 *avatars* and 1 virtual *camera*.[17] Suppose that container space

[17] The third entity in the simulation is a virtual camera that records the other two entities in the scene, and hence is not visible within the three-dimensional illustration of Figure 9.6.

is modelled as a discrete grid world together with relative orientation relationships among the entities as per the partitioning scheme of the Single-Cross Calculus (Freksa, 1992). For this discussion, further suppose that the *camera* is the only entity that is able to move, that is, change location from one grid-cell to another.

For a scenario such as this, spatio-temporal abduction serves as a basis of *scenario and narrative completion*, and for this particular example, the derivation of ideal *camera placements* serves as a side-effect of the abduction process. Figure 9.19 consists of a *narrative* (completion) from time points t_1 to t_{12}, denoting an *abduced* evolution of the system, as represented by the sequence of qualitative state descriptions for *two* stationary and *one* moving entity. For clarity, images from a three-dimensional simulation are included together with the relational illustrations for each of the time points. From an initial narrative description consisting of information about only some of the time points,[18] the narrative completion has been abduced on the basis of available *camera actions—pan, zoom, move*—and pre-specified knowledge or heuristics, referred to as *film idioms*, about desired camera placements, for example *establishing shot, external shot, mid-shot, close-up*, and so forth. In this example, the resulting narrative may be used by a virtual reality or an automatic cinematography system to generate automatic visualizations for a script.

With respect to the automatic cinematography example, it is easy to intuitively infer the general structure of causal explanation (by abduction) within spatial information. Consider the illustration in Figure 9.20 for a hypothetical (e.g. branching) situation space that characterizes the complete evolution of a system. In Figure 9.20—the situation-based history $< s_0, s_1, \ldots, s_n >$ represents one path, corresponding to an actual time line $< t_0, t_1, \ldots, t_n >$, within the overall branching-tree structured situation space. Given incomplete narrative descriptions, for example corresponding to only some ordered time points (such as in Figure 9.19) in terms of high-level spatial (e.g. topological, orientation) and occurrence information, the objective of causal explanation is to derive one or more paths from the branching situation space, that could best-fit the available narrative information. Of course, the completions that bridge the narrative by interpolating the missing spatial and action/event information have to be consistent with domain-specific and domain-independent rules (e.g. pertaining to dynamics).

A formalization of the above stated problem of scenario and narrative completion by spatio-temporal abduction in the Event Calculus may be consulted in Bhatt and Flanagan (2010). The motivations and broad research questions underlying the approach may be referred to in Bhatt (2010).

[18] These are, for instance, (implicitly) available from linguistic descriptions about *acts* and *scenes* within a drama or film script. Here, the progression of the script can be thought of as an imaginary evolution of the system.

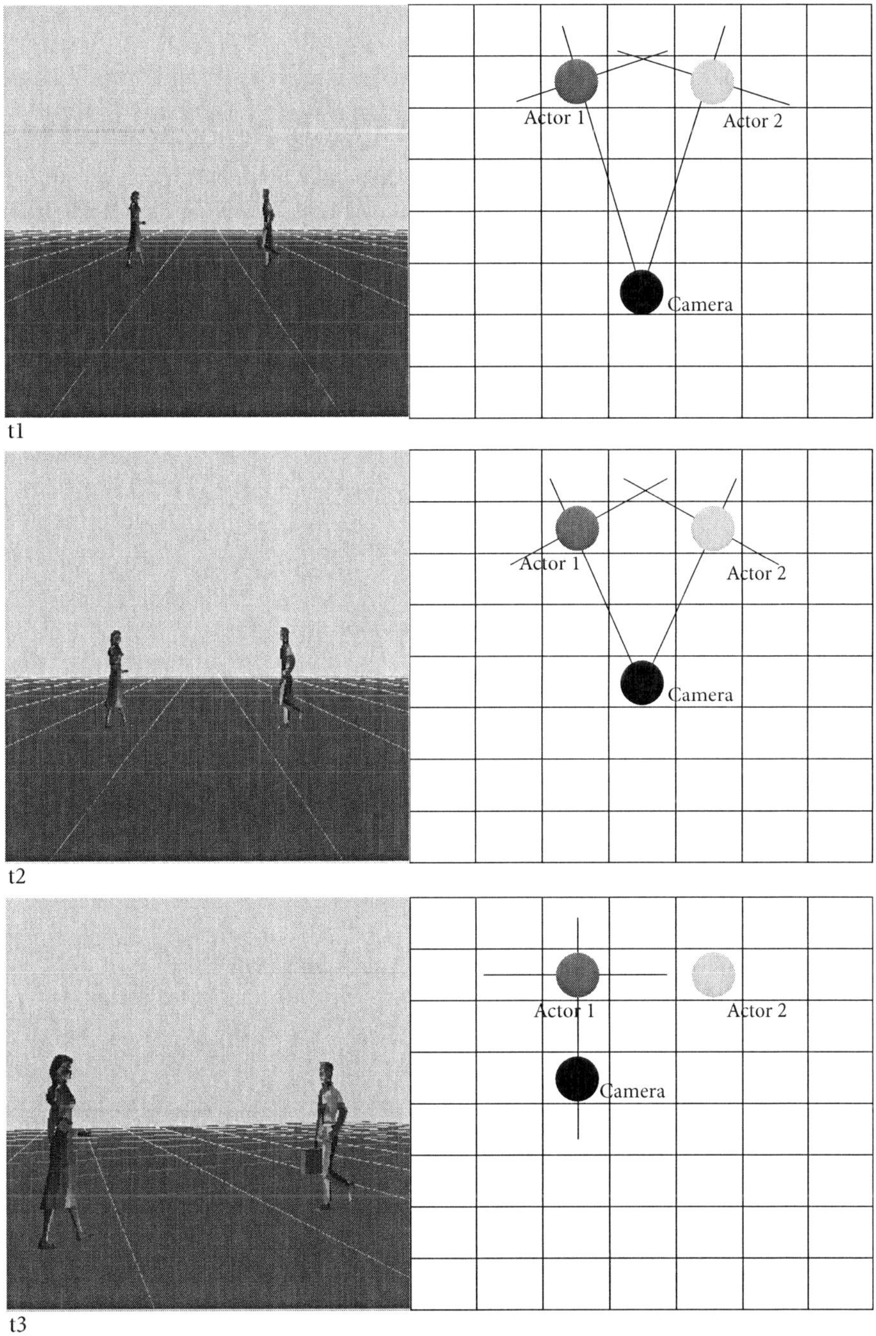

FIGURE 9.19 Scenario and narrative completion by abduction. (Source: Bhatt and Flanagan, 2010)

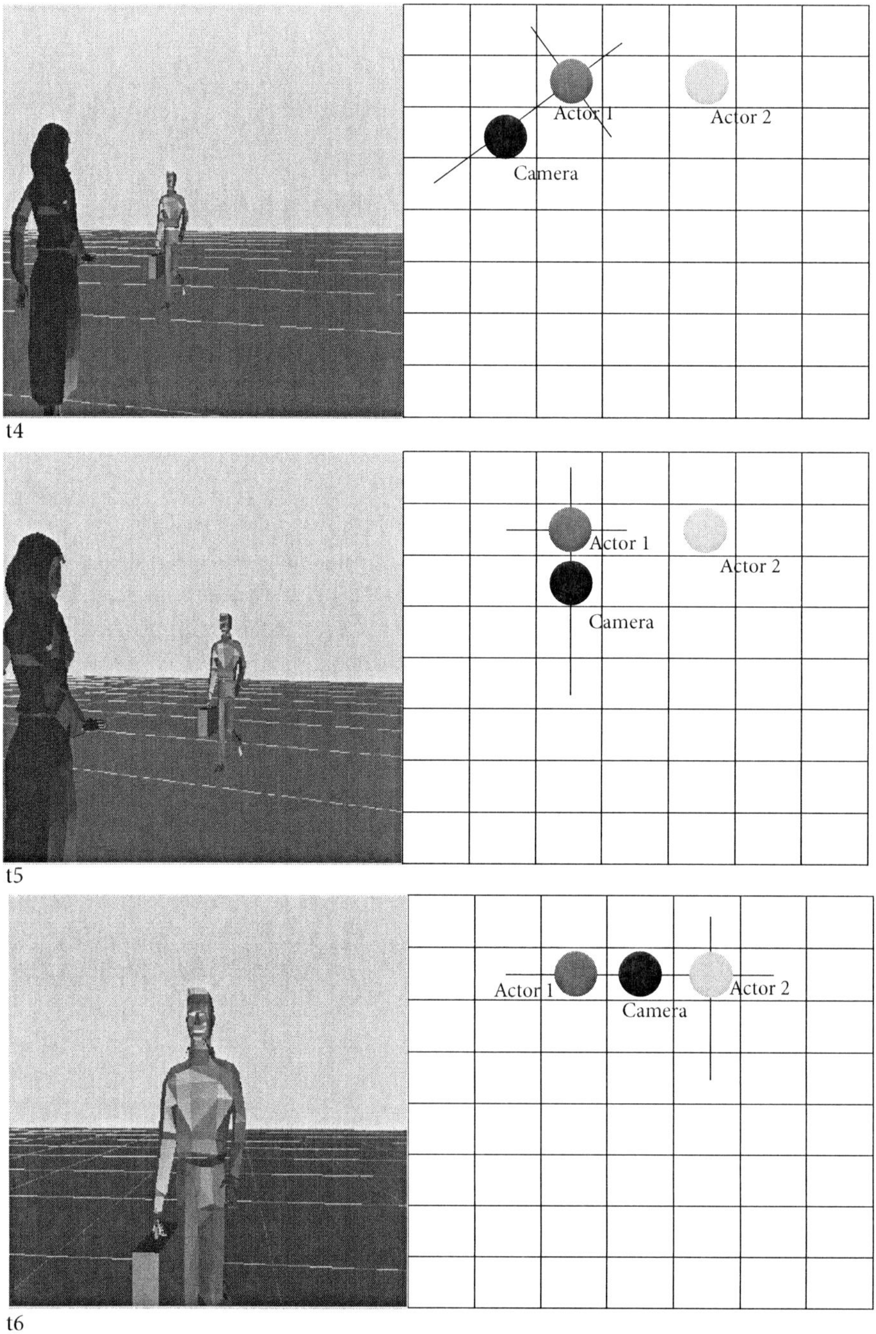

FIGURE 9.19 Continued

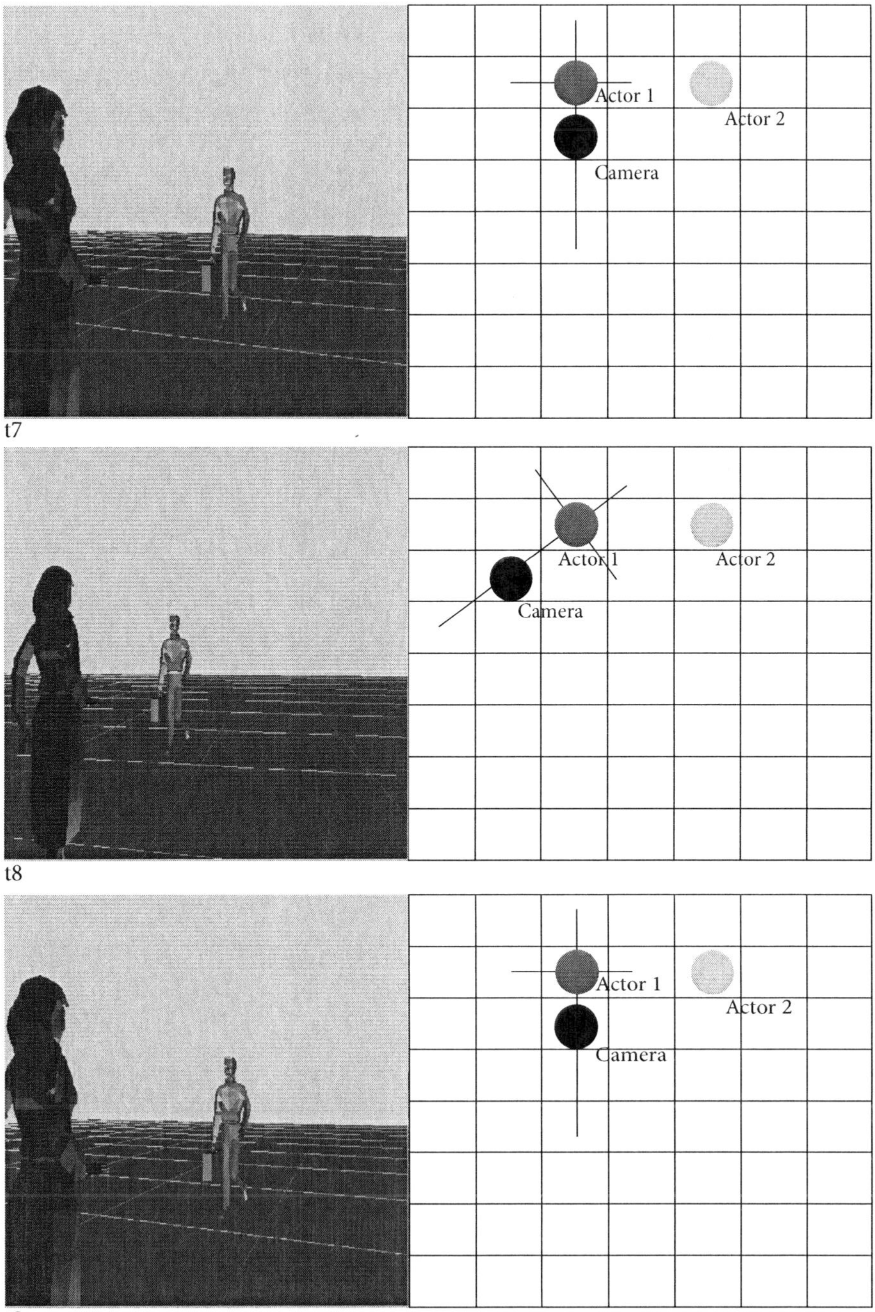

FIGURE 9.19 Continued

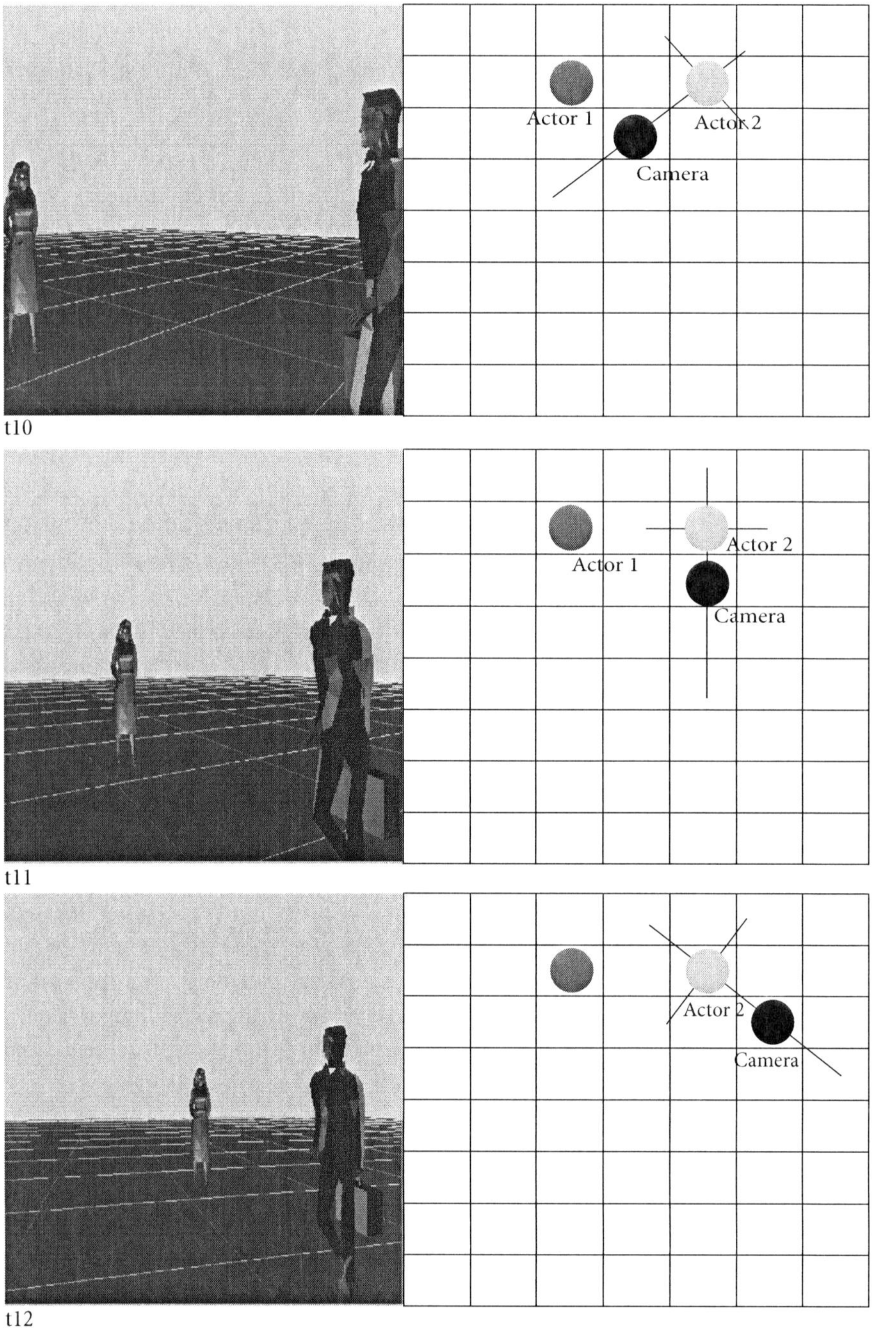

Figure 9.19 Continued

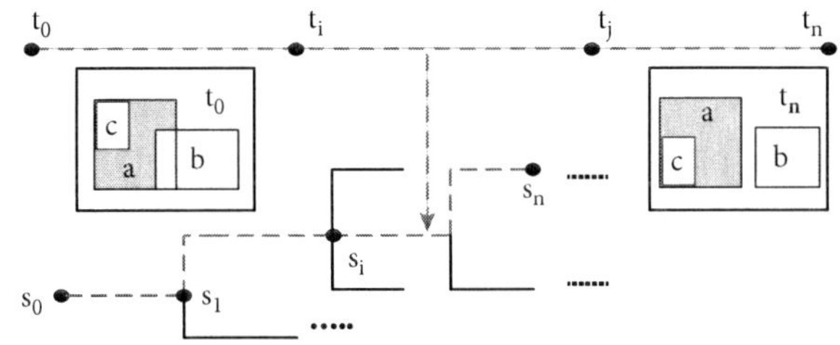

FIGURE 9.20 Branching/hypothetical situation space

9.6 Summary and discussion

Spatial assistance systems aim to transfer the cognitive stress involved in a human analytical activity onto a system, by externalizing and operationalizing the decision-making processes involved therein. From the viewpoint of human *assistance*, *assurance*, and *empowerment*, this chapter considered a range of application domains, including architecture design assistance, creative media assistance, and navigation assistance (e.g. emergency scenarios). This chapter focused on the spatial informatics concerned with the conceptual, formal, and computational aspects of 'space' within this range of *spatial assistance systems*. Specifically, the main contributions of this chapter are:

1) identification of the nature of human-centred spatio-linguistic conceptions that occur within a specific class of application domains;
2) development of a formal framework that may be used as an interfacing mechanism between the spatio-linguistic conceptualization of humans and its qualitative abstraction as per our interpretation of structural form;
3) enabling fundamental reasoning patterns, as identifiable with respect to a particular philosophy of general spatial reasoning within systems that creatively or constructively, assist, assure, and empower humans.

Whereas the above were illustrated in the context of diverse application scenarios, the discussions focused on the domain of architecture design assistance in particular. This is because spatial design as a problem-solving activity involves rich, real-world problem characterizations and solutions: the domain also enables concrete specification and illustration of problems and their solutions. The other two domains, involving creative media design, and real-time emergency assistance were chosen because in addition to being practically relevant, they offered significant breadth in order to illustrate the generality and transferability of the basic ideas concerning structural forms, multi-perspective, and multi-modal abstractions, and the computational problems addressed in the chapter.

9.6.1 Further pointers to literature

Much has been left out of the chapter: a precise definition and formalization of structural form—each constituent component thereof—and a formal treatment of the computational aspects concerning checking for functional design consistency using rule-based specifications has been omitted. For example, rule-based specifications, which we have implemented within the Constraint Logic Programming (CLP) framework (Jaffar and Maher, 1994), present their own set of fundamental challenges involving the realization of a 'spatial semantics' such that rules may refer to spatial entities and relationships (Bhatt et al., 2011b). Similarly, an elaboration of the rather contrived notion of (logic-based) spatio-temporal abduction has not been included. Spatio-temporal abduction, which can be implemented (Bhatt and Flanagan, 2010) using the Event Calculus formalism (Kowalski and Sergot, 1986), has only been described with respect to the computational structure that it acquires; much more may be elaborated on in this regard, for example, with respect to its precise formalization, the integration of the semantics of a qualitative spatial calculus within the event calculus, and the general applicability of spatio-temporal abduction in domains outside of the one discussed here (e.g. activity abduction in smart environments, geospatial dynamics). For the interested reader these aspects, and the general agenda that underlies and inspires this line of research may be consulted in Bhatt (2010), Bhatt and Freksa (2010), and Bhatt and Loke (2008).

Finally, although we presented the concept of multi-perspective semantics and multiple modalities in sufficient detail, the industrial underpinnings and relevance of this work were not presented in the chapter: at a practical level, we have developed a multi-perspective, multi-modal spatial data access framework (Schultz and Bhatt, 2010) designed to serve the informational and computational requirements of architectural design assistance systems that are intended to provide intelligent spatial decision-support and analytical capabilities. In this context, we ensure interoperability with commercial tools by utilizing the stipulations of the Building Information Model (BIM) (Eastman et al., 2008) and the Industry Foundation Classes (IFC) (Froese et al., 1999). Exemplary prototypes are also in progress as a part of the design assistance tool *DSim* (Bhatt et al., 2010) that is being designed to offer reasoning, simulation, and visualization capabilities.

10

Cognitive operations in tour planning*

INESSA SEIFERT AND THORA TENBRINK

10.1 Tour planning

Tour planning is a difficult task that people have to deal with when preparing a journey to a foreign city or a country. Travellers have to make decisions on what they are going to do, they must specify a set of *activities*. Along with the question *what*, they have to decide *where* the activities will take place. Since most of the interesting places and attractions are distributed around a country or a city, the corresponding locations have to be grouped together, to reduce the time of travelling from one place to another (Brown, 2006).

Journeys are usually constrained in time, so that tour planners have to fix the durations of their activities and put them into a feasible temporal order. The information on *what*, *when*, and *where* has to be mentally arranged from fragmented, loosely coupled knowledge derived, for example, from geographic maps, travelling guides, or conveyed by friends. A tour planning task implies finding appropriate locations where the planned activities should take place, and putting them into an appropriate temporal order which is constrained by the temporal scope of a journey (Hägerstrand, 1975).

In some respects, tour planning is similar to the abstract mathematical problem known as the 'Travelling Salesman Problem' (TSP). The TSP involves identifying the shortest path that connects a number of previously defined points with each other before returning to the starting point. This well-researched problem is known to be NP-hard (Karp, 1972), and poses a range of problems for humans (Wiener and Tenbrink, 2008) as well as machines (Applegate et al., 2007). Tour planning involves the additional challenge that locations need to be identified before the selected locations can be connected with each other. The constraints, furthermore, differ in

* Special thanks go to Holger Schultheis for his support with the inferential statistics. We wish to thank two anonymous reviewers who provided valuable comments that substantially improved the quality of the manuscript.

that the journey typically has temporal restrictions that need to be accounted for, but path optimization is not necessarily required.

Since many everyday problems resemble the TSP in one way or another, human problem solving strategies that are used across various tasks have attracted the attention of artificial intelligence (AI) researchers and cognitive psychologists. The TSP is a useful research paradigm to address and formalize spatial planning in a controlled way while varying crucial aspects, such as the number and distribution of points. The primary advantage of this paradigm is that it can be used to reduce the complexity of spatial relationships to a simple, schematized scenario, corresponding to basic human representation (Chown, this volume). Studies conducted by MacGregor and Ormerod (1996) revealed that people produce nearly optimal solutions to TSPs that include up to 20 locations. These solutions exposed a small number of crossing lines. Graham et al. (2000) showed that the processing time required for solving an instance of the TSP was a linear function of the problem size (computational complexity of O(n)). This finding leads to the assumption that people utilize mechanisms that require a low computational effort, which can be explained by the limitations of human cognitive capacity.

While human performance on well-defined spatial optimization tasks such as the TSP is a well-studied area of research, the more mundane problem solving processes taking place during holiday tour planning tasks have received less attention. In striking contrast to this lack of a substantial cognitive basis, a broad range of navigation and tour planning assistance systems are available on the market. Insights about how people solve such naturalistic problems using touristic maps would support the development of systems that provide cognitively adequate, and thus more user-friendly, assistance in holiday planning. This idea parallels the chapters by Gallay et al. and Bhatt et al. in this volume; insights from spatial cognition research are essential for the improvement of assistance software for challenging spatial tasks.

To shed light on the cognitive processes taking place during tour creating tasks, we reviewed the literature related to the TSP and further relevant problem solving tasks, and additionally conducted an explorative study that involved a natural setting, namely a two-week journey on the well-known holiday island of Crete in Greece. This work aims to understand:

- how people structure the problem space;
- what problem solving operators (Ericsson and Simon, 1984) are involved during tour planning;
- what kinds of spatial problem solving strategies help people solve the task more efficiently;
- whether the resulting tours expose small numbers of crossing lines similar to the TSP solutions; and
- how people communicate the resulting solutions.

This work focuses on problem solving operators and spatial strategies as well as conceptual schemes that people employ to communicate the resulting tours. The problem solving operators are outlined in an exhaustive encoding vocabulary that involves insights from the existing empirical findings and models of human performance on the TSP. This vocabulary was used to analyse the retrospective verbal protocols collected during the tour planning study.

10.2 Mental processing of spatial knowledge

In this section, we review empirical findings and cognitive models that provide valuable insights into how people process and manage the complexity of spatial planning tasks.

10.2.1 Cognitive Model of Planning

The computational Cognitive Model of Planning (CMP) was proposed by Hayes-Roth and Hayes-Roth (1979) in order to explain, implement, and reproduce cognitive processes taking place during a resource-limited scheduling task. To observe the cognitive processes, these authors employed the method of verbal think-aloud protocol analysis. Participants were instructed to solve a resource-limited scheduling problem: the planning of a day's errands according to the given priorities as well as spatial and temporal constraints, using a map representing an unfamiliar city environment. The instruction contained more errands than could be accomplished within a given temporal limit.

The errand sequencing task shares a number of characteristics with a tour planning task, and so we draw upon the CMP model as one of our main resources for current purposes. Errands can be regarded as activities that have to be planned in an unfamiliar city environment. Similar to the partially specified activities in tour planning, some of the errands could be accomplished at different optional locations. However, a major difference concerns specific additional temporal assignments and constraints that had to be taken into account.

The analysis of the verbal protocol data by Hayes-Roth and Hayes-Roth (1979) showed that participants structured the problem space into different levels of abstraction. During the planning task, they detected and pursued opportunities at a detailed level that were not necessarily in line with a currently instantiated abstract plan. In that sense, their planning behaviour seemed rather chaotic, since coherent decisions were followed by less coherent opportunities that induced changes at different levels of abstraction.

The processes at a higher level of abstraction encompassed the grouping of errands into clusters, and sequencing the clusters of errands in a specific temporal order. The spatial clusters were associated with the cardinal directions including the (diagonal) combinations such as the southeast (SE) or northwest (NW) corner of the map. The particular sequencing of errands took place at a detailed level of granularity. During the errand sequencing phases, participants sometimes modified the high-level plan by incorporating those errands which were close to the errands currently included in the detailed plan, even if they belonged to a different spatial cluster. Thus, they observed a *closest errand next* principle while ignoring the high-level plan. Furthermore, there were individual differences; while some of the participants treated the errand-sequencing task primarily as a *scheduling* problem, others saw it as a Travelling Salesman Problem. The scheduling problem model determines the order of errands according to their specific features, such as priorities. In contrast, the TSP model optimizes the distances between the locations where the errands are accomplished. Unfortunately, there are no results regarding which of the problem models contributed to better results, in terms of the length of the produced paths.

10.2.2 Human performance on the Travelling Salesman Problem

One of the first systematic studies (MacGregor and Ormerod, 1996) on human performance given small TSPs (10 and 20 nodes) revealed that humans could handle this task exceptionally well; the deviation from the optimal path was not greater than 0.75 per cent. Furthermore, only very few of the human solutions had crossing lines. These findings triggered major debates about the underlying mechanisms that allow people to reduce cognitive effort while solving the TSP. Two major views are prominent in the literature (see Tenbrink and Wiener, 2009, for a review). The first view assumes that people instantiate an overall shape resembling a convex hull that encompasses the boundary points in the problem space. The second view argues for hierarchical spatial problem solving strategies.

10.2.2.1 Convex-hull problem solving strategy The first view is based on the assumption that people structure the problem space by applying a convex hull, which encompasses the boundary points (MacGregor et al., 2000). Using the convex-hull problem solving strategy, people sketch, but do not connect, arcs among the boundary points and instantiate new spatial relations that allow for guiding their search. MacGregor et al. (2000) assume that people randomly select a starting point and direction of travel (clockwise or counter-clockwise). Then they subsequently insert points into the solution path using different insertion criteria that are employed as heuristic methods for solving the TSP (similar to findings in Operations Research, cf. Golden et al., 1980). However, one strong argument against the cognitive adequacy of this model is the computational complexity of the insertion-based heuristics (i.e. $O(n^2 * \log(n))$), which is much higher than the computational effort of the human cognitive mechanisms ($O(n)$).

10.2.2.2 Hierarchical problem solving strategies The other view is based on the assumption that human spatial cognition is hierarchical (Stevens and Coupe, 1978; Tversky, 1993; Hirtle, 1998). To solve TSP problems efficiently, people employ spatial clustering mechanisms, which allow for reducing the amount of information processed in each problem solving step. For example, the participants of the study conducted by Hayes-Roth and Hayes-Roth (1979) utilized cardinal directions for structuring the schematized two-dimensional representation of the environment into the four clusters NE, NW, SE, and SW. These spatial clusters resemble sectors of the Internal Cardinal Direction Model (ICM-9) proposed by Lyi et al. (2005). In addition to eight compass-based directions, the ICM-9 incorporates a middle sector 'M' in the centre of an environment. The ICM-9 represents the mental conceptualization of depicted environments. In the following, we summarize research on hierarchical TSP strategies.

10.2.2.3 Hierarchical Pyramid TSP model Graham et al. (2000) proposed a *Hierarchical Pyramid* model consisting of different layers that represent different levels of granularity. The procedure for solving a TSP using a hierarchical model is based on a *coarse-to-fine* planning principle adopted from AI spatial systems (Leiser and Zilbershatz, 1989; Kuipers, 2000; Chown, 2000; Kuipers et al., 2003). To begin with, nodes that represent clusters at the highest level of granularity are connected with each other. Then the procedure moves to a lower level by using the *cheapest insertion* criterion regarding the distance between the corresponding super-ordinate nodes, and obtains a partial solution. The partial solutions produced in this way are then integrated, and the high-level nodes are subsequently removed. The procedure is repeated for each layer of granularity. The solutions produced by the hierarchical model exposed the same characteristics as those identified by MacGregor and Ormerod (1996). In this case, the run-time complexity of the *cheapest insertion* heuristics is much higher than the effort of the human cognitive mechanism. Furthermore, the straightforward *coarse-to-fine* approach is not consistent with the chaotic and opportunistic planning behaviour detected by Hayes-Roth and Hayes-Roth (1979), as described above.

10.2.2.4 Heuristics change model Another recent computational model of human performance on the TSP (Cutini et al., 2005) also involves hierarchical clustering principles for structuring the problem space. The approach resembles opportunistic planning behaviour by searching in different directions according to the perceptual input of the problem solving agent. It employs *nearest neighbour* (NN), *direction down* (DD), and *direction right* (DR) heuristics to choose the next location to be visited. The model implements a control mechanism, which is responsible for changing the heuristic according to the perceived clusters of locations.

Although this model implements the opportunistic planning behaviour reported by Hayes-Roth and Hayes-Roth (1979), the run-time complexity of the nearest neighbour (NN) search procedure ($O(n^2)$) is still higher than $O(n)$.

10.2.2.5 Fine-to-coarse TSP models Pizlo et al. (2006) modified the Hierarchical Pyramid model by proposing a *Foveating Pyramid* model. The new model identifies, at the beginning of the problem solving process, clusters of locations on many levels of resolution, and establishes spatial relations between the clusters. In contrast to the previous *coarse-to-fine* approach, this model implements a *conceptual fovea*, which represents the focus of attention. To simulate the limitations of the human cognitive capacity, the fovea maintains only a small amount of information on the finest level of granularity. At the same time, it maintains a representation of distant locations at higher levels of granularity. The search procedure is described as moving the fovea from the current location on the tour to the next one, using a gradual transition between the resolutions. The new clusters, resulting from movement of the simulated fovea to the next locations, are incorporated in the representation and are gradually updated using the *cheapest insertion* criterion.

The new model reduced the amount of information processed during each problem solving step, but still employed the computationally complex *cheapest insertion* search procedure. The characteristics of the solutions produced by the new model are similar to the preceding Hierarchical Pyramid model and the characteristics of the TSP solutions produced by humans (MacGregor and Ormerod, 1996).

A cognitive model of *region-based* planning proposed by Wiener and Mallot (2003) operates on a hierarchical knowledge representation which consists of places and regions. The connectivity relations discovered among places as well as regions are incorporated into a *fine-to-coarse* planning heuristic. The heuristic operates on a focal egocentric representation of the environment that contains information about nearby locations at a fine level of granularity, as well as information about distant locations at a coarser level of granularity.

In each problem solving step, the representation is iteratively updated, such that information for the immediate next move is always available. The *fine-to-coarse* planning heuristic simultaneously employs different levels of granularity, similar to the approach pursued in the *Foveating Pyramid* model (Pizlo et al., 2006) as just described. The content of the working memory is retrieved from spatial long-term memory according to the current ego positions and the corresponding target locations.

This hierarchical spatial planning strategy allows for the complexity of the task to be reduced, since it requires less memory in comparison to coarse-to-fine approaches and therefore less cognitive effort. A similar strategy was shown to be employed in a navigational version on human performance in a task similar to the TSP in regionalized environments (Wiener et al., 2009). Tenbrink and Wiener

(2009) re-used this design in a paper-and-pencil version showing path trajectories, and furthermore collected retrospective reports in which participants reported their strategies. In addition to the connectivity relations between the perceived regions, these verbalizations included mentioning direction relations that guided the search for an optimal path, and revealed a number of intricate strategies that were combined to reduce task complexity for the participants.

10.2.3 Summary of problem solving models All of these presented models of human performance in TSP tasks share the same characteristic of resulting in a small number of crossing lines. The models, however, use different approaches to structure the problem space, such as by conceptualizing regions or imagining a convex hull. Later in this chapter, we will compile the cognitive processes mentioned in the literature into an extensive encoding vocabulary. This vocabulary will be used as an annotation scheme to detect problem solving steps and strategies during a tour creating task, given to participants in an explorative study. Details of this study (together with a different focus of analysis) were published in Tenbrink and Seifert (2011); here our focus lies on the association of mental operators known from the earlier literature. To set the stage for such an association, we now turn to a brief outline of our study.

10.3 Tour planning study

Our tour planning study aimed at exploring the cognitive problem solving processes during a weakly specified tour planning task such as creating an individual journey to a foreign country. The objectives of the study encompassed the following research questions:

- Do people structure a large-scale environment into regions (further denoted as high-level regions) when creating a holiday tour?
- Do people employ a convex-hull or a hierarchical spatial planning strategy during weakly specified tour planning tasks?
- Do the solution paths to a weakly specified tour planning task contain a small number of crossing lines similar to TSP solutions?
- What concepts do people use for communicating the resulting tours?

Since it can be assumed that retrospective verbal protocols include information about how people solve tasks given to them (Ericsson and Simon, 1984), the verbalization of the spatial planning processes should contain mention of mental operators such as:

- Structuring large-scale environments into high-level regions;
- Definition of activities;

- Spatial planning strategies such as:
  - Hierarchical strategy,
  - Convex-hull strategy.

In order to observe human problem solving behaviour during underspecified spatial planning in a natural setting, we designed a task that was relatively familiar and easy to understand. In particular, the aim was to plan an individual journey to a foreign country that should include various tourist activities. For this purpose, we needed a compact environment that offered a large number of opportunities for tourist activity. The well-known holiday island of Crete in Greece turned out to be ideal for this purpose, since Crete spans only 260 km from east to west, and 60 km from north to south. We obtained a suitable map of Crete that displays diverse potential tourist activity types (Figure 10.10 below shows an annotated fraction of such a map, providing an impression of the design).

We reasoned that the oblong shape of the island of Crete (Figure 10.1, left) might be a factor that potentially influences some aspects we were interested in, such as the number of crossing lines in the tours produced by the participants, or the cognitive planning processes. Travellers can easily cross the island by car at least twice a day from north to south. To compare the island of Crete with a more regularly shaped environment, we cut the map of Crete into several pieces and reassembled them to an imaginary island that we internally called Cretopia. In this way, we created another map (Figure 10.1, right) that was more balanced regarding its width and length, but provided the same amount of information to the participants of the study. The size of the map of Crete was approximately 83 cm × 27 cm, and that of Cretopia was 65 cm × 57 cm.

10.3.1 Design

The tour planning study encompasses both verbal and behavioural data. The verbal data include a tour plan produced by the participants and a retrospective protocol with a description of the decisions made during the task. The behavioural data are tours drawn on a geographic map of a large-scale environment.

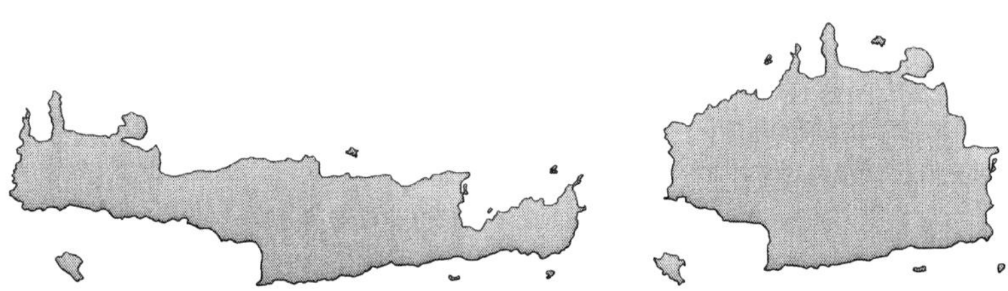

FIGURE 10.1 Shapes of Crete (left) and Cretopia (right)

10.3.2 Participants

The participants of the study were 26 school students (17–18 years, 12th grade) and one university student (23 years old). Thirteen of them (8 female and 5 male) were (randomly) assigned to the Crete condition, and 14 (8 female and 6 male) to the Cretopia condition. Each participant solved the tour planning task just once. In the Cretopia condition, one female participant created a tour with eight crossing lines, demonstrating that not only a circular trip, but also other types of solutions were possible. However, since this solution is obviously exceptional, we only analysed the remaining 13 data sets in the Cretopia condition further. The study was conducted in German.

10.3.3 Materials

The particular interest of this study was to provide empirical support for the operators that invoke the definition of new activities, and a comparison of the characteristics of the resulting tours to TSP solutions. The map of Crete, and consequently the new island Cretopia, contained information about various locations, such as cities, towns, villages, and a street network that encompassed high ways, roads, and hiking paths. The locations were labelled with specific symbols denoting the activity types that could be performed there (see Figure 10.1 for an impression). The activity types encompassed sightseeing attractions such as museums, archaeological excavations, caves, castles, harbours, and beaches. The maps contained topographic information about different landscapes such as gorges, mountains, rivers, valleys, and seacoasts. Altogether, the employed representations of the natural environment incorporated factors that should invoke mental structuring processes, such as the natural distribution of locations associated with symbols, and geographic regions representing landscapes.

In the top right corner of the map of Crete, and in the top left corner of the map of Cretopia, there was a legend that contained the meanings of the symbols, and another one that provided information about distances in kilometres between the major cities of Crete, and likewise of Cretopia. The names of the major cities of Cretopia were changed in order to avoid an association with the potentially familiar cities of the island of Crete. The resemblance of our fantasy island Cretopia to a natural one was so convincing that, after the study, several students asked whether Cretopia was one of the islands in Greece, and where this island was situated.

10.3.4 Procedure

Each participant was provided with written instructions about the tour planning task, a coloured map printout, a pencil, and several sheets of paper. They were asked to plan a trip to Crete (or, in the Cretopia condition, simply *the island*) for two imaginary friends, who wanted to travel about the island and accomplish various activities during their trip. A journey had to start and end at the same destination, namely one of the three main airports seen on the map. It should include diversified activities accomplished at different locations, and cover 14 days. However, it was left to the participants to decide about the number and types of activities to be accomplished. Participants were allowed to make notes and annotations, and to draw the resulting tours on their individual maps. The participants were told to assume a car as the means of transportation to travel from one destination to another. The instructions included three tasks:

1) Draw the tour on the map.
2) Describe on a piece of paper a tour plan of a holiday trip, which incorporates a list of activities, overnight stays, and different destinations.
3) Provide a written retrospective report. The exact (translated) instruction for this task was as follows: 'When you have finished the list (for example after half an hour), we would like to know how you did this! What did you start with? And what was the next step of consideration? How did you deal with the map? Please describe in a text what was important and what thoughts and decisions played a role. This text can help your friend to make plans on his own the next time!'

Participants were given about 20 minutes to accomplish the first and second tasks, and were then encouraged to finish these and start with the last task, which took about 30 minutes. This last task was decisive for our purposes of accessing the high-level cognitive problem solving processes. In the next section, we will motivate our extended encoding vocabulary in detail, which was used to analyse these reports, and which also served as the basis for an implementation of a tour planning assistance model in Seifert (2009).

10.4 Encoding vocabulary for the analysis of verbal reports

The investigation of verbal representations in other areas of human spatial cognition, such as wayfinding, has proved to be a particularly valuable way of accessing cognitive problem solving processes (Denis, 1997; Denis et al., 1999; Tversky and Lee, 1999; Soh and Smith-Jackson, 2004). Therefore, we considered verbalizations to be a reliable data source for the investigation of problem solving behaviour in underspecified spatial planning tasks. This approach to analysing spatial language complements analyses of route descriptions (Taylor and Brunyé, this volume; Denis and Fernandez, this volume) and related verbalizations of spatial concepts (Avraamides et al., this volume; Brosset et al., 2008; Rehrl et al., 2009).

Ericsson and Simon (1984) introduced a method for determining possible problem solving processes from verbal protocols. The methodology involves the specification of an encoding vocabulary, which is used to identify and describe mental operators. Ericsson and Simon (1984) proposed the following theoretical assumptions about the general structure of cognitive problem solving processes:

- Human problem solving behaviour is viewed as a search though a problem space. During this search, people accumulate knowledge about the problem domain.
- Each problem solving step involves an application of a task-relevant operator that brings new knowledge and allows for moving to a new state in the problem space.

The original method of retrospective protocol analysis proposed by Ericsson and Simon (1984) aims at accessing people's thoughts directly after performing a given problem solving task. After finishing the task, people have to recollect and describe what they were thinking during the task. In doing so, this method allows the remaining content of human working memory to be retrieved. In the scope of the tour planning study, we employed a slightly different retrospective verbalization method that aimed to access high-level cognitive processes such as spatial problem solving strategies. The participants of the study were asked to describe the decisions they made while creating an individual journey, with their friends as possible addressees of this description in mind. Although verbal data collected from such a retrospective description exposes less direct step-by-step information than think-aloud protocols could potentially provide, we expected that the reports should contain more verbalizations of conscious high-level cognitive processes and problem solving strategies.

An encoding vocabulary needs to be designed in alignment to the structure of the problem space. Ericsson and Simon (1984) suggest defining task-relevant operators by considering all theoretically possible elementary actions that invoke changes of entities and relations belonging to the problem space. The resulting actions can then be compared to the cognitive processes reported in the literature on related tasks. In this way, the encoding vocabulary provides a solid basis for the development of new theories, models, and research questions. The empirical evidence of new hypotheses can be validated by determining the characteristics of occurrences of the encoded task-relevant operators in verbal data collected during empirical studies.

In the following, the problem domain is broken down into basic entities, relations, and actions.

10.4.1 Basic entities

The basic entities and relations of the spatio-temporal planning domain are the following:

- **Activities:** Spatio-temporal planning involves the sequencing of activities similar to the errand sequencing task. Therefore, the basic entity that is instantiated and manipulated during the task is an *activity*, which incorporates an *activity type*, a relation to a *spatial assignment*, and a *duration*. The basic relation between activities is the temporal order: one activity instance follows *after* another activity instance.

- **Spatial assignments**: A spatial assignment is defined as a *location* or a *region*. Locations are regarded as point-based objects, and regions are regarded as extended objects that contain a set of locations. Regions represent spatial areas that contain a set of spatially proximate locations. The mentioning of such areas can be found in the descriptions of tourist behaviour observed by Brown (2006), and also in the transcribed verbal protocols published in Hayes-Roth et al. (1979) and Hayes-Roth and Hayes-Roth (1979).
- **Routes**: Activities should be performed at different places; people have to define *routes* between the corresponding spatial assignments. Problem solving processes indicating the specification of routes were mentioned in the Cognitive Model of Planning (CMP) (Hayes-Roth and Hayes-Roth, 1979).
- **High-level regions**: The CMP, as well as empirical studies focused on the human performance on the TSP, report problem solving processes that contribute to structuring the problem space into spatial clusters, also called regions. Such spatial clusters and regions are further specified as *high-level regions*.
- **Current plan**: The temporal order of activities, spatial assignments, and routes between them are encompassed in a *current plan* which undergoes various manipulations during the planning task.

10.4.2 Basic relations

Based on the mental spatial relations proposed by McNamara (1986), a mental representation structure instantiated during spatial planning tasks can be visualized as shown in Figure 10.2.

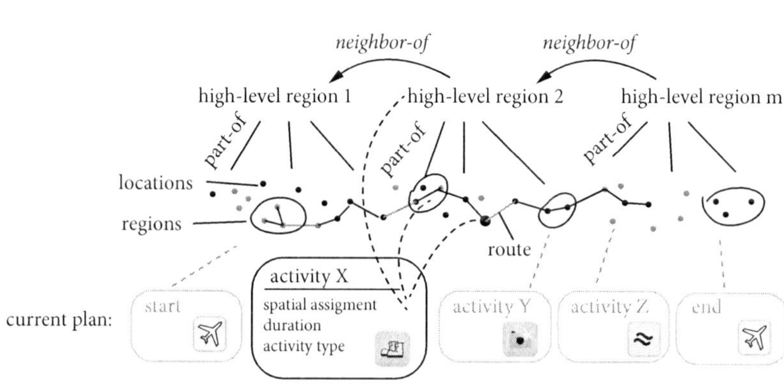

FIGURE 10.2 Mental spatial entities and relations instantiated during spatial planning

Mental structuring processes of a large-scale environment result in the definition of spatial clusters or high-level regions. The high-level regions belong to the highest level of granularity of the problem space.

- **Neighbour-of**: The neighbour-of relations represent directed region connectivity relations, which were observed during navigation tasks in regionalized environments (Wiener and Mallot, 2003). Similarly, the model of mental spatial relations proposed by McNamara (1986) represents adjacent relations that incorporate cardinal directions between high-level regions.
- **Part-of**: The high-level regions contain both locations and regions. According to the fine-to-coarse models of human performance on the TSP, the containment relations are represented as *part-of* relations.
- **Temporal order**: Activities have to be put into an appropriate order under consideration of the overall temporal scope of a tour.

The spatial assignments of the instantiated activities can be associated with each of the spatial entities that belong to the hierarchical spatial representation structure (illustrated as dashed lines in Figure 10.2).

10.4.3 Basic actions

An intermediate problem solving state can be viewed as a state of a current plan. It involves a sequence of activities with partially specified attributes. It can also be a complete feasible tour plan, which may fail to fulfil the personal preferences and, therefore, require additional modifications. The basic actions encompass the following categories:

- Manipulating activities and their attributes;
- Relating activities to spatial assignments;
- Sequencing activities;
- Defining routes;
- Structuring the representation of an unfamiliar environment into high-level regions;
- Assigning activities to high-level regions;
- Sequencing of high-level regions;
- Selecting a problem model.

The connection between an operator and the corresponding empirical basis is distinguished by the following labels: CMP for the Cognitive Model of Planning, and HP-TSP for the models of human performance on the Travelling Salesman Problem.

10.4.4 Manipulation of activities

As already mentioned, an activity involves three attributes: activity type, duration, and spatial assignment. Defining a tour plan requires the creation of new instances of activities and the specification of their attributes. The literature on spatial planning considers specific pre-defined problems that should be understood and accomplished by the participants of the studies. The following operators are derived from the theoretical assumptions regarding the basic actions that are necessary for the creation of a new activity.

- Create activity:
  - define activity type;
  - define duration.

It has to be possible to remove tentative activities from the current plan. The following operators are responsible for removing activities, if the situation requires it, together with all attributes from the current plan, or removing one of the attributes of an activity.

- Remove activity;
- Remove activity type.

The attributes of an instantiated activity can be changed during the planning process.

- Change activity type;
- Change duration.

The operators introduced so far are responsible for the instantiation of new activities. Since tour creation tasks have not been addressed in this way in the literature yet, there is no connection to the empirical basis.

10.4.5 Relating activities to spatial assignments

A spatial assignment involves a location or region containing a set of locations. During the errand-sequencing tasks, the errands were arranged according to the proximity of appropriate locations termed as the *closest errand next* strategy. The region-based model of route planning (Wiener and Mallot, 2003; see Section 10.2.2.5) operates on a hierarchical region-based representation. Here, the planning processes involve the definition of the next region to be visited.

- Define relation to spatial assignment (CMP, HP-TSP);
- Remove relation to spatial assignment (CMP);
- Change relation to spatial assignment (CMP).

The cognitive processes described in the CMP have a connection to the operators responsible for the manipulation of the relations between spatial assignments and activities (*define*, *remove*, and *change*).

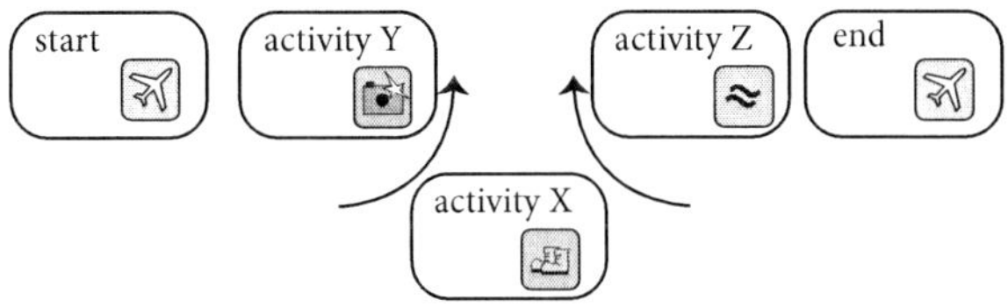

FIGURE 10.3 Inserting an activity

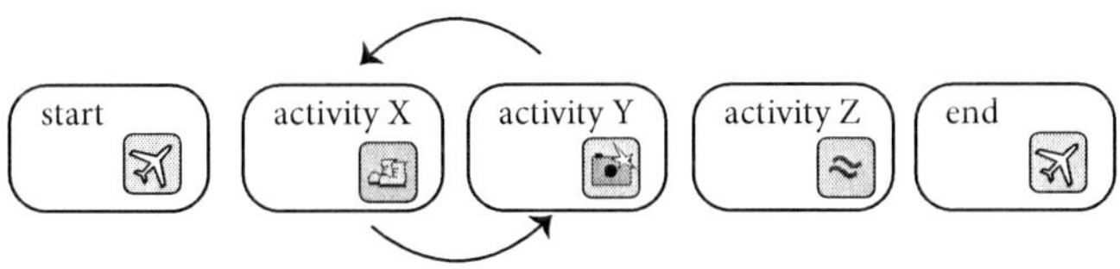

FIGURE 10.4 Swapping activity X and activity Y. It is possible to put an activity before or after any other activity

10.4.6 *Sequencing of activities*

Various changes of the temporal order of activities were observed during the errand sequencing task (CMP; Hayes-Roth and Hayes-Roth, 1979). Figure 10.3 illustrates an example sequence of activities with different activity types where a new activity is inserted, and Figure 10.4 illustrates a change of temporal order.

- Insert activity (CMP);
- Change activity order (CMP).

10.4.7 *Definition of routes between spatial assignments*

As described above, the models of human performance on the TSP demonstrate different strategies for defining routes that connect locations. The *convex hull* strategy involves defining imaginary arcs between boundary points, and connecting the remaining points between the locations by inserting them between the boundary points according to the insertion criteria described in Section 10.2.2.1.

- Define route from location to location (HP-TSP).

The hierarchical models of human performance on the TSP as well as the fine-to-coarse route planning heuristic operate on hierarchical knowledge representations that contain connectivity relations not only between locations (places), but also between locations and regions. Therefore, we assume that people define routes at different levels of granularity; for example between a region and a location, or between a region and another region (Figure 10.5):

- Define route from location to region (HP-TSP);
- Define route from region to region (HP-TSP);
- Define route from region to location (HP-TSP).

10.4.8 Structuring the problem space into high-level regions

The CMP contains spatial clusters denoted as SW, NW, SE, and NW, which divide the map used in the study into four sectors according to the internal cardinal direction relations (Hayes-Roth and Hayes-Roth, 1979).

- Define spatial structure (CMP, HP-TSP).

Similar structuring processes were also implemented in the hierarchical models of human performance on the TSP. The models employed clusters of nodes, also denoted as regions, to structure the spatial dimension into different levels of granularity.

- Define high-level region (CMP, HP-TSP).

The definition of high-level regions is the result of the structuring process. Participants of the errand sequencing task grouped several errands together and assigned them to a spatial cluster. Accordingly, the operators that implement these basic actions are defined as:

- Grouping activities (CMP);
- Assign activity to high-level region (CMP).

Figure 10.6 illustrates a group of activities with different activity types assigned to a high-level region. It also shows a single activity assigned to a high-level region. Activities related to a specific region or a location at the detailed level of granularity 'inherit' the corresponding relation of the high-level region.

10.4.9 Sequencing of high-level regions

Sequencing of high-level regions was observed during the errand sequencing task (Hayes-Roth and Hayes-Roth, 1979). Also, hierarchical models of the human

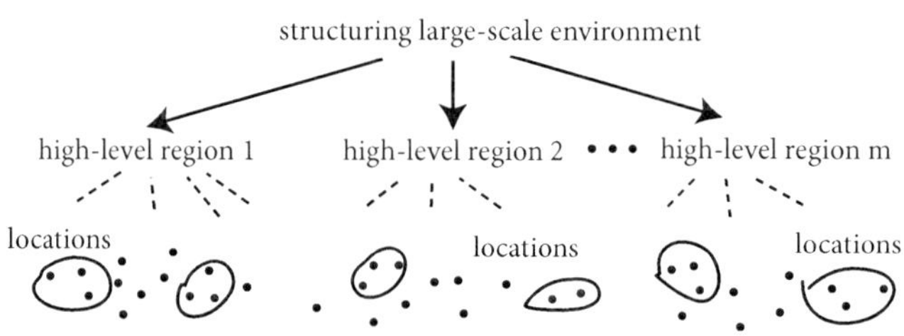

FIGURE 10.5 Activity instances associated with specific spatial assignments and routes defined at different levels of granularity

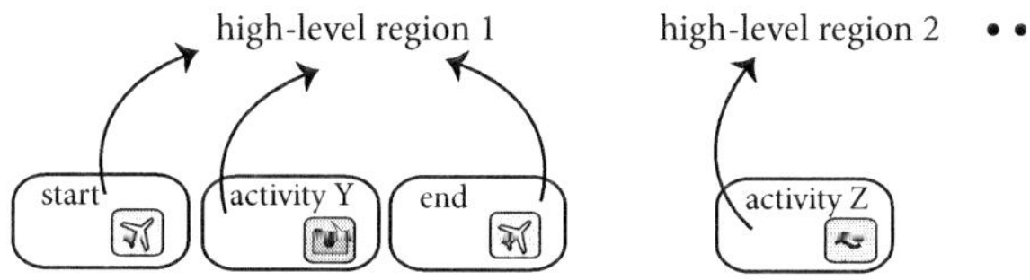

FIGURE 10.6 Assigning activities to high-level regions

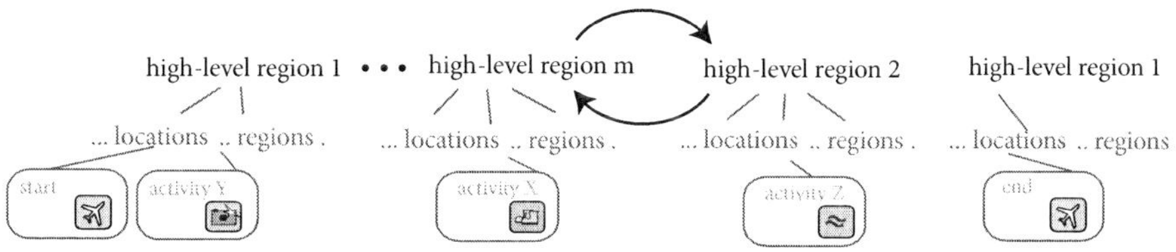

FIGURE 10.7 Changing the order of high-level regions

performance on the TSP perform operations at the highest level of granularity, by putting high-level nodes into a specific order to reduce the complexity of the problem solving task. Therefore, we assume that there must exist operators which are responsible for the sequencing of high-level regions.

- Set order of high-level regions (CMP, HP-TSP);
- Change order of high-level regions (CMP, HP-TSP).

These operators allow for specifying or changing the order of high-level regions. Figure 10.7 illustrates the basic action that invokes changing the order of high-level regions.

10.4.10 Selecting a problem model

The CMP implemented two problem models, which allowed for the sequencing of errands according to the optimization criteria selected by the participants. The first problem model was denoted as a scheduling model, which involved sequencing errands according to their specific properties (such as priorities). The second problem model was termed a TSP model. It involved the optimization of an overall route connecting the spatial assignments of errands.

Our spatio-temporal planning task encompasses activities that are featured by specific activity types. However, it considers no priorities. Accordingly, we propose to define a problem model for the sequencing of activities according to activity types, and another problem model for sequencing activities according to the optimality of the corresponding routes.

- Select problem model (CMP).

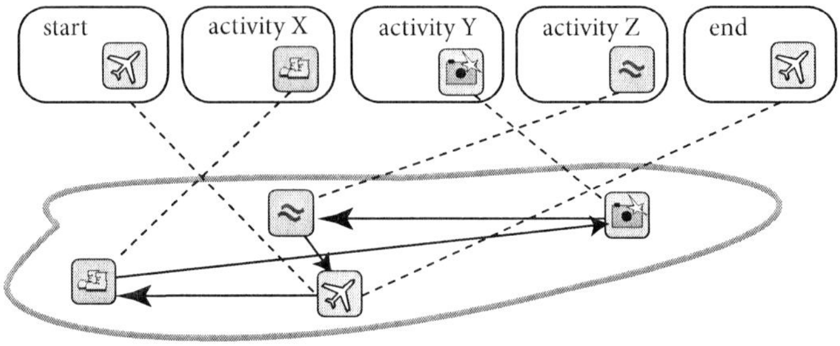

FIGURE 10.8 Scheduling model approach

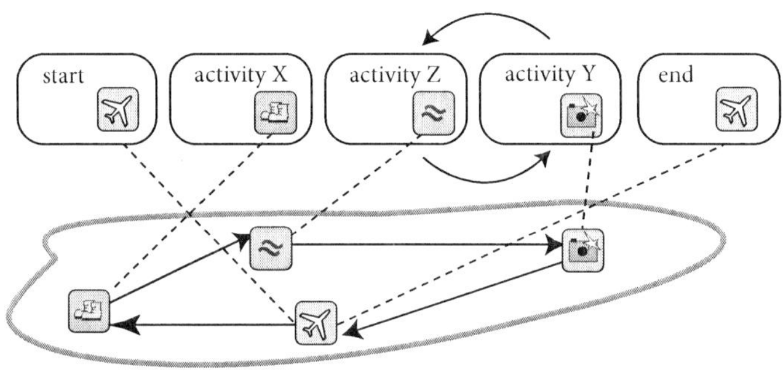

FIGURE 10.9 TSP problem model approach

Figures 10.8 and 10.9 illustrate the two different methods for approaching the example spatio-temporal planning task. Figure 10.8 depicts an example environment which includes four different locations. The activities assigned to these locations are ordered according to the activity type without consideration of the optimality of the underlying route. Figure 10.9 illustrates changing the order of activities according to the optimal route that connects the related spatial assignments.

10.5 Results

Having introduced the basic operators used to analyse the verbal reports collected in the explorative study described in Section 10.3, we are now in a position to report the study results. We begin with an illustration of an example data set (Section 10.5.1) and will then describe the results of the analysis for each part of the collected data sets (Sections 10.5.2, 10.5.3, and 10.5.4).

10.5.1 Example data set

The example data set includes a travel route drawn on the map (Figure 10.10), a tour plan with the proposed activities and destinations (Table 10.1), and a retrospective

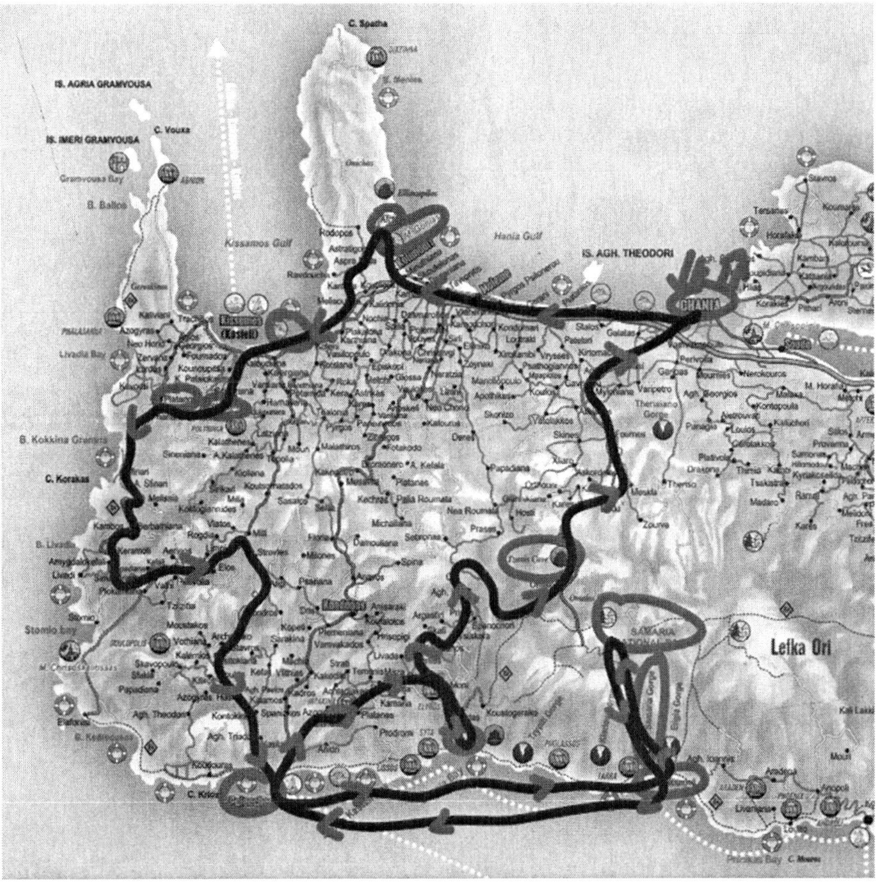

FIGURE 10.10 Example tour drawn on the map of Crete

report describing the problem solving steps and decisions made during the planning task (Table 10.2). We provide the verbal data in the original German together with an English translation.

10.5.2 *Task 1: Characteristics of the tours produced*

We compared the features of the tours drawn by the participants on the maps to the characteristics of the solutions to the Travelling Salesman Problem produced by people. According to our hypothesis, the tours produced by the participants should involve a small number of crossing lines.

Results showed that most of the collected tours represented round trips. The majority of solutions exposed a small number of crossing lines and can be regarded as circular tours. Only two routes in the Crete condition and four routes in the

TABLE 10.1 **Example tour plan**

English (translation)	German (original)
Chania Airport	Chania Airport
Chania—1 night, city tour,	Chania—1 Nacht, Stadtaufenthalt,
M. Gonias, monastery sightseeing	M. Gonias, Kloster Besichtigung
Overnight stay in Afrata	Übernachtung in Afrata
Campground, from there go to the beach, enjoy an easy day	Campingplatz, von dort aus zum Strand, einen ruhigen Tag erleben
On the next day, go to Polyrinia, visit an archaeological site	Am nächsten Tag nach Polyrinia, Ausgrabungsstätte besuchen
Overnight stay in Planatos	Übernachtung in Planatos
To Paleochora, city tour, overnight stay	Nach Paleochora, Stadtbesichtigung, Übernachtung
Go by ferry to Aghia Roumeli, overnight stay	Mit der Fähre nach Aghia Roumeli, Übernachtung
Visit of Samaria Gorge	Besuch der Samaria Gorge
Overnight stay in Aghia Roumeli	Übernachtung in Aghia Roumeli
Next day visit the Samaria National Park	Am nächsten Tag in den Samaria National Park
Overnight stay in an Alpine hut	Übernachtung in Berghütte
Back to Aghia Roumeli	Zurück nach Aghia Roumeli
By ferry to Paleochora	Mit der Fähre nach Paleochora
By car to Sougia beach	Mit dem Auto nach Sougia Strand
To Tzanis Cave, sightseeing in the cave	Nach Tzanis Cave, Höhlenbesichtigung
Back to Chania	Zurück nach Chania
Day trip: visit the folklore museum near Sivaras	Tagesausflug: Besuch des Heimatkundemuseums bei Sivaras

Cretopia condition had a single crossing line each, leading to an overall shape resembling a horizontally aligned eight (∞). As stated earlier, there was one additional solution with eight crossing lines, which is not further analysed here. Routes sometimes included a detour (see the example tour in Figure 10.10), possibly induced by the underlying road network that had to be taken into consideration when travelling by car. Table 10.3 lists the number of detected crossing lines for each tour drawn on the map.

In contrast to our earlier assumptions, the Cretopia condition contained on average more crossing lines than the Crete condition. The inferential statistic is calculated using the binomial distribution. If participants had no preferences regarding the crossing lines, the probability of creating a tour either with or without crossing lines would each be 0.5. The probability of making crossing lines in the

TABLE 10.2 **Example description of the problem solving steps and decision**

English (translation)	German (original)
First, I read the text carefully and marked the important passages. I then looked at the map as a whole. Since 14 days is not enough time to tour the whole island, I've broken the island up into three parts in my mind, and assigned an area to each of the airports. Then, I looked at the legend and the scale. The area, which I would let the tourists visit, needed to have as many different points of interest described in the legend, that could be easily reached, as possible. Therefore I decided on the area around Chania. I wanted the tourists to cover all of the western part of the island and accomplish each activity at least once. So, I didn't pay attention to the types of roads, only making sure that they did not drive along hiking paths.	Zuerst habe ich mir den Text aufmerksam durchgelesen und entscheidende Stellen markiert. Dann habe ich die Karte als ganzes betrachtet. Da in 14 Tagen nicht die gesamte Insel in Ruhe zu besichtigen ist, habe ich die Insel in meinem Kopf in drei Teile geteilt, und jedem Flughafen einen Bereich zugeteilt. Dann habe ich mir die Legende und den Maßstab angeschaut. Der Bereich den ich das Paar bereisen lassen würde musste möglichst viele verschiedene Punkte, die in der Legende genannt werden aufweisen, die aber auch gut erreichbar waren. So habe ich mich für den Bereich um Chania entschieden. ich wollte das Paar einmal um den westlichen Inselteil reisen lassen und sie dabei jede Aktivität einmal erleben lassen. Dabei habe ich kaum auf die verschiedenen Straßenarten geachtet, lediglich darauf, die beiden nicht über Fußwege fahren zu lassen.

TABLE 10.3 **Number of crossing lines in the Crete and Cretopia conditions**

Condition	Crossing lines (average)	Standard deviation	Probability
Crete (n=13)	2 (0.15)	0.36	0.0017 < 0.05
Cretopia (n=13)	4 (0.31)	0.61	0.011 < 0.05
Total (n=26)	6 (0.23)	0.50	0.00026 < 0.05

resulting tours is 0.0026 for both conditions, with 0.0017 in the Crete condition and 0.011 in the Cretopia condition. These results are significant.

The length of the route segments between the selected locations revealed that the students overestimated the time required for travelling between the destinations, although the map contained information about the scale and the distance costs between major cities (3 cm in the map corresponds to 10 km in the physical environment).

10.5.3 Task 2: Conceptual choices in describing tour plans

We analysed the structure of the tour plans to identify the conceptual choices used by the participants to convey the solutions. The analysis of the data sets revealed two conceptual schemes: a location-based scheme and a day-based scheme.

The location-based scheme represents a sequence of sub-tours (trips) from one location to another. Each trip included a target location, the description of planned activities together with their locations, and the duration of the trip in days.

The day-based scheme contained an explicit temporal order of days. Each day was filled with at least one activity. In some cases the schemes were merged: each day or pair of days covered a small trip, or in contrast several days were put together due to the same kind of activity during the days.

The example tour given in Table 10.1 represents a location-based scheme, since the description included different destinations in which the participant planned overnight stays without explicitly conveying the temporal structure of the journey.

Table 10.4 shows that most participants conveyed the resulting tour plans using the day-based scheme. To calculate the inferential statistics for the conceptual choices, we used the binomial distribution under an assumption of equal probability of preferences for day-based and location-based schemes. The probability of choosing the location-based scheme is 0.13, which is not significant in the Crete condition. The probability of the day-based scheme in the Cretopia condition, as well as for both conditions combined, is significant.

Most tour plans represent a sequence of 14 days with corresponding activities that could be accomplished each day. The participants planned on average two activities per day and used terms such as 'in the morning' or 'in the evening' to convey temporal assignments. Some of the participants assigned precise points in time, such as starting a trip at 10:00 and visiting a monastery at 12:00, and also precise durations, like 3 hours for relaxing on the beach. This kind of precision, however, can be regarded as an exception.

10.5.4 Task 3: Analysis of the retrospective reports

The paper-based verbal protocols were made available in a digital format. Each report was divided into separate, informationally meaningful units that contained an

TABLE 10.4 **Conceptual choices of the structure of tour plans**

Condition	Location-based scheme	Day-based scheme	Probability
Crete, # tours	3	10	$0.13 > 0.05$
Cretopia, # tours	1	12	$0.00012 < 0.05$
Total # tours	4	22	$0.0012 < 0.05$

occurrence of at least one task-relevant operator. Some of these units included the application of more than one operator. However, there were never more than two annotations per unit. Table 10.5 summarizes the task-relevant operators together with examples of related units.

10.5.4.1 Coding schemes The coding schemes encompass a subset of the operators outlined in the encoding vocabulary (Section 10.4). This subset contains only those operators that were mentioned in the verbal protocols. As already mentioned, mental problem solving processes involve different levels of granularity and abstraction.

Table 10.5 shows the operators in the spatial dimension detected in the collected data sets. The table includes operators that perform the mental structuring of a

TABLE 10.5 **Operators in the spatial dimension detected in the verbal descriptions**

Mental operator	Label	Gran. & Abstr.	Example: English (translation)	Example: German (original)
Define spatial structure	SS (A)	abstract	I've mentally broken the island up into three parts, and assigned an area to each airport.	dann habe ich die Insel in meinem Kopf in drei Teile geteilt, und jedem Flughafen einen Bereich zugeteilt.
Define route from region to region	R (C)	coarse	From the national park to the coast, by ferry to an island, from there to the mainland, then to the mountains, along a riverbank, etc.	Vom Nationalpark an die Küste, mit der Fähre auf eine Insel, von dort aus ans Festland in gebirgiges Gelände, an einem Fluss entlang etc.
Define route from region to location (or location to region)	R (M)	mixed	I've decided to cross Crete along the coast west of Heraklion.	ich habe mich dafür entschieden Kreta an der Küste westlich von Heraklion zu 'durchschreiten'.
Define route from location to location	R (D)	detailed	(Pashia Ammos)... and from there further to Goutnia.	(Pashia Ammos)... und von dort aus geht es dann weiter nach Goutnia.
Define relation to spatial assignment	SA (C)	coarse	Then, I decided about the airport and considered all the highlights of that area + what would be interesting for myself.	Dann für FH entschieden, und alle Besonderheiten der Umgebung gesucht + was mich interessieren würde.
Define relation to spatial assignment	SA (D)	detailed	Here, I paid particular attention to visiting Heraklion once...	Hier speziell darauf geachtet, einmal nach Heraklion zu kommen...

large-scale environment into high-level regions, the definition of routes, and the specification of new activities by relations to spatial assignments, durations, and activity types. The coding schemes do not contain the manipulation of activities, that is, changing the order of activities; this may be related to the fact that the protocols were written in retrospect. There was no explicit mention of which problem model was applied; no participant reported which of the two dimensions guided the search for a solution, the activities or the optimization of the route connecting locations. However, we detected specific properties of verbal protocols that indicated the problem models.

Tables 10.6 and 10.7 show the operators in the temporal and activity dimensions. The original operators derived from the problem description encompass definitions of durations and activity types only at a detailed level of granularity. In the retrospective protocols, however, we also found the utterances that mention activity types and durations at a coarse level of granularity. This difference was subsequently included in the coding schemes. The operators employed for sequencing activities at a detailed level of granularity are not included, since they were not mentioned by the participants in the retrospective descriptions. The coding schemes were applied by two different coders, with an intercoder agreement of 81 per cent (94 correspondences of 115 units randomly selected from different conditions).

10.5.4.2 Frequency of operators in the verbal reports Figures 10.11 and 10.12 give the relative frequency of operators per condition within all units. It is clear from a visual inspection that the conditions do not differ in any obvious respects with regard to the assignment of operators. The specification of activity types is most frequent in both conditions, which may be due to the fact that the participants were explicitly asked to plan different activities. The definition of activity types was communicated

TABLE 10.6 **Operators in the temporal dimension detected in the verbal descriptions**

Mental operator	Label	Gran. & Abstr.	Example: English (translation)	Example: German (original)
Define duration	D (C)	coarse	Since I can't see the whole island easily in 14 days,	da in 14 Tagen nicht die gesamte Insel in Ruhe zu besichtigen ist,
Define duration	D (D)	detailed	I've planned one day for each activity, or staying at a camp ground for 2–3 days, to make the trip more varied.	ich habe für jede Aktivität einen Tag, bzw. für den Campingplatz 2–3 Tage berechnet um die Fahrt möglichst abwechslungsreich zu gestalten.

TABLE 10.7 **Operators in the activity type dimension detected in the verbal descriptions**

Mental operator	Label	Gran. & Abstr.	Example: English (translation)	Example: German (original)
Define activity type	AT (C)	coarse	Not to plan everything in advance, only the main things—spontaneity!	nicht schon alles vorher planen, nur Hauptsachen—Spontaneität!
Define activity type	AT (M)	mixed	Whenever possible, I tried to avoid three things of the same kind (for example 3 castles).	Möglichst nicht drei gleiche Sachen versucht zu finden (z.B. 3 × Burg).
Define activity type	AT (D)	detailed	Stay at beautiful beaches, if possible, not overcrowded with tourists.	Aufenthalte an schönen Badestränden, nach Möglichkeit nicht zu sehr touristisch belagert.

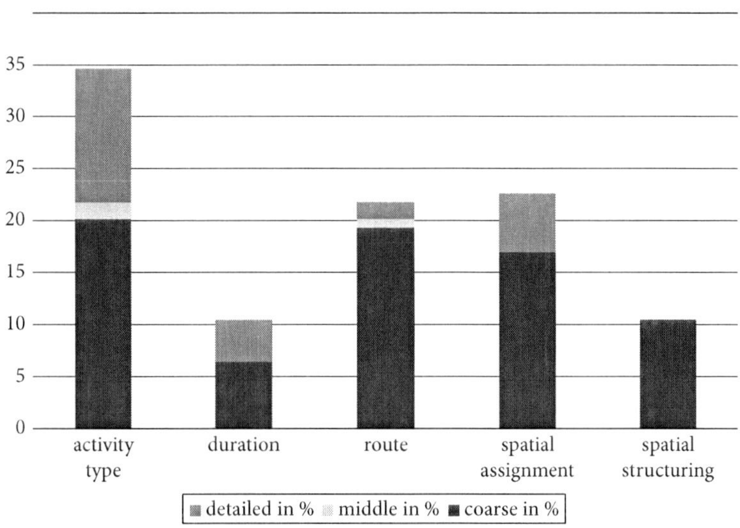

FIGURE 10.11 Crete condition: usage of mental operators in %

at different levels of granularity. For example, one of the participants expressed the diversity of activity types by categories like 'Natur' (nature), followed by 'Kultur' (culture). Sometimes explicit criteria were given for the selection of specific activity types, such as 'Whenever possible, I tried to avoid three things of the same kind (for example 3 castles)' (see Table 10.7). We assigned this kind of utterance to the mixed level of granularity, since the participant provided a concrete example for the

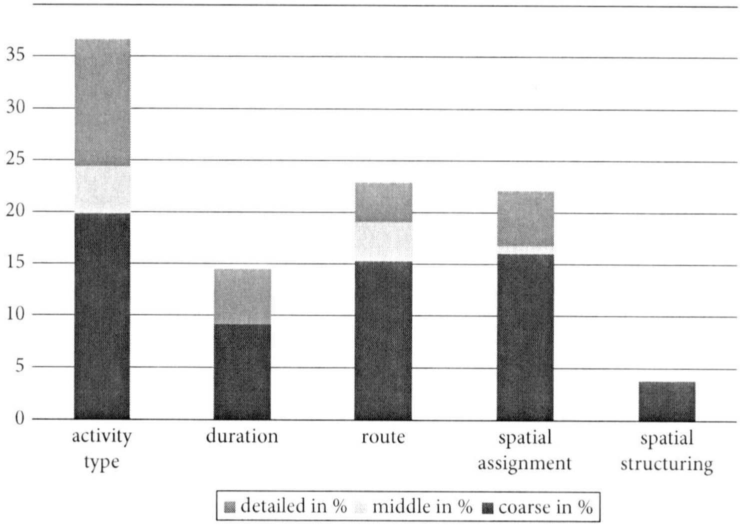

FIGURE 10.12 Cretopia condition: usage of mental operators in %

selection criterion. The routes and spatial assignments were communicated roughly equally often, mostly at a coarse level of granularity.

10.5.4.3 Individual differences During the analysis of the verbal data, we detected three groups of participants with individual preferences regarding the frequency of mentioning specific operators. The first group ('activity type planners') primarily described the kinds of activities the travellers were going to accomplish, whereas the second group ('route planners') focused on the routes. The third group ('location planners') preferred to mention how and why they selected different locations. To specify this observation we calculated how many times each operator was used by each subject. Then we determined the most frequent operator reported by each participant's verbal description. Some of the operators were mentioned equally often within the verbal report. We decided to add these operators as a most frequently employed operator to the overall result (which leads to a sum of more than 100 per cent, since some participants fell into more than one preference group).

The most prominent group is the 'activity type planners', represented by 53.8 per cent of participants in the Crete and 53.8 per cent in the Cretopia condition. The 'route planners' represent 30.8 per cent of the Crete condition, and 38.5 per cent of the Cretopia condition. The third group are 'location planners', 38 per cent of the Crete condition and only 23.1 per cent of the Cretopia condition. The Crete condition encompasses more 'location planners' than 'route planners,' while the Cretopia condition exposes more 'route planners' than 'location planners'.

10.5.4.4 Spatial planning strategies Along with determining the occurrences of operators in the retrospective verbal protocols, we were interested in the spatial planning strategies employed by the participants in order to solve the task more efficiently. According to our hypothesis, we considered two possible strategies: the hierarchical strategy and the convex-hull strategy. To determine the hierarchical strategy, we searched the utterances that mentioned geographic regions such as landscapes, or high-level regions structured according to the internal cardinal directions (e.g. the western part of the island). Three of 13 descriptions in the Crete condition (23 per cent), and 2 of 13 in the Cretopia condition (15.4 per cent) mentioned this strategy, as in the example given in Table 10.8.

To determine indications of the convex-hull strategy, we searched the collected data for occurrences of the terms 'circle', 'ellipse', or a 'circular trip'. We found 5 of the 13 data sets in the Crete condition (38.5 per cent), and 5 of the 13 data sets in the Cretopia condition (38.5 per cent) that mentioned one of the terms associated with the convex-hull strategy.

The fragment of a retrospective report shown in Table 10.9 demonstrates both strategies coming into operation concurrently. The participant describes the identification of clusters of locations that shared specific content, such as landscapes and gorges, as well as the mental image of a line and then a circle going through these clusters of locations. This was the only participant who used both strategies. All other reports pointed to either a hierarchical or convex-hull strategy.

10.6 Summary and Discussion

The retrospective protocols collected during the study contain utterances that indicate mental structuring processes. Some of the participants mentioned the variety of different activity types as a criterion for selecting specific parts of the island. Others utilized the cardinal direction relations, such as the western part of the island, for structuring the environment into high-level regions.

We detected different categories of verbal descriptions regarding the frequency of mentioning activity types, locations, or routes. Based on these findings, we defined different groups of travel planners: 'activity type planners', 'route planners', and

TABLE 10.8 **Hierarchical spatial planning strategy**

English (translation)	German (original)
From the national park on the coast, by ferry to an island, from there to the mainland, then to the mountains, along a riverbank.	vom Nationalpark an die Küste, mit der Fähre auf eine Insel, von dort aus ans Festland in gebirgiges Gelände an einem Fluss entlang.

TABLE 10.9 **Concurrent usage of the convex-hull and the hierarchical strategy**

English (translation)	German (original)
I then searched for a line leading to the beach, since good weather dominates the island. This line should go through a varied landscape of activities. After arriving at the beach, I tried to make a large circle out of the line in order to be able to see as much as possible. The travel routes were mostly drawn through the landscapes to avoid city traffic. During the design of this circle, locations that exposed many items of one specific activity type were important (e.g. 5 gorges).	Dann habe ich eine Linie zum Strand gesucht, da auf d. Insel gutes Wetter herrscht. Diese Linie sollte durch eine abwechslungsreiche Landschaft an Aktivitäten gehen. Am Strand angekommen versuchte ich aus der Linie noch einen möglichst großen Kreis zu machen, damit man möglichst viel sieht. Die Fahrtrouten sind größtenteils durch Landschaften gezogen, um Stadtverkehr zu meiden. Bei d. Erschließung d. Kreises waren mir vor allem die Stellen wichtig, an denen es viel von einem gab (z.B. die 5 Schluchten).

'location planners'. The 'activity type planners' reported the kind of activities they planned, the 'route planners' focused on describing routes, and the 'location planners' conveyed locations they targeted. This finding indicates the selection of a problem model, similar to the scheduling and the TSP model observed by Hayes-Roth and Hayes-Roth (1979) during the errand sequencing task. The TSP model might be applied by the 'location' and 'route planners', whereas the scheduling model might be preferred by the 'activity planners'. This interesting finding could be a starting point for a further study to determine the individual preferences and differences in how to approach a tour planning problem.

Similar to earlier TSP solutions found in the literature, the tours drawn on the maps by the participants exposed a small number of crossing lines. Due to the peculiar oblong shape of the island of Crete, we assumed that people might create tours with many crossing lines. However, the participants of the Cretopia condition (using a rounder-shaped island) created solution paths with more crossing lines on average than the participants of the Crete condition. One further person in the Cretopia condition produced a solution path with eight crossing lines (considered an outlier). However, the influence of the shape of an unfamiliar environment on the characteristics of the tours needs further elaboration.

The verbal reports of the study pointed to both the hierarchical as well as the convex-hull strategy. However, the convex-hull strategy is represented more often in the collected data sets than the hierarchical strategy.

The tour plans produced by the participants could be classified into two conceptual schemes. The location-based scheme includes a set of sub-tours that cover small temporal intervals (for example, several days). The day-based scheme involves a list of temporally ordered days associated with locations and planned activities.

The majority of the participants of the tour-planning study conveyed the resulting tours using a day-based scheme.

The present tour planning study indicates ways in which the formal models of human performance surveyed in this chapter that were based on more schematic versions of the TSP could be modified to incorporate aspects that relate to holiday tour planning, regarding the strategies and the underlying representation that they are operating on. In this way, our study provides an example of using empirical performance results along with formal mechanisms (similar to those explored in this volume, for instance, by Chown, by Nguyen and Wachsmuth, and by Barclay and Galton, modelling schematization aspects other than those explored here) so as to inform spatial assistance systems. The existing models implement either a convex-hull or hierarchical problem solving strategy to manage the complexity of TSP related planning processes. In our tour planning study, the collected verbal data reflect both strategies coming together in operation. Therefore, each model of human performance on the TSP on its own could not explain the observed behaviour. This finding corresponds to recent results by Tenbrink and Wiener (2009) with respect to the complexity of conceptual strategies involved in a TSP related task, likewise reflected in verbal protocols.

The most interesting and important open question is the mental representation of the spatial environment and its structure that people use for solving spatial planning tasks. The TSP models operate on a metric Euclidean plane, using a metric *cheapest insertion* heuristics. In contrast, mental representations are known to be qualitative and schematic (Chown, this volume). The insights of the current study were used by Seifert (2009) to implement a qualitative region-based model of map-based tour planning that allows for applying a region-based direction heuristic. The heuristics exposes a lower run time complexity of $O(n * \log(n))$ than the insertion-based heuristics that is $O(n^2 * \log(n))$ and thus provides a substantial basis for the development of intuitive assistance systems.

The compiled encoding vocabulary provides a more complete description of the mental operators and problem solving strategies involved in this type of spatial planning problem than has previously been available. It establishes a strong connection between the existing empirical studies of cognitive models of human performance on the TSP and the tour planning problem addressed here. This vocabulary can be used as a starting point for further empirical studies as well as enhanced models of human performance on spatial planning problems.

The exploratory tour planning study presented here raises questions for further research that could be pursued in more controlled experiments, such as: Why do people choose a particular problem model? Does a particular problem model actually explain particular tour planning behaviour at a more fine-grained level? What can problem models and conceptual schemes reveal about cognitive tour planning in terms of competing models?

11

Navigation assistance for blind pedestrians: guidelines for the design of devices and implications for spatial cognition*

MATHIEU GALLAY, MICHEL DENIS, AND MALIKA AUVRAY

11.1 Introduction

Moving independently in order to explore unknown areas of an urban environment is a cognitively complex task that requires many different capacities. Navigation from point A to point B can be seen as a three-step process. Step 1 is the *preliminary planning phase*: Before any physical action occurs, there is a need to plan the journey. To do so, travellers have to take into account their point of departure and their destination, compare the different routes linking these points and decide on the basis of criteria of distance, travel time, and safety, which is the best itinerary to follow. This process results in a mental route to follow that is made of turning points interconnected by road segments. After the preliminary planning phase, there is navigation itself, involving two distinct behavioural levels. Step 2 is *global navigation*, which consists in reaching various turning points composing the itinerary, remembering the direction to take at intersections and keeping that direction to stay on the road. Finally, step 3 is *fine navigation*, a short-term task which consists in avoiding unforeseen obstacles and overcoming difficulties (zebra crossings, traffic lights, cars, etc.). It should be mentioned that Step 3 could be considered as a sub-process of Step 2, as it is obvious that

* This chapter was prepared as part of the research project NAVIG, a project funded by the French National Research Agency. The research consortium includes three CNRS-supported laboratories specializing in interaction and disability (IRIT-IHCS, Toulouse), human and artificial vision (CERCO, Toulouse), and human–machine communication (LIMSI-CNRS, Orsay); two companies active in the fields of artificial vision (SpikeNet Technology) and geolocalization (NAVOCAP); and an institute of special education for visually-impaired persons (INJA-CESDV).

someone who does not successfully manage fine navigation will not be able to fulfil the global one. However, these two behaviours are clearly distinguished by the different perceptual, attentional, and motor processes involved in them.

These three components of pedestrian navigation, which will be described in more detail in the subsequent sections, are selectively hampered by a lack of vision. Indeed, sighted people usually deal with the preliminary planning phase using a map, whereas global navigation is achieved through distal information made available by vision, which also allows the identifying of obstacles to avoid on the road. An effective navigation aid device for blind people on the move is therefore expected to overcome the lack of vision for each of these three components, in indoor as well as in outdoor environments.

Offering blind people alternative ways to have access to the spatial information normally conveyed by vision relates to some of the most exciting challenges in the field of spatial cognition. For instance, developers have to address the role of prior visual experience on the spatial cognition abilities of early and late blind individuals when aiming to provide them with guiding tools that are efficient in compensating for their respective impairments (Pasqualotto and Proulx, 2012). More specifically, offering a verbal description of the spatial environment as an alternative to the traditional map-based preliminary planning phase raises the issue of the characteristics of linguistic materials, which—at least for some—are considered to be modality-specific (e.g. Klippel and Montello, 2007; see also Avraamides, Mello, and Greenauer in this volume). The development of electronic guiding devices is also closely related to the progress made in the field of formal models, as spatial environments have to be recorded and efficiently described by computers in order to provide users with relevant spatial information (see Barclay and Galton, and Kordjamshidi, Hois, van Otterlo, and Moens in this volume).

11.2 Commercialized systems

11.2.1 Preliminary planning phase

It is particularly important for a blind person during the preparation of an errand, to have information on the general shape of the itinerary and on the landmarks involved in it. The availability of this kind of information can reduce fear and feelings of anxiety due to navigation in an unknown environment and can therefore allow a better use of available cognitive resources, which may then be allocated to the navigation task itself. The most complete alternative to the traditional visual maps conventionally used by sighted people are audio-tactile maps (Parkes, 1988; Miele et al., 2006).

For some time now, it has been possible for blind people to generate from home, via specific software, the purely tactile map of a new place they want to explore. It is possible to emboss this map at home with the right equipment. Such a map contains relief streets at its centre and their corresponding Braille labelling at the periphery. Tactile maps have been proved to be efficient in environment learning by several studies. Ungar et al. (1994) compared the performance of visually-impaired children (aged 5–11) who were asked to learn an environment made of toys displayed randomly across the floor of a large hall either by direct exploration or by exposure to the tactile map of the environment. Blind children learned more accurately from the tactile map than from direct exploration. An advantage for tactile map learning was also reported in an ecological environment by Espinosa et al. (1998). Blind adults were asked to learn an approximately two-kilometre route in downtown Madrid and to reproduce it later while making direction estimations between locations on the route. The participants who learned the route with a tactile map and direct exploration performed better than those who resorted only to direct exploration, and this was the case for both route knowledge and direction estimations.

The problem with a purely tactile map is that the space available to include Braille information is limited. As a consequence, not all street names are listed, and in addition there is a lack of information about points of interest. This is the reason why a lot of effort is made on the implementation of audio-tactile maps at home. Audio-tactile maps require, in addition to a specific printer, the acquisition of a graphic tablet. As the tactile relief map of a desired geographical area is superimposed on the map displayed on the tablet, it becomes possible to pinpoint a street segment to get its name (multiple clicks may be used to provide additional information such as segment length or the precise spelling of a street's name). These cards are considered to be a valuable tool for the planning phase, but overall, audio-tactile maps have important financial and temporal costs.

A study by Golledge et al. (2004) on blind people's wishes about the functionalities of a navigation aid device shows that even if audio-tactile maps are considered as being acceptable tools for preparing an itinerary, they are requested by fewer interviewees than simple verbal information about the route. Indeed, those interviewed by the authors asked for vocal information prior to displacement, instructions on length and the number of segments composing the itinerary, the number of turning points it contains, as well as the size of the angles formed at reorientation points. More detailed information such as landmarks, potential obstacles, and audio and podo-tactile cues present on the itinerary are also considered useful. When they were asked about their usual way of having access to such information, blind people answered that they obtain it mainly by asking other people. Still regarding preferences, Golledge et al.'s study also revealed that blind people would like a system in which requests are entered using vocal communication. Instructing the system verbally was preferred to any other proposed means (such as phone keypad, computer keyboard, or Braille keyboard). For output information, the desired means of transmission were also verbal. These systems were, in order of preference, a small loudspeaker mounted in a necklace or attached to the shoulder, bone conduction headphones, or a simple headset. The most undesired output is a Braille keyboard or conventional headphones covering both ears.

11.2.2 Global navigation

GPS-based guidance systems for blind people have been available on the market (especially in the United States) for a few years now. Rather than providing an exhaustive market survey, this chapter focuses on the presentation of existing systems' overall functioning (see Roentgen et al., 2008, for a complete review) and suggests means for their potential improvement. In addition, this chapter focuses on GPS-based technologies and not on mere visuo-tactile or visuo-auditory conversion systems (for a review of these sensory substitution devices, see Auvray and Myin, 2009; for a discussion on their implications on spatial cognition, see Auvray et al., 2005). Among the best-known systems, it is worth citing BrailleNote GPS and Trekker, both built by the HumanWare Group. These products have been subject to scientific evaluation by an American foundation, the National Federation of the Blind (NFB's Access Technology Staff, 2006) and by Denham et al. (2004).

The BrailleNote GPS system consists of a Braille PDA connected to satellite receptors (Figure 11.1). Several CDs containing maps of various regions of the United States are provided with a flash card on which the software is installed. It allows users to easily create, store, and erase both pedestrian and road trips. It offers a certain number of options: it can compute the way back to the departure point and allows users to choose between clockwise or more traditional left/right directions or even between a detailed presentation of all the crossroads along a defined path and a simpler presentation of the directions to follow. A storage system for categorically sorted points of interest (e.g. hotels, restaurants, hospitals) offers the possibility to use

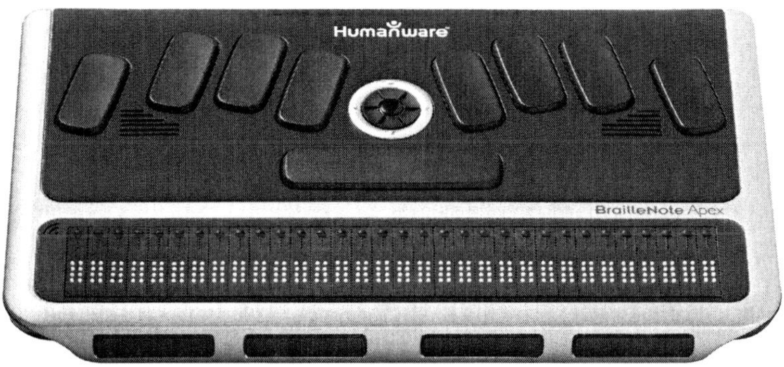

FIGURE 11.1 Illustration of the commercial device 'BrailleNote GPS' by the Humanware Group

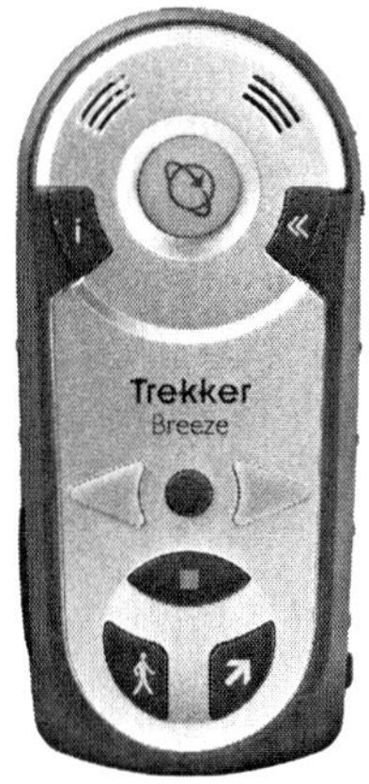

FIGURE 11.2 Illustration of the commercial device 'Trekker BREEZE' by the Humanware Group

these points as additional information for navigation or to enter one of them as a destination. Finally, these points of interest and street names are available both during real-time navigation and the home trip-planning phase through an 'exploratory mode'. The NFB underlined in its assessment that BrailleNote is a system providing good flexibility and ease of use but regretted that it is not currently possible to print or emboss a created route at home. Denham et al. (2004) reported great variability in the accuracy of information given in order to access points of interest, but stressed that it is a valuable tool for obtaining information on a given geographical area.

Another well-known navigation aid device, Trekker, is composed of a PDA containing geographical maps, software to install, a wireless GPS receiver, an external speaker and a Velcro belt to hold together all these elements (Figure 11.2). Like BrailleNote, it allows users to find their current position and search streets and points of interest as they move, and it also allows them to choose between pedestrian and driving modes, and between online and home modes. It provides detailed information about crossroads (e.g. number of segments, type of intersection). Input interaction is made through contextual buttons and a Braille keyboard. Output interaction is purely verbal through the external speaker that provides clear sound perception and does not block environmental sounds. This device was judged by NFB as having good pedestrian guiding qualities, a very useful option that describes streets and intersections, and as being easy to use due to the similarities with widely-used Microsoft Windows menus. However, the wide variety of menus and options available were said to make menu browsing tedious. In addition, the use of Braille for the inputting of instructions was not acknowledged as being the most practical or the fastest way to interact with the system.

NFB's evaluation of navigation aid devices also included a section dedicated to the limitations common to all GPS-based systems. Indeed, there are inherent problems associated with the use of this technology. For example, a GPS receiver does not work 100 per cent of the time, but only when satellite coverage is satisfactory (the receiver needs to communicate with multiple satellites to provide an optimal signal). Climatic conditions and user's location can also affect the quality of geolocalization as the signal is not easily accessible when the weather is cloudy or when the user is indoors or trapped between tall buildings. The maps used by the system must be regularly updated to avoid becoming obsolete, and even if they are updated, it is very difficult to integrate information about roadworks or traffic conditions. Finally, the authors highlight the problem of energy dependence: The onboard batteries are heavy, have limited life duration, and must be recharged quite often in order to accompany users until the end of their course. These limitations have a detrimental effect on blind people's autonomy and the subsequent sections of this chapter will highlight relevant ways of overcoming them.

11.2.3 Fine navigation

In addition to guide dogs and white canes, electronic systems can help users to avoid potential obstacles littering the route. For instance, Farcy et al. (2006) presented two of those systems, developed by their research team, 'Tom Pouce' and 'Télétact' (Figure 11.3).

The Tom Pouce technology allows its users, through an infrared light emission/reception system, to be warned by a vibration of the presence of an obstacle in the distance with about 20 per cent error. The device is fixed on the user's white cane and has a horizontal angular resolution of 20° (corresponding to an average shoulder width) and 50° vertical (to protect from the knees to the head, under-knee protection

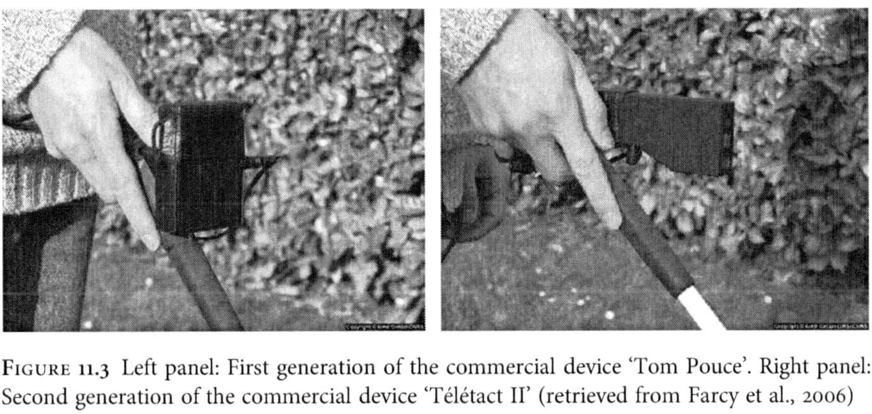

FIGURE 11.3 Left panel: First generation of the commercial device 'Tom Pouce'. Right panel: Second generation of the commercial device 'Télétact II' (retrieved from Farcy et al., 2006)

being ensured by the cane itself). The device is adjustable to detect obstacles located at 50 cm, 1.5 m, or 3 m, depending on the user's preference. Being based on light reflection, the limits of this device are late detection of small black objects (that do not reflect enough light) and too early detection of reflective road signs.

Télétact uses laser beams to detect obstacles. This system is able to provide the distance to an obstacle with 1 per cent error in an area ranging from 10 cm to 10 m from the user, the distance being encoded in more or less intense vibrations, or with a melody whose pitch and rhythm are based on distance. The main limitation of this system lies in its difficulty of use, because the detection of obstacles requires a manual scan of the environment that has to be learned by the user.

There are many other obstacle detection devices, running on the same principle (emission/reception of ultrasounds or light signal and tactile communication) such as the 'Bat K' sonar (Bay Advanced Technologies; Kay, 1964) but it is acknowledged that none of them can replace the white cane or dog. Besides, many of these devices are fixed directly on the stick, as a detection complement.

11.3 Systems under development, global navigation, and guidance means

So far, navigation aid devices handle the planning phase on their own, based on points of departure and arrival triggered by users. Moreover, to avoid obstacles, blind people prefer to use a white cane or a guiding dog rather than one of the available fine navigation aid devices. Research teams in recent studies on assistance for blind pedestrians are, therefore, mainly focused on investigating global navigation. Technology is making progress in this direction and new means of improving global navigation are under investigation in several research laboratories, with the aim of developing more optimal devices, adapted to the needs of blind people.

There are different ways to guide a person to a predetermined point in space such as RFID (Radio Frequency Identification) or RIAS systems (Remote Infrared Audible Signage) which allow users to hold in their hand a signal-sensitive device which will guide them to specific points, such as a bus stop, a cab phone, or a store, by emitting a signal. However, these systems require a heavy pre-installation of the areas in which navigation will take place. Most of the progress made in global navigation therefore relies on the development of satellite guidance technology. The idea of using GPS navigation to help blind people find their way has been independently suggested by Collins (1985) and by Loomis (1985). In particular, Loomis and his colleagues have been very active in this research area, especially with regard to the development of an efficient guidance interface for blind people called Personal Guidance System (PGS) (Figure 11.4).

To evaluate these devices, the tests usually consist in asking blind participants to follow a route made of segments and shift points while experimenters measure travel time, walking speed, and the distance travelled. Together with post-experimental

FIGURE 11.4 Left panel: Illustration of the version of the Personal Guidance System developed between 1992 and 1995. Right panel: Miniaturized version of the system developed between 1999 and 2000. (Retrieved from Loomis et al., 2001. © The American Academy of Optometry 2001.)

questionnaires about ease of use, comfort, and the feeling of safety provided, these measures allow the assessment of the potential of several guidance interfaces based on GPS technology.

11.3.1 Audio-based guidance systems

A number of studies have been conducted by Loomis and his colleagues to assess the effectiveness of their device, the Personal Guidance System. With a binaural headset, it is possible to emit spatialized sounds in the user's ears, so that they are perceived as coming from a precise location in the environment. The idea is that this type of signal is a direct perceptual signal that does not require additional cognitive processing as opposed to speech signals.

Loomis et al. (1998) compared the virtual sound guidance mode to three more classical speech-based guidance modes with tests performed on planned routes in open areas of the campus of the University of California in Santa Barbara. The common basis of these guidance modes were the following: the system included a GPS module able to transmit and receive satellite signals, a back-packed computer containing a Geographic Information System (GIS) to locate the users in space, and an orientation sensor (electronic compass) to determine at any one time in which direction the users are moving. In the 'virtual' mode, the compass was fixed to a helmet and was therefore located on the users' heads, allowing them to hear verbal information (the number corresponding to the next turning point) coming from the turning point they needed to reach. The sound's intensity gradually increased as users got closer to their goal. When users entered a 5 m-radius virtual circle around the turning point, they heard a virtual sound coming from the next turning point. In the 'left / right' mode, the compass was fixed on the users' chests and allowed the path followed by them to be corrected by sending verbal signals through the headset ('left', 'right', or 'straight'). The 'bearing' mode was the same as the one described above, with additional verbal information concerning the angle of rotation to be made by users to face the next turning point ('left 80°'). Finally, the 'no compass' mode was the same as the bearing mode except that the information on the direction taken by users did not result from compass data, but from the extrapolation of two successive GPS position points (which entails that if the user stops moving, the system is no longer able to send information).

Recordings of the time taken to complete the route indicated that the best guidance mode among those tested here was the 'virtual' mode, which gave a shorter travel time than the other three systems. Moreover, the preference orderings showed that 5 out of the 8 participants from whom rankings were obtained preferred the 'virtual' mode. The guidance mode causing the longest travel times and worst subjective judgements was the 'no compass' mode. This experiment highlighted the need to use a compass to deliver information regarding the direction taken by users and the interest of using virtual sounds to guide pedestrians.

A second experiment conducted by Loomis et al. (2005) introduced a new guidance interface called the Haptic Pointer Interface (HPI), based on the RIAS system. Users carried in their hands a box that was connected to an electronic compass. When the user's hand made an angle of less than 10° around the next turning point, the computer emitted beeping sounds through a loudspeaker attached to the chest. This experiment compared five guiding systems, two of them based on virtual sounds (using a binaural headset) and the other three based on the HPI (using speakers). The 'virtual speech' mode was the same as the virtual mode of the previous experiment except that users heard the distance that separated them from the next turning point instead of the corresponding number. The 'virtual tone' mode guided users by emitting a spatialized beep. The 'HPI tone' emitted beeps when the hand pointed to the right direction (less than 10° from the turning point). The 'HPI speech' system gave verbal instructions ('straight' if the hand pointed in the right direction, 'left' or 'right' when the hand deviated by more than 10° from the turning point). Finally, the 'body pointing' mode was similar to the 'HPI tone' but instead of the hand, it is the body that was taken into account to calculate the direction. The distance to the next turning point was given orally every 8 seconds for all systems (additionally to other information types), except for the 'virtual speech' which gave this information continuously.

The efficiency of virtual sound guidance was thus confirmed by this experiment, since the 'virtual speech' mode induced the shortest travel times. Coming next is 'virtual tone', then 'body pointing', 'HPI speech', and finally 'HPI tone'. If different guidance modes prompted different travel times, the distances covered were similar with all modes, indicating that the systems were equally efficient when it comes to following a trajectory. Therefore, turning points were the precise location where time was saved thanks to virtual systems. Indeed, whereas with a pointing system (hand or body), users needed some time to find the direction of the next turning point, it was immediately available to their ears with a virtual audio guidance system. Subjective judgements made by participants suggested that each system has its advantages and disadvantages. Manual pointing was considered easier than body pointing, but had the disadvantage of occupying a hand (knowing that most of the time the other hand is already occupied with a white cane or guide dog). The virtual sounds had the advantage of being quickly informative and leaving the hands free, but the requirement to use a headset for binaural rendering (while other systems use speakers) disrupted the perception of environmental sounds, which was perceived as being a real issue by blind people. Finally the presence of verbal information was appreciated because it is informative, but it required more attention than hearing simple beeps.

With respect to the use of such devices in ecological conditions, a study by Marston et al. (2006) aimed at moving the tests from controlled environments (campus) to everyday structured environments (building blocks structured by walls and sidewalks versus the city park). The study also aimed at comparing the HPI mode to a virtual sound mode improved by using an air-tube headset, which does not block environmental sounds. In addition, the experimenters gave their participants the opportunity to choose when to enable audio information. The two systems (HPI and virtual sound) provided satisfactory results in both types of environment, with, however, a non-significant advantage in favour of the virtual sound mode. The authors underlined that in the building block environment, which is structured by roads and sidewalks, participants tended to use less sound information than inside the park, which is less structured. Participants evaluated very positively the sound produced by air-tube headsets, they rated it as 'not blocking external sounds' (mean of 4.5 on a 5 point scale where 1 = strongly disagree and 5 = strongly agree) and considered that a commercialized guidance system should offer this kind of sound reproduction (mean of 4.65 on the same scale).

In order to determine how attentional resources are allocated, Klatzky et al. (2006) highlighted what appears to be a considerable advantage for virtual sounds over language. In 'normal' navigating conditions (the only task is to reach a point B from a point A), a verbal information guidance mode ('left', 'right') resulted in similar performance to the one obtained with a spatial sound mode. However, the latter turned out to be superior to the verbal mode when the navigation interfered with a distracting task (here, *N*-back task: Report in a scrolling list, the appearance of the same item with *N*-intervals, the values of *N* varying in order to increase or reduce the cognitive load). Indeed, travel times and travelled distances were shorter and the distracting task performance was higher with spatialized sounds than with speech guidance. Therefore, it appears that spatial sounds are being processed at a perceptual level, whereas language requires additional cognitive mediation. Due to lower cognitive load, spatial sounds allow some cognitive resources to be freed and that can therefore be allocated to another task, such as talking with someone else along the way, keeping a shopping list in mind, or getting ready for an appointment.

A study conducted by Walker and Lindsay (2006) investigated the effect of the type of sound used for guidance (sonar pulse versus pure sine wave versus burst of pink noise), the effect of the size of the virtual circle around the turning points (0.5 m, 1.5 m, or 15 m), and the learning effect (analysis of progress made along three different paths). The recordings of the time taken and the distance travelled led to the following conclusions. First, this study confirmed that a virtual sound efficiently guides pedestrians, as all participants managed to go to the end of the three paths. It also showed a significant improvement in performance with experience, as participants (blind people without previous visual experience) were faster and covered less distance as successive tests occurred. The results revealed a significant improvement between the first and the second test and between the second and the third one, which indicates that the threshold for optimal learning was not achieved during the three sessions and thereby that performance can still improve. As expected (based on the sound spectrum), the pink noise brought about the best performance, but this effect did not reach the significance threshold. Finally, the size of the virtual circle around the turning point seemed to have an influence on the time taken and the distance travelled. The authors found that a medium-sized circle (about the size of a human step) is the safest solution (even if it is not the one leading to the shortest travel time). Indeed, it is easy to walk outside of the circle if it is too small, which may lead the user to turn around and waste time. In addition, even if the existence of a big circle saves time because it is easier to find, making turns too early in an urban context can lead to walking straight into a wall or crossing the street in the wrong place. In conclusion, it seems that learning effects and the effects of the circle size have a greater influence on performance than the type of sound used for guidance.

11.3.2 Tactile guidance systems

Audition is not the only sensory modality that has been tested as a means of guidance. Marston et al. (2007) have tested the efficiency of two guidance systems, one based on sound, the other on touch (via a vibrating motor attached to the waist). Both had a binary functioning: they sent a stimulus when users were moving toward the correct direction ('on-course cue') or otherwise when they moved away from the correct direction ('off-course cue'). This binary guidance system was efficient because it gave travel times that were similar to those obtained with previously tested systems, with equal performance for the tactile system and the sound system. Moreover, it was well accepted by users who agreed (mean of 4.9 on a 5-point scale) that this type of guidance should be available as an option on a commercial device. However, this binary guidance mode cannot be used as the only means of communication between the system and its user as the addition of verbal information is sometimes necessary (such as street names), but it has the advantage of being very simple and not requiring too many attentional resources.

The use of tactile stimulation for navigation aids has also been investigated by a team of German researchers who first presented the functioning principles of their system (Henze et al., 2006). The latter consists of a GPS navigation module added to a belt equipped with an electronic compass and six vibrating motors that can be activated individually or simultaneously, with variable intensities of vibration for greater accuracy. The computing of the user's location and itinerary is supported by the system, which reduces the cognitive load associated with map information processing. The use of a tactile device has the advantage of providing a directly interpretable stimulation (like virtual sounds) without clogging the ear canals and remaining relatively unobtrusive.

Heuten et al. (2008) subsequently presented the results of an experimental test of their tactile device. The authors measured the accuracy in the perception of direction and the efficiency of assistance provided by the device in terms of guidance in a real environment. The tests revealed that participants perceived the direction indicated by the system with an average error of 15°. Acuracy was strongly dependent on the difference between the angle of body orientation and the position of the vibration motor providing stimulation. Indeed, the six motors were placed around the waist 60° away from each other, which led to good performance when the orientation to perceive was close to one of the six motors and to lower accuracy when it occurred between two motors (in this case, the respective intensity of the vibration of the motors adjacent to the direction had to be taken into account, which complicated the interpretation of direction). Humans' ability to perceive tactile stimulations has to be taken into account. It turns out that the tests showed greater accuracy when the to-be-collected stimuli were in front of the body rather than behind, which is a happy coincidence, since it is very unusual to need to move backwards in a navigation task.

Heuten et al. (2008) evaluated the efficiency of their device with blindfolded sighted participants in a tracking task on a predefined itinerary, in an open environment (no path to follow or specific landmarks) in which were recorded the average deviation from the original itinerary and the total time taken over the course. Two different itineraries were tested, one with angular corners, the other with long curves. The walking speeds were, respectively, 3.0 km/h and 3.3 km/h and deviation averages with respect to predefined paths were 6.57 m and 7.21 m, respectively. The authors underlined the fact that this precision could be improved. First, tactile stimulation corresponding to the next point of reorientation was given when users entered a circle of 15 m around a point. Then, there were no landmarks like walls or sidewalks along the path tested. Reducing the diameter of the circle and adding landmarks to the itinerary could therefore help to guide blind people in a more efficient way. However, these studies revealed that a purely tactile stimulation is sufficient to transmit GPS information to a pedestrian in an intelligible way.

Table 11.1 summarizes the main characteristics, advantages, and disadvantages of the navigation devices discussed in this section.

11.4 GPS, spatial representations, and frames of reference

A new generation of navigation devices should be soon available to the public and provide assistance to blind people in their daily activities. However, from a fundamental point of view, it seems necessary to consider the long-term impact that the use of a navigation assistance device may have on mental representations of the environment. While it is generally acknowledged that the use of GPS helps people form better spatial representations of their environment, the opposite effect can sometimes be observed. Placing undue reliance on a guidance system that computes an itinerary itself may well result in following instructions in a passive way without a proper encoding of the spatial relations linking the various geographical elements. This would result in strengthening the spontaneous tendency of blind people to adopt an egocentric perspective, in contrast to an allocentric perspective.

11.4.1 Egocentric and allocentric reference frames in navigation tasks

Two well contrasted frames of reference are distinguished in spatial cognition: *egocentric* and *allocentric* reference frames (O'Keefe and Nadel, 1978; see also Tversky, 2003). The way an itinerary is described (route versus survey description) is expected to orient an individual towards an egocentric or an allocentric reference frame (see Avraamides, Mello, and Greenauer, and Taylor and Brunyé in this volume). Route descriptions consist in sequences of instructions specifying the changes in direction that people must make to reach their destination (e.g. 'turn left', 'walk 150 metres', 'turn right'). It is characterized by a serial aspect whose direct consequence is that if users omit one instruction, they will be lost and therefore unable to reach their destination. These route descriptions are not flexible and cannot be reorganized. They are based on an egocentric reference frame, as opposed to survey descriptions, which are based on an allocentric reference frame. The latter involves knowledge of spatial relations that unite the various elements of the environment regardless of the observer's position.

The chapter by Taylor and Brunyé in this volume offers an extensive overview of the differences between route and survey perspectives. The authors crucially highlight

TABLE 11.1 **Summary of the main characteristics of the navigation devices**

	Name	Sensory modality	Information received by users	Transmission device	Advantages	Disadvantages
Fine navigation aid devices	Tom Pouce	Touch	Manual vibration	Vibrator fixed on a white cane	Easy to use	Numerous detection errors
	Télétact	Touch / Audition	Manual vibration / Musical sounds	Vibrator / Speaker fixed on a white cane	Precise obstacle detection	Difficult to use, requires important learning
Existing devices	BrailleNote GPS	Audition / Touch	Verbal language / Braille language	Speaker / Keyboard	Complete environmental description	High level of attentional resources needed, draws attention to user
	Trekker	Audition	Verbal language	Speaker	Complete environmental description	High level of attentional resources needed, draws attention to user
Developing devices	Auditive PGS (Loomis et al., 2005)	Audition	Binaural sounds	Binaural headset	Effective guidance, low level of attentional resources needed	Covers both ears, transmits limited information
	Tactile PGS (Marston et al., 2007)	Touch	Binary vibration	Simple vibrator	Effective guidance, low level of attentional resources needed	Transmits limited information
	Tactile Wayfinder (Heuten et al., 2008)	Touch	Hips vibration	Vibrating belt with 6 motors	Effective guidance, low level of attentional resources needed	Transmits limited information
	NAVIG (Katz et al., 2010)	Audition	Verbal language / Binaural sounds	Binaural headset	Effective guidance, low level of attentional resources needed, complete environmental description	Covers both ears, heavy equipment

that perspective switching and spatial inferences generation are critical skills for spatial problem solving; two behaviours that are allowed by the use of a survey perspective. An efficient allocentric representation of space would allow one to mentally explore an environment, as opposed to physically explore it, and would consequently allow a reorganization of itineraries such as alternative routes or shortcuts. Benefits from taking that kind of perspective have been experimentally confirmed (Brunyé and Taylor, 2008a; Brunyé et al., 2008).

The construction of an allocentric representation, of a city for example, requires full integration of the various paths known in this city, and the inclusion of such knowledge in a geographic perspective using the cardinal directions. This construction is a cognitively difficult process and is, in sighted people, greatly facilitated by the use of the visual modality, as it allows multiple and simultaneous perception of spatial elements, even if they are distant. According to Millar (1994), it is the perception of distal elements that allows encoding spatial relations between elements of the environment, and therefore allows the construction of an allocentric reference frame.

11.4.2 Blindness and reference frames

In cases of partial or total lack of vision, the construction of surrounding areas is performed by integrating information from other sensory modalities. Blind people mainly use podo-tactile (e.g. walking on a lawn, then a road, then on pavement), kinaesthetic (such as air brushing the skin and vestibular information), and auditory information (such as sounds coming from the cars or traffic lights) to visualize mentally the paths they follow. This multisensory integration, since it is based on sequentially collected information, involves a high cognitive and temporal cost.

The lack of distal information associated with multisensory integration cost leads blind people to rely on egocentric navigation strategies rather than on allocentric ones. This tendency explains why blind people are more likely to organize spatial information in a route form rather than in a survey form. Noordzij et al., (2006) tested the qualities of mental representation constructed by blind and sighted people on the basis of a verbal description. The environment (a zoo) was described either from an egocentric (route description) or an allocentric point of view (survey description). The results of this study suggest that blind people's mental representations constructed on the basis of a verbal description share the same metric properties as those constructed by sighted people. In a priming task, participants were presented with two animal names that were either present or absent from the original description. The participants' task consisted in judging as fast as possible if the second animal was actually present in the zoo. Just like the sighted, when both elements were really parts of the initial description, a priming effect occurred: the closer the two elements were geographically, the faster the participants responded. Moreover, and again as in sighted people, this proximity effect disappeared when the animals to be judged were close in text but far in space. The authors therefore concluded that blind people have a real spatial (and not only verbal) mental representation of verbally described environments. Another task consisted in comparing distances between elements of the environment. Again, the same effect was shown: the larger the difference between the two distances to be compared, the shorter the response times. The quality of the mental representation constructed by blind and sighted people appeared to be similar, that is, they share the same metric properties. However, and importantly, the analysis of the effects of the different types of description revealed that blind participants constructed a more robust spatial representation from a route description, whereas sighted people performed better with a survey description.

Other findings from experiments on the acquisition of spatial knowledge via locomotion are in line with this conception. They confirm that the route perspective imposed by locomotor experience is especially well integrated by blind people. Two recent studies used the mental scanning paradigm to assess the metric validity of the representation of distances based on locomotor learning. In the first study (Afonso et al., 2010), participants learned the positions of objects in a room. Then, they were invited to mentally scan the distances separating pairs of objects. This task is classically used to assess whether or not the time required to scan a given distance is correlated with the corresponding actual distance. When such a correlation is found, it is taken as indicating that metric distances are effectively represented in the mind in an analogical format. Sighted people who have learned configurations of objects visually typically generated response times which reflect such positive correlation. The same was found when they learned these configurations from verbal descriptions. But when they learned the same set of inter-object distances from locomotor exploration in actual space while being blindfolded, they did not give any sign of time/distance correlation, which shows that they have not constructed any consistent internal representation in which distances would be represented. On the contrary, blind people's responses did show a strong pattern of time/distance correlation, which attests that their motor experience has helped them build a realistic metric representation of configurations learned from an egocentric route perspective. Furthermore, the difficulty experienced by blindfolded sighted people was reflected by scanning times, which were consistently longer than those of blind people.

The same was found by Iachini and Ruggiero (2010), who also investigated the strategies used by people while performing mental scanning following locomotor learning of an environment. One strategy was described by the participants as placing a focus on a map-like representation of the learned configuration, without any reference to the experience of movement which was involved during learning. Another strategy was described as being based on the reconstruction of an image of the participants' own moving body along the paths and the revival of the kinaesthetic feeling associated with it. The relevant point here is that the kinaesthetic strategy—which is inherently associated with the route perspective—is the dominant strategy of blind people when they evoke the paths via mental scanning, and even still more markedly in congenitally blind people. This finding, again, is consistent with Millar's (1994) hypothesis that the egocentric perspective is privileged by blind people when they encode spatial relations among objects. As a result, blind people are strongly biased towards forming route-based sequential representations when processing spatial information.

The use of an egocentric perspective is sufficient to succeed in numerous spatial tasks, which explains why sometimes similar performance was found between sighted and blind individuals. To give some examples, Rieser et al. (1986), and Loomis et al. (1993) compared spatial memory performance in blind and blindfolded sighted participants. With respect to the task, Loomis et al. (1993) accompanied the participants in their study along a straight or curved path, which people were subsequently asked to reproduce. Rieser et al. (1986) guided the participants in an unfamiliar environment, from a departure point to six other places marked by the presence of objects that people would later have to point at with their finger. In these two experiments, no significant difference was found between the performance of blind and sighted people. These findings suggest that visual deprivation, whether congenital or acquired, has little or no effect on the estimation of directions and distances between elements of an area that has been physically explored. Indeed, the reproduction of a physically experienced itinerary is based on kinaesthetic and vestibular memory, corresponding to an egocentric frame of reference. However, when the task involves performing a change of perspective or spatial inferences, differences between sighted and blind people appear. In their experiment, Rieser et al. (1986) also tested the quality of the encoding of spatial relations that had not been physically explored. In the absence of any possibility to reproduce a path through vestibular and kinaesthetic memory, congenitally blind people did not perform as well as sighted and late blind individuals.

Blind people are oriented to the use of an egocentric encoding of space, a strategy sometimes poorly suited to solving complex spatial problems. If the egocentric representation of space is the perspective blind persons adopt by default, this does not mean that they are unable to adopt an allocentric perspective. Tinti et al. (2006) compared the performance of blind and blindfolded sighted people on several tasks involving an allocentric reference frame (path completion, directions and distance estimation, map drawing). In this particular experimental context (explored environments were restricted in size and sighted participants had to use a white cane because they were blindfolded), the authors observed better performance in blind people as compared to sighted people. While this finding should not be interpreted as indicating that blind people have better spatial representations than sighted people, it appears from this experiment that early visual experience is not a necessary condition for the construction of complex spatial representations.

11.4.3 Virtual sounds to represent distal elements

It has been suggested that one of the main constraints leading blind people to the use of egocentric strategies is the lack of distal information (Millar, 1994). Subsequently, studies tested the efficiency of environmental sounds as auditory landmarks (i.e. a functional equivalent to Millar's visual distal object) in navigation tasks. Morrongiello et al. (1995) evaluated spatial knowledge in 5 to 9-year-old blindfolded sighted and congenitally blind children. Participants had to move in a large-scale environment containing four distinct objects and were asked to remember the objects' position in order to reach them when they were asked to. Participants performed the task in two conditions: without, and then with an auditory landmark (a metronome placed next to one of the four objects). As performance was better (in terms of both position accuracy and path efficiency) with the auditory signal setting at every tested age, the authors concluded that auditory landmarks can serve as a point of reference to help people keep track of their position while moving, allowing them to form a better survey representation. However, it should be mentioned that, in this study, the landmark condition was always performed after the no-landmark condition (in order to avoid learning effects in the no-landmark condition). Thus, this conclusion has to be taken with a pinch of salt. Nevertheless, the sharp increase in performance levels following the introduction of the landmark suggests that sounds can be used as distal elements in the construction of a survey perspective.

A set of experiments using binaural sounds in order to represent distal cues in a virtual auditory environment revealed that allocentric strategies are used more by blindfolded sighted participants in the presence of an auditory distal cue (Gallay et al., 2010). Although this result has yet to be extended to other situations, it suggests that this kind of option could be implemented in a navigation aid device designed for blind users. Thus, by using sound spatialization technology to compensate for the absence of vision, a navigation system can offer blind users the opportunity to 'attach' virtual sounds to specific places in the environment (e.g. home, workplace, or even favourite hairdresser). These sounds can be activated whenever the users want to get an idea of the direction and distance to these points they have chosen to bookmark. In the context of long-term use, this option would allow blind people to connect the many fragmented pieces of information they have (several egocentric representations of the different paths known) for forming a more robust and more consistent spatial representation of the environment in which they live.

It is worth mentioning that it also seems necessary to investigate the effects of the presence of auditory distal cues in a longitudinal approach. Indeed, the strengthening of the spatial mental representation of a complex environment such as a city requires a number of interactions with the environment, and can therefore only be observed in the long term. If this hypothesis proves to be true, the implementation of such an option in navigation aid devices for blind people could not only help users perform their daily trips, but also eventually allow them to avoid the trap of GPS dependence, by getting rid of the guidance system and constructing a better, self-sufficient mental representation of their environment.

11.5 Guidelines for developing aid devices

The observations on existing and developing systems described in the previous sections allow us to draw some recommendations for the development of new guidance systems that are more feature-complete and closer to blind users' needs. To be more complete and more useful than a simple sequence of instructions, a navigation aid device for blind people should not only help in global navigation, but it should also help in route planning and fine navigation while using an intuitive and effective interaction system.

11.5.1 Preliminary planning phase

In order to have access to information about navigation several means are possible. Among them, audio-tactile maps are very useful tools for planning a journey. They provide a great amount of spatial information and allow users to have a good mental representation of the environment. On the other hand, the use of such maps requires buying heavy and expensive materials and devoting time to the manual exploration of maps. A lighter and simpler solution can be considered to prepare journeys, such as a simple presentation of the overall directions and the names of the streets to be taken, but this aspect of navigation should not be neglected.

The other aspect of route planning concerns the content of the information to be displayed. In order to better understand blind people's habits and needs when performing pedestrian navigation in an unfamiliar environment as well as the influence of route planning on subsequent navigation, an ergonomic study was conducted in our laboratory (Brunet, 2010; Brunet et al., submitted). Several blind people were followed by the experimenter during the execution of an urban navigation task, from the preliminary planning phase to reaching the final point. They were asked to explain each step of their usual trip planning process and then to walk along the way, using the information they had previously gathered and asking the experimenter if any information was missing. The main result, concerning route planning, was that people do not plan trips in the same way. Three dominant strategies were identified: (1) individual planning with computer help; (2) planning with human help; (3) no planning at all, but asking other pedestrians for the way while navigating.

These results suggest that, in order to fit different users' needs and habits, a navigation aid system should allow them to choose between travel with or without a home planning phase. In addition, the information delivered on the go should differ depending on whether or not there was any home planning performed by the user. For users willing to plan their trip, the system should provide clear information and users should be given the option of setting the level of detail they want in the instructions. It has been shown that minimal descriptions (e.g. 'take Danton Street', 'take St-Germain Boulevard on the right') are better stored by individuals than excessively detailed descriptions (e.g. 'take up Danton Street till the second traffic lights and then turn right at the pharmacy on St-Germain Boulevard'; see Denis et al., 2006). However, it may be useful for users to have additional information about points of interest present on the path ('May I stop at the cash dispenser on my way to the store?'), the length of the segments ('Should I keep going for 20 m or 200 m?'), or the angular size of the turns they are to take.

Concerning the grammar of verbal materials, it should be noted that route directions produced by visually-impaired people have features that differentiate them sharply from those of the sighted. Thus, some principles should guide their design. Undoubtedly, the most 'friendly' verbal information that blind people may wish to have assist them in spatial orientation is information provided by other blind people. The analysis of route directions generated by sighted people shows that they closely combine prescriptive and descriptive statements, that is, instructions to move and descriptions of the visual landmarks signalling the points where a reorientation must take place (e.g. Denis, 1997; Michon & Denis, 2001; Daniel et al., 2003; Denis and Fernandez in this volume). The descriptive pattern is quite distinct for blind persons, whose route directions are essentially procedural, alternating progression and reorientation prescriptions, and referring to the limited subset of environmental features that are accessible to their perception. Interestingly, references to the segments' length are typically more frequent in blind than in sighted people, and this of course must be taken into account when designing navigation aids for blind users. Even more crucial are the reorientation instructions. While sighted people rarely refer to angular measures of the reorientations, blind people need detailed information on the geometrical configuration of an environment at turning points. Lastly, proprioceptive and vestibular information is conveyed in the blind's verbal route directions, as well as informative auditory patterns (such as on which side is the traffic in the street). Route instructions for blind people include features that are so specific that the navigational performance of sighted people using such instructions is poorer than that of blind people (Denis, 2006).

In summary, taking these specificities into account would make for efficient route planning, adapted to blind people's needs. Such route planning will allow users, in the short run, to be reassured and thereby decrease the anxiety aroused by moving in an unfamiliar environment. In the longer run, it will also allow users to form a better mental image of their environment.

11.5.2 Global navigation

Regarding output interaction, language is the means that is most requested by users. However, as we saw earlier, several works (Loomis et al., 2005; Heuten et al., 2008) have revealed that there are many other existing means of guidance, such as virtual sounds or tactile guidance, each of them having specific advantages. Thus, it is for guidance itself that the choices to be made concerning the modality to use will be the most delicate. Indeed, one of the few points of agreement among developers is the following: the system must be able at any moment to know the orientation of the user's head, which requires the presence of an electronic compass. Otherwise, studies on pedestrian guidance have revealed that each guidance mode has its own advantages and disadvantages. The conventionally-used guidance mode (speaker giving verbal information) has the advantage of being complete in terms of describing the environment and not to block the ear canal for good perception of external sounds. It has the disadvantage of cognitively saturating the user (verbal information involves a cost in time of presentation and a cognitive cost in processing; see Klatzky et al., 2006) and to draw other people's attention to him or her. These problems would be solved by using a binaural headset and spatial sounds to help reach turning points, because this combination represents a lower cognitive cost and provides quite effective guidance while significantly reducing the number of instructions to be given to achieve the goals and sub-goals. However, the use of virtual sounds necessarily involves the use of a headset covering both ears, something that blind people do not seem willing to accept (thereby raising the importance of testing bone conduction and air-tube headphones). Finally, it is possible to consider a purely tactile guidance mode, since it leads to a quite similar performance to other systems, but it appears difficult to overcome verbal instructions, as many details such as street names or descriptions of points of interest will necessarily require the use of language.

In general, it seems that a good compromise would involve the use of virtual sounds for guidance (provided by a headset offering satisfactory levels of external sound perception) and the use of language for more complete descriptions. Developers have to keep in mind the goal of minimizing the amount of data provided and give only strictly necessary information, in order not to oppress the users, who should still have the opportunity to request more complete descriptions if they need to.

11.5.3 Fine navigation

With respect to fine navigation, aid devices are not yet able to replace the white cane or dog, but it is possible to combine them with the use of visual-to-auditory conversion systems (Auvray et al., 2007; Hanneton et al., 2010) or by obstacle detection equipment (Farcy et al., 2006). Nevertheless, the systems currently developed should take into account the fact that one of the user's hands is occupied by the white cane or dog and must therefore keep the use of manual controls to the bare minimum. Instead of using ultrasound and light beams, it is now possible to use artificial vision and object recognition technology to detect the presence of unexpected elements along the path. Two miniature cameras inserted in a pair of glasses can provide stereoscopic vision. Associated with object recognition software, such a device can allow object detection as fast as 10 milliseconds. In addition to detecting the presence of unexpected objects in the immediate vicinity, it should allow users to detect environmental targets such as mailboxes or walkways as well as helping indoor navigation by detecting walls, doors, and doorknobs (see the chapter by Barclay and Galton in this volume for further information on object recognition and description by machines).

Furthermore, the use of artificial vision is expected to help reduce the negative effects of GPS use. One of the most recurrent complaints made by users about GPS technology is that it often instructs the user to 'turn right in 10 metres' when the target street has just been passed. Crossing information from GIS on the layout of streets and presence of points of interest (e.g. telephone booths, poles, bus stops) with environmental information received by cameras will, in certain cases of poor satellite reception, increase the accuracy of the user's location by double checking data. It should be mentioned that the aim of combining the use of GPS, GIS, and object recognition technologies to help blind people with their everyday autonomy is at the core of the French project NAVIG (see Dramas et al., 2008; Gallay et al., 2010; Katz et al., 2010; Brilhault et al., 2011; for follow-up research on the project, see Figure 11.5 for an illustration of the device). However, it is also worth mentioning that a concern raised by such technology is to find a good compromise between available features and everyday use of the system. Indeed, a system consisting of a GPS chip and an object recognition camera connected by a laptop will offer great prospects for the autonomy development of blind people but will raise the problem of powering all these devices and of course intrusiveness and discretion problems.

11.5.4 Input interaction and interface

Concerning user/system interaction and considering that Braille keyboard input interaction offered by several existing systems involves a high cognitive and temporal cost, other means of instructing the system have to be found. As arising from Golledge et al.'s (2004) study and from reports of blind users (Brunet, 2010), it seems more desirable to use a box with some simple buttons along with contextual speech recognition software for effective input interaction. In addition, system flexibility and the customization level are important points for ease of use. The system should offer many features such as the opportunity to save, reverse, and share itineraries created by users. It should also contain information about public transport networks in cities where it will be used (as well as any disturbances) and must provide access to a list of predefined points of interest (such as stores and monuments) while offering the possibility of creating new ones.

Finally, developers have to keep in mind that all potential users have different needs and the system will have to adapt to these individual needs. Indeed, an already independent user having a good knowledge of the city will not need the same

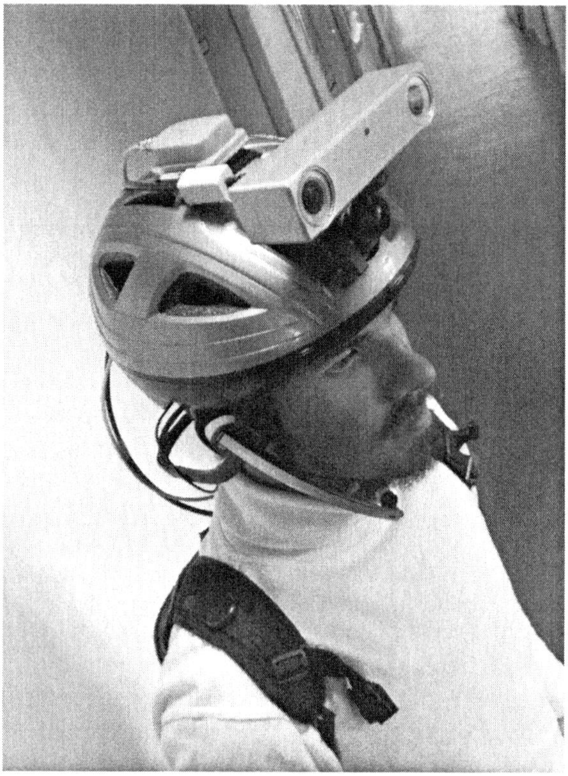

FIGURE 11.5 NAVIG prototype including a GPS receiver, a stereoscopic camera, and a head motion tracking device mounted on a helmet. Microphone and headphones are used to transmit speech and audio information

amount of information as an inexperienced user making a trip for the first time. Similarly, as suggested by Walker and Lindsay (2006), this type of aid device will lead to significant learning effects, as people get used to it. This means that after a year, users will be able to get rid of pieces of information they no longer consider helpful. In general, as Golledge et al. (2004) pointed out, blind people seem to expect a high degree of customization of this kind of device. The virtual sound guidance, for instance, can be considered as particularly effective by some and very unpleasant by others who should then be able to replace the binaural headset and virtual sounds by a loudspeaker and verbal information.

11.6 Conclusions

This chapter has described a variety of navigation aid devices developed for blind people. Such devices are useful for planning a trip, for retrieving information concerning the route to follow, or for identifying points of interest while navigating in outdoor environments, or even to avoid obstacles along the way. Taken together, technology advances and innovative research projects may soon offer blind people new opportunities to improve their autonomy in their daily life.

The use of personalized navigation planning phases and audio-tactile maps would allow blind people to learn about their physical environment while providing safety and decreasing the level of anxiety associated with navigation in unfamiliar surroundings. More investigations have to be conducted on the possibilities offered by virtual sounds, for instance, on the eventual benefits of the binaural description of a crossroad using virtual trajectory sounds, but this technology already provides new guidance opportunities, as effective as speech, with the advantage of involving a lower cognitive load. With this new means of guidance, the possibility arises to reduce the amount of information delivered to users, allowing them to focus on something else as well as having a better perception of environmental sounds. Fine navigation is made easier by new technologies, such as object-recognition cameras that allow the detection of obstacles along the way, as well as object-grasping and indoor navigation.

Finally, GPS-based navigation devices that were originally designed for car navigation are more and more oriented to the various needs of specific users. Pedestrians do not need the same guidance information as car drivers and blind pedestrians do not need the same guidance information as sighted ones. Blind people, albeit able to adopt an allocentric perspective, are naturally directed towards an egocentric perspective. They are therefore more likely than the sighted to fall into the trap of passive satellite guidance. Daily use of an aid device by blind persons may end up trapping them in the use of a route perspective and lead them to ignore survey representations. Further empirical studies are thus needed to assess whether the introduction of virtual sounds as distal elements can help blind people to construct a more robust allocentric representation of their environments. In every instance, to make sure that a system built in order to give greater autonomy to blind people does not (paradoxically) result in the opposite effect, that is, making users more and more dependent on the device, it is important to take into account the long-term effects of the use of such devices.

References

Note: Entries with just one or two authors appear in alpha order by name, but three or more authors are treated as et al. and appear in chronological order of works.

Afonso, A., Blum, A., Katz, B. F. G., Tarroux, P., Borst, G., and Denis, M. (2010). Structural properties of spatial representations in blind people: Scanning images constructed from haptic exploration or from locomotion in a 3-D audio virtual environment. *Memory and Cognition*, 38, 591–604.

Aiello, M., Pratt-Hartmann, I. E., and v. Benthem, J. F. (2007). *Handbook of Spatial Logics*. Berlin: Springer.

Alexander, C., Ishikawa, S., and Silverstein, M. (1977). *A Pattern Language: Towns, Buildings, Construction*. New York: Oxford University Press.

Allen, G. L. (1997). From knowledge to words to wayfinding: Issues in the production and comprehension of route directions. In S. C. Hirtle and A. U. Frank (eds), *Spatial Information Theory: A Theoretical Basis for GIS* (pp. 363–72). Berlin: Springer.

Allen, G. L. (2000). Principles and practices for communicating route knowledge. *Applied Cognitive Psychology*, 14, 333–59.

Allen, G. L. (2003). Gestures accompanying verbal route directions: Do they point to a new avenue for examining spatial representations? *Spatial Cognition and Computation*, 3, 259–68.

Alpaydin, E. (2004). *Introduction to Machine Learning*. Cambridge, MA: MIT Press.

Anderson, A. H., Bader, M., Bard, E., Boyle, E., Doherty, G., Garrod, S., Isard, S., Kowtko, J. McAllister, J. Miller, J., Sotillo, C., Thompson, H., and Weinert, R. (1991). The HCRC Map Task Corpus. *Language and Speech*, 34(4), 351–66.

Applegate, D. L., Bixby, R. E., Chvátal, V., and Cook, W. J. (2007). *The Traveling Salesman Problem: A Computational Study*. Princeton, NJ: Princeton University Press.

Appleyard, D. (1970). Styles and methods of structuring a city. *Environment and Behavior*, 2, 100–17.

Arijon. D. (1976). *Grammar of the Film Language*. New York: Hastings House.

Arkin, R. (1992). Behavior-based robot navigation for extended domains. *Adaptive Behavior*, 1, 201–25.

Ashmead, D. H., Davis, D. L., and Northington, A. (1995). Contribution of listeners' approaching motion to auditory distance perception. *Journal of Experimental Psychology: Human Perception and Performance*, 21, 239–56.

Auvray, M. and Myin, E. (2009). Perception with compensatory devices. From sensory substitution to sensorimotor extension. *Cognitive Science*, 33, 1036–58.

Auvray, M., Hanneton, S., Lenay, and O'Regan, J. K. (2005). There is something out there: Distal attribution in sensory substitution, twenty years later. *Journal of Integrative Neuroscience*, 4, 505–21.

Auvray, M., Hanneton, S., and O'Regan, J. K. (2007). Learning to perceive with a visuo-auditory substitution system: Localization and object recognition with The Voice. *Perception*, 36, 416–30.

Avraamides, M. N. (2003). Spatial updating of environments described in texts. *Cognitive Psychology*, 47, 402–31.

Avraamides, M. N. and Kelly, J. W. (2008). Multiple systems of spatial memory and action. *Cognitive Processing*, 9, 93–106.

Avraamides, M. N. and Kelly, J. W. (2010). Multiple systems of spatial memory: Evidence from described scenes. *Journal of Experimental Psychology: Learning, Memory, and Cognition*, 36(3), 635–45.

Avraamides, M. N., Loomis, J. M., Klatzky, R. L., and Golledge, R. G. (2004). Functional equivalence of spatial representations derived from vision and language: Evidence from allocentric judgements. *Journal of Experimental Psychology: Learning, Memory, and Cognition*, 30(4), 801–14.

Avrahami, J. and Kareev, Y. (1994). The emergence of events. *Cognition*, 53, 239–61.

Baddeley, A. D. (1992). Working memory. *Science*, 255, 556–59.

Baddeley, A. D., Emslie, H., Kolodny, J., and Duncan, J. (1998). Random generation and the executive control of working memory. *Quarterly Journal of Experimental Psychology*, 51A, 818–52.

Baldwin, D., Andersson, A., Saffran, J., and Meyer, M. (2008). Segmenting dynamic human action via statistical structure. *Cognition*, 106, 1382–407.

Barclay, M. J. (2010). *Reference object choice in spatial language: machine and human models.* PhD thesis, University of Exeter, UK.

Barclay, M. J. and Galton, A. P. (2008a). A scene corpus for training and testing spatial communication systems. *Proceedings of the AISB convention: Communication, Interaction and Social Intelligence*, 10, 26–9.

Barclay, M. J. and Galton, A. P. (2008b). An influence model for reference object selection in spatially locative phrases. In *Proceedings of the Sixth International Conference on Spatial Cognition* (pp. 216–32). Berlin: Springer.

Barsalou, L. W. (1999). Perceptual symbol systems. *Behavioral and Brain Sciences*, 22, 557–660.

Barsalou, L. W. (2005). Situated conceptualization. In H. Cohen and C. Lefebvre (eds), *Handbook of Categorization in Cognitive Science* (pp. 619–50). St. Louis: Elsevier.

Baskaya, A., Wilson, C. and Özcan, Y. Z. (2004). Wayfinding in an unfamiliar environment. *Environment and Behavior*, 36(6), 839–67.

Bateman, J. A. (2010a). Situating spatial language and the role of ontology: Issues and outlook. *Linguistics and Language Compass*, 4(8), 639–64.

Bateman, J. A. (2010b). Language and space: A two-level semantic approach based on principles of ontological engineering. *International Journal of Speech Technology*, 13(1), 29–48.

Bateman, J. A., Tenbrink, T., and Farrar, S. (2007). The role of conceptual and linguistic ontologies in interpreting spatial discourse. *Discourse Processes*, 44, 175–212.

Bateman, J. A., Hois, J., Ross, R., and Tenbrink, T. (2010). A linguistic ontology of space for natural language processing. *Artificial Intelligence*, 174(14), 1027–71.

Beauchemin, S. S. and Barron, J. L. (1995). The computation of optical flow. *ACM Computing Surveys*, 27(3).

Beeson, P. (2008). *Creating and utilizing symbolic representation of spatial knowledge using mobile robots*. PhD Thesis, The University of Texas at Austin.

Beeson, P., Dodayil, J., and Kuipers, B. (2010). Factoring the mapping problem: Mobile robot map-building in the hybrid spatial semantic hierarchy. *The International Journal of Robotics Research*, 29(4), 428–59.

Benedikt, M. L. (1979). To take hold of space: isovists and isovist fields. *Environment and Planning B: Planning and Design*, 6(1), 47–65.

Bhatt, M. (2010). Reasoning about space, actions and change: A paradigm for applications of spatial reasoning. In S. Hazarika (ed.), *Qualitative Spatial Representation and Reasoning: Trends and Future Directions*. USA: IGI Global.

Bhatt, M. and Flanagan, G. (2010). Spatio-temporal abduction for scenario and narrative completion (a preliminary statement). In M. Bhatt, H. Guesgen, and S. Hazarika (eds), *International Workshop on Spatio-Temporal Dynamics, European Conference on Artificial Intelligence (ECAI-10)* (pp. 31–36). ECAIWorkshop Proceedings., and SFB/TR 8 Spatial Cognition Report Series, August 2010. Available at <http://www.sfbtr8.spatial-cognition.de/papers/SFB_TR8Rep023-06_2010.pdf>, last accessed 29 April 1013.

Bhatt, M. and Freksa, C. (2010). Spatial computing for design: An artificial intelligence perspective. In NSF *International Workshop on Studying Visual and Spatial Reasoning for Design Creativity* (SDC'10). Available at <http://mason.gmu.edu/~jgero/conferences/sdc10>, last accessed 29 April 2013.

Bhatt, M. and Loke, S. (2008). Modelling dynamic spatial systems in the situation calculus. *Spatial Cognition and Computation*, 8(1), 86–130.

Bhatt, M., Dylla, F., and Hois, J. (2009). Spatio-terminological inference for the design of ambient environments. In K. S. Hornsby, C. Claramunt, M. Denis, and G. Ligozat (eds), *Conference on Spatial Information Theory* (COSIT'09) (pp. 371–91). Berlin: Springer.

Bhatt, M., Ichim, A., and Flanagan, G. (2010). Dsim: A tool for assisted spatial design. In *Proceedings of the 4th International Conference on Design Computing and Cognition* (DCC'10). Stuttgart.

Bhatt, M., Guesgen, H., Woelfl, S., and Hazarika, S. (2011a). Qualitative spatial and temporal reasoning: Emerging applications, trends and directions. *Journal of Spatial Cognition and Computation*, 11(1).

Bhatt, M., Lee, J. H., and Schultz, C. (2011b). CLP(QS): A declarative spatial reasoning framework. In *Proceedings of the 10th International Conference on Spatial Information Theory (COSIT 11)* (pp. 210–30) Berlin, Heidelberg: Springer.

Bilge, A. R. and Taylor, H. A. (2010). Where is 'here' in nested environments? Spatial updating from different sources. *Spatial Cognition and Computation*, 10(2–3), 157–83.

Biocca, F., Tang, A., Owen, C., Mou, W., and Fan, X. (2005). Mobile infospaces: Personal and egocentric space as psychological frames for information organization in augmented reality environments. In D. Schmorrow (ed.), *Foundations of Augmented Cognition* (Proceedings of HCI International 2005, Vol. 11, pp. 154–63). Mahwah, NJ: Lawrence Erlbaum Associates.

Bitgood, S. (1992). Visitor orientation and circulation: Some general principles. *Visitor Behavior*, 7(3), 15.

Bitgood, S. (1995). Visitor circulation: Is there really a right-turn bias? *Visitor Behavior*, 10(1), 5–6.

Bloom, P., Peterson, M. A., Nadel, L. and Garrett, M. F. (eds) (1996). *Language and Space*. Cambridge, MA: The MIT Press.

Bosco, A., Filomena, S., Sardone, L., Scalisi, T. G., and Longoni, A. M. (1996). Spatial models derived from verbal descriptions of fictitious environments: The influence of study time and the individual differences in visuo-spatial ability. *Psychologische Beiträge*, 38, 451–64.

Branigan, E. (1984). *Point of View in the Cinema: A Theory of Narration and Subjectivity in Classical Film*, Berlin and New York: Mouton.

Branigan, E. (2006). *Projecting a Camera: Language-Games in Film Theory*. Abingdon: Routledge.

BremLBO. (2003). Bremische Landesbauordnung. Available at: <http://www.bauordnungen.de/html/bremen.html>, last accessed 24 March 2013.

Brentano, F. (1874). *Psychology from an Empirical Standpoint*. Leipzig: Duncker und Humbolt. (Reprint: New York: Humanities Press, 1973).

Brilhault, A., Kammoun, S., Gutierrez, O., Truillet, P. and Jouffrais, C. (2011). Fusion of artificial vision and GPS to improve blind pedestrian positioning. *International Conference on New Technologies, Mobility and Security*. 7–10 February, Paris, France.

Brooks, R. (1987). Intelligence without representation. *Artificial Intelligence*, 47, 139–59.

Brooks, R. (1991). New approaches to robotics. *Science* 13, 253:(5025), 1227–32.

Brosset, D., Claramunt, C., and Saux, E. (2008). Wayfinding in natural and urban environments: a comparative study. *Cartographica*, 43(1), 21–30.

Brown, B. (2006). Working on problems of tourists. *Annals of Tourism Research*, 34, 364–83.

Brunet, L. (2010). *Etude des stratégies et besoins des personnes non-voyantes lors de la navigation pour la conception d'un dispositif d'aide performant et accepté [Study of blind persons' navigation needs and strategies for the development of an efficient and accepted aid device]*. Master's Thesis, Department of Ergonomics, Université Paris-Sud 11, Orsay, France.

Brunet, L., Gallay, M., Darses, F., and Auvray, M. (submitted). Etude des stratégies et des besoins des piétons non-voyants lors de la navigation en environnement urbain [Study of the strategies and needs of blind pedestrians in urban environment].

Brunyé, T. T. and Taylor, H. A. (2008a). Extended experience benefits spatial mental model development with route but not survey descriptions. *Acta Psychologica*, 127(2), 340–54.

Brunyé, T. T. and Taylor, H. A. (2008b). Working memory in developing and applying mental models from spatial descriptions. *Journal of Memory and Language*, 58(3), 701–29.

Brunyé, T. T. and Taylor, H. A. (2009). When goals constrain: Eye movements and memory for goal-oriented map study. *Applied Cognitive Psychology*, 22, 1–16.

Brunyé, T. T., Taylor, H. A, Rapp, D. N., and Spiro, A. B. (2006). Learning procedures: The role of working memory in multimedia learning experiences. *Applied Cognitive Psychology*, 20, 917–40.

Brunyé, T. T., Taylor, H. A, and Worboys, M. (2007). Levels of detail in descriptions and depictions of geographic space. *Spatial Cognition and Computation*, 7(3), 227–66.

Brunyé, T. T., Rapp, D. N. and Taylor, H. A. (2008). Representational flexibility and specificity following spatial descriptions of real-world environments. *Cognition*, 108(2), 418–43.

Brunyé, T. T., Ditman, T., Mahoney, C. R., Walters, E. K., and Taylor, H. A. (2010a). You heard it here first: Readers mentally simulate described sound. *Acta Psychologica*, 135(2), 209–15.

Brunyé, T. T., Mahoney, C. R., and Taylor, H. A. (2010b). Moving through imagined space: Mentally simulating locomotion during spatial description reading. *Acta Psychologica*, 134, 110–24.

Bryant, D. J. (1997). Representing space in language and perception. *Mind and Language*, 12, 239–64.

Bryant, D. J., Tversky, B., and Franklin, N. (1992). Internal and external spatial frameworks for representing described scenes. *Journal of Memory and Language*, 31, 74–98.

Burgess, N. (2006). Spatial memory: How egocentric and allocentric combine. *Trends in Cognitive Sciences*, 10, 551–7.

Burgess, N. and Hitch, G. J. (1999). Memory for serial order: A network model of the phonological loop and its timing. *Psychological Review*, 106(3), 551–81.

Burnett, G., Smith, D., and May, A. (2001). Supporting the navigation task: Characteristics of 'good' landmarks. In M. A. Hanson (ed.), *Contemporary Ergonomics 2001* (pp. 441–6). London: Taylor and Francis.

Byron, D., Koller, A., Striegnitz, K., Cassell, J., Dale, R., Moore, J., and Oberlander, J. (2009). Report on the first NLG challenge on generating instructions in virtual environments (GIVE). In ENLG '09: *Proceedings of the 12th European Workshop on Natural Language Generation* (pp. 165–73). Morristown, NJ: Association for Computational Linguistics.

Caggiano, V., Fogassi, L., Rizzolatti, G., Thier, P., and Casile, A. (2009). Mirror neurons differentially encode the peripersonal and extrapersonal space of monkeys. *Science*, 324, 403–6.

Carlson L. A. and Hill, P. L. (2008). Processing the presence, placement, and properties of a distractor in spatial language tasks. *Memory and Cognition*, 36, 240–55.

Carlson, L. A. and Van Deman, S. R. (2004). The space in spatial language. *Journal of Memory and Language*, 51, 418–36.

Carlson, L. A. and van der Zee, E. (eds) (2005). *Functional Features in Language And Space: Insights from Perception, Categorization and Development*. Oxford: Oxford University Press.

Carlson-Radvansky. L. A. (1996). Constructing spatial templates: The influence of reference frame selection. Available at <ww.cs.indiana.edu/event/maics96/Proceedings/Radvansky/radvansky.html>, last accessed 2 May 2013.

Carlson-Radvansky L. A. and Irwin, D. (1994). Reference frame activation during spatial term assignment. *Journal of Memory and Language*, 33, 646–71.

Carlson-Radvansky L. A. and Radvansky, G. A. (1996). The influence of functional relations on spatial term selection. *Psychological Science*, 7(1), 56–60.

Cattaneo, Z., Vecchi, T., Cornoldi, C., Mammarella, I., Bonino, D., Ricciardi, E., and Pietrini, P. (2008). Imagery and spatial processes in blindness and visual impairment. *Neuroscience and Biobehavioral Reviews*, 32, 1346–60.

Chabanne, V., Peruch, P., Denis, M., and Thinus-Blanc, C. (2003). Mental scanning of images contructed from visual experience or verbal descriptions: The impact of survey versus route perspective. *Imagination, Cognition, and Personality*, 23(2–3), 163–71.

Choi, Y. K. (1999). The morphology of exploration and encounter in museum layouts. *Environment and Planning B: Planning and Design*, 26(2), 241–50.

Chown, E. (1999). Making predictions in an uncertain world: Environmental structures and cognitive maps. *Adaptive Behavior*, 7(1), 1–17.

Chown, E. L. (2000). Gateways: An approach to parsing spatial domains. In *International Conference on Machine Learning (ICML)*. Workshop on Machine Learning of Spatial Knowledge (pp. 1–6).

Chown, E. and Boots, B. (2008). Learning cognitive maps: Lessons for robotics. In M. Jefferies, and W. K. Yeap (eds), *Robotic and Cognitive Approaches to Spatial Mapping*. Berlin, Heidelberg: Springer.

Chown, E., Kaplan, S., and Kortenkamp, D. (1995). Prototypes, location and associative networks (PLAN): towards a unified theory of cognitive mapping. *Cognitive Science*, 19, 1–52.

Chown, E., Fishman, J., Strom, J., Slavov, G., Hermans, T., Dunn, N., Lawrence, A., Morrison, J., and Krob, E. (2009). The Northern Bites 2008 Standard Platform Robot Team, <http://www.tzi.de/spl/bin/view/Website/Teams2009>, last accessed 2 May 2013.

Christianson, D. B., Anderson, S. E., Wei He, L., Salesin, D. H., Weld, D. H., and Cohen, M. F. (1996). Declarative camera control for automatic cinematography. *AAAI '96 Proceedings of the thirteenth national conference on Artificial intelligence* Vol. 1, (pp. 148–55), Palo Alto, CA: AAAI Press.

Clark, A. (1989). *Microcognition: Philosophy, Cognitive Science, and Parallel Distributed Processing*. Cambridge, MA: MIT Press.

Clayton, K. and Habibi, A. (1991). Contributions of temporal contiguity to the spatial priming effect. *Journal of Experimental Psychology: Learning, Memory, and Cognition*, 17(2), 263–71.

Cohn, A. G. and Renz, J. (2007). Qualitative spatial reasoning. In F. van Harmelen, V. Lifschitz, and B. Porter (eds), *Handbook of Knowledge Representation* (pp. 551–83). Amsterdam, Oxford: Elsevier.

Collins, C. (1985). On mobility aids for the blind. In D. Warren and E. Strelow (eds), *Electronic Spatial Sensing for the Blind* (pp. 35–64). Dordrecht: Martinus Nijhoff.

Couclelis, H. (1996). Verbal directions for way-finding: Space, cognition, and language. In J. Portugali (ed.), *The Construction of Cognitive Maps* (pp. 133–53). Dordrecht: Kluwer.

Couclelis, H., Golledge, R., Gale, N., and Tobler, W. (1987). Exploring the anchor-point hypothesis of spatial cognition. *Journal of Experimental Psychology*, 7, 99–122.

Coventry, K. R. and Garrod, S. C. (2004). *Saying, Seeing and Acting: The Psychological Semantics of Spatial Prepositions*. Hove and New York: Psychology Press.

Coventry, K. R., Cangelosi, A., Rajapakse, R., Bacon, A., Newstead, S., Joyce, D., and Richards, L. V. (2005). Spatial prepositions and vague quantifiers: Implementing the functional geometric framework. In C. Freksa, M. Knauff, B. Krieg-Brückner, and B. Nebel (eds), *Spatial Cognition IV. Reasoning, Action and Interaction* (pp. 98–110). Berlin: Springer.

Coventry, K. R., Tenbrink, T. and Bateman, J. (eds) (2009a). *Spatial Language and Dialogue*. Oxford: Oxford University Press.

Coventry, K. R., Tenbrink, T., and Bateman, J. (2009b). Spatial language and dialogue: Navigating the domain. In K. R. Coventry, T. Tenbrink, and J. Bateman (eds), *Spatial Language and Dialogue* (pp. 1–7). Oxford: Oxford University Press.

Curiel, J. M. and Radvansky, G. A. (1998). Mental organization of maps. *Journal of Experimental Psychology: Learning, Memory, and Cognition*, 24, 202–14.

Cutini, S., Di Ferdinando, A., Basso, D., Bisiacchi, P., and Zorzi, M. (2005). A computational model of planning in the traveling salesman problem. In M. B. B. G. Bara L. Barsalou (ed.), *Proceedings of the Twenty Seventh Annual Conference of the Cognitive Science Society*. Mahwah, NJ: Erlbaum.

Dale R. and Reiter, E. (1995). Computational interpretations of the Gricean maxims in the generation of referring expressions. *Cognitive Science*, 19, 233–63.

Daniel, M.-P. and Denis, M. (1998). Spatial descriptions as navigational aids: A cognitive analysis of route directions. *Kognitionswissenschaft*, 7, 45–52.

Daniel, M.-P. and Denis, M. (2004). The production of route directions: Investigating conditions that favour conciseness in spatial discourse. *Applied Cognitive Psychology*, 18(1), 57–75.

Daniel, M.-P., Tom, A., Manghi, E., and Denis, M. (2003). Testing the value of route directions through navigational performance. *Spatial Cognition and Computation*, 3, 269–89.

Dasen, P. R. and Mishra, R. C. (2010). *Development of Geocentric Spatial Language and Cognition: An Eco-Cultural Perspective*. Cambridge, UK: Cambridge University Press.

Davies, C. and Peebles, D. (2010). Spaces or scenes: Map-based orientation in urban environments. *Spatial Cognition and Computation*, 10(2), 135–56.

De Beni, R., Pazzaglia, F., Gyselinck, V., and Meneghetti, C. (2005). Visuospatial working memory and mental representation of spatial descriptions. *European Journal of Cognitive Psychology*, 17, 77–95.

de Berg, M., van Kreveld, M., Overmars, M., and Schwarzkopf, O. (2000). *Computational Geometry: Algorithms and Applications* (2nd edn). Berlin: Springer.

De Raedt, L., Frasconi, P., Kersting, K., and Muggleton, S. (eds) (2008). *Probabilistic Inductive Logic Programming*. LNCS, Springer.

De Vega, M. and Rodrigo, M. J. (2001). Updating spatial layouts mediated by pointing and labelling under physical and imaginary rotation. *European Journal of Cognitive Psychology*, 13, 369–93.

de Waal, F. B. M. (2002). Evolutionary psychology: The wheat and the chaff. *Current Directions in Psychological Science*, 11(6), 187–91.

Denham, J., Leventhal, J., and McComas, H. (2004). Getting from point A to point B: A review of two GPS systems. *AccessWorld*, 5.

Denis, M. (1996). Imagery and the description of spatial configurations. In M. de Vega, M. J. Intons-Peterson, P. N. Johnson-Laird, M. Denis, and M. Marschark, *Models of Visuospatial Cognition* (pp. 128–97). New York: Oxford University Press.

Denis, M. (1997). The description of routes: A cognitive approach to the production of spatial discourse. *Current Psychology of Cognition*, 16(4), 409–58.

Denis, M. (2006). Spécifications pour un système d'aide verbale au déplacement pour des piétons aveugles dans des environnements urbains complexes [Specifications for a system of verbal assistance to the navigation of blind pedestrians in complex urban environments]. In E. Bajolet, M.-F. Mattei, and J.-M. Rennes (eds), *Quatre ans de recherche urbaine 2001–2004: Action Concertée Incitative Ville, Ministère de la Recherche* (vol. 2, pp. 208–11). Tours, France: Presses Universitaires François-Rabelais.

Denis, M. and Cocude, M. (1997). On the metric properties of visual images generated from verbal descriptions: Evidence for the robustness of the mental scanning effect. *European Journal of Cognitive Psychology*, 9, 353–79.

Denis, M. and de Vega, M. (1993). Modèles mentaux et imagerie mentale [Mental models and mental imagery]. In M.-F. Ehrlich, H. Tardieu, and M. Cavazza (eds), *Les Modèles Mentaux: Approche Cognitive des Représentations* (pp. 79–100). Paris: Masson.

Denis, M. and Denhière, G. (1990). Comprehension and recall of spatial descriptions. *European Bulletin of Cognitive Psychology*, 10, 115–43.

Denis, M. and Zimmer, H. D. (1992). Analog properties of cognitive maps constructed from verbal descriptions. *Psychological Research*, 54(4), 286–98.

Denis, M., Pazzaglia, F., Cornoldi, C., and Bertolo, L. (1999). Spatial discourse and navigation: An analysis of route directions in the city of Venice. *Applied Cognitive Psychology*, 13, 145–74.

Denis, M., Michon, P.-E., and Tom, A. (2006). Assisting pedestrian wayfinding in urban settings: Why references to landmarks are crucial in direction-giving. In G. L. Allen (ed.), *Applied Spatial Cognition: From Research to Cognitive Technology* (pp. 25–51). Mahwah, NJ: Erlbaum.

Deschacht, K. and Moens, M. F. (2009). Semi-supervised semantic role labeling using the latent words language model. In *Empirical Methods in Natural Language Processing*, EMNLP'09, pp. 21–9.

Deutscher, G. (2005). *The Unfolding of Language: An Evolutionary Tour of Mankind's Greatest Invention*. New York: Henry Holt and Company.

Deyzac, E., Logie, R. H., and Denis, M. (2006). Visuospatial working memory and the processing of spatial descriptions. *British Journal of Psychology*, 97, 217–43.

Dixon, P. (1987). The structure of mental plans for following directions. *Journal of Experimental Psychology: Learning, Memory, and Cognition*, 13, 18–26.

Dramas, F., Oriola, B., Katz, B. F. G., Thorpe, S. J., and Jouffrais, C. (2008). Designing an assistive device for the blind based on object localization and augmented auditory reality. *Proceedings of the 10th international ACM SIGACCESS conference on Computers and Accessibility* (pp. 263–4). Halifax, Nova Scotia, Canada. New York: ACM.

Drucker S. M. and Zeltzer, D. (1995). Camdroid: A system for implementing intelligent camera control. In *I3D '95: Proceedings of the 1995 symposium on Interactive 3D graphics*, (pp. 139–44). New York: ACM.

Duncan, J., Emslie, H., Williams, P., Johnson, R., and Freer, C. (1996). Intelligence and the frontal lobe: The organization of goal-directed behavior. *Cognitive Psychology*, 30, 257–303.

Duwe, I., Kessler, K., and Strohner, H. (2002). Resolving ambiguous descriptions through visual information. In K. R. Coventry and P. Olivier (eds), *Spatial Language. Cognitive and Computational Perspectives* (pp. 43–67). Dordrecht: Kluwer Academic.

Eastman, C., Teicholz, P., Sacks, R., and Liston, K. (2008). *BIM Handbook: A Guide to Building Information Modeling for Owners, Managers, Designers, Engineers and Contractors*. Frontiers in Artificial Intelligence and Applications. Hoboken, NJ: Wiley, 2008.

Egenhofer, M. (2009). A reference system for topological relations between compound spatial objects. In C. Heuser and G. Pernul (eds), *3rd International Workshop on Semantic and*

Conceptual Issues in GIS (SeCoGIS 2009)—Advances in Conceptual Modeling—Challenging Perspectives, Lecture Notes in Computer Science, Vol. 5833.

Ehrich, V. and Koster, C. (1983). Discourse organization and sentence form: The structure of room descriptions in Dutch. *Discourse Processes*, 6, 169–95.

Ehrlich, K. and Johnson-Laird, P. N. (1982). Spatial descriptions and referential continuity. *Journal of Verbal Learning and Verbal Behavior*, 21, 296–306.

Elias, B. (2003). Extracting landmarks with data mining methods. In W. Kuhn, M. F. Worboys, and S. Timpf (eds), *Spatial Information Theory: Foundations of Geographic Information Science* (pp. 375–89). Berlin: Springer.

Elias B. and Brenner, C. (2004). Automatic generation and application of landmarks in navigation data sets. In Peter F. Fisher (ed.), *Developments in Spatial Data Handling: 11th International Symposium on Spatial Data Handling* (pp. 469–80). Berlin, Heidelberg: Springer.

Elmogy, M., Habel, C., and Zhang, J. (2009). Spatial language for route-based humanoid robot navigation. *Cognitive Processing*, 10 (Suppl. 2), S208–S211.

Emmorey, K., Tversky, B., and Taylor, H. A. (2000). Using space to describe space: Perspective in speech, sign, and gesture. *Journal of Spatial Cognition and Computation*, 2, 157–80.

Ericsson, K. A. and Simon, H. A. (1984). Protocol analysis: Verbal reports as data. Cambridge, MA: MIT Press.

Eschenbach, C. (1999). Geometric structures of frames of reference and natural language semantics. *Spatial Cognition and Computation*, 1(4), 329–48.

Espinosa, M.-A., Ungar, S., Ochaíta, E., Blades, M., and Spencer, C. (1998). Comparing methods for introducing blind and visually impaired people to unfamiliar urban environments. *Journal of Environmental Psychology*, 18, 277–87.

Estevez, A. and Calvo, M. G. (2000). Working memory capacity and time course of predictive inferences. *Memory*, 8(1), 51–61.

Evans, V. and Chilton, P. (eds) (2010). *Language, Cognition and Space: The State of the Art and New Directions*. London: Equinox Publishing.

Farah, M. (1988). Is visual imagery really visual? Overlooked evidence from neuropsychology. *Psychological Review*, 95(3), 307–17.

Farcy, R., Leroux, R., Jucha, A., Damaschini, R., Grégoire, C., and Zogaghi, A. (2006). Electronic travel aids and electronic orientation aids for blind people: Technical, rehabilitation and everyday life points of view. In M. A. Hersh (ed.), *Proceedings of the Conference & Workshop on Assistive Technology for Vision and Hearing Impairment (CVHI 2006)*.

Farmer, E. W., Berman, J. V. F., and Fletcher, Y. L. (1986). Evidence for a visuo-spatial scratchpad in working memory. *Quarterly Journal of Experimental Psychology*, 38(4), 675–88.

Ferguson, E. L. and Hegarty, M. (1994). Properties of cognitive maps constructed from text. *Memory and Cognition*, 22, 455–73.

Fernandez, G. (2000). *Processes cognitifs mis en oeuvre dans la compréhension et la mémorisation de descriptions d'itinéraires: Une approche différentielle.* [Cognitive processes in the comprehension and memory of route directions: A differential approach.] Unpublished doctoral thesis, Université René-Descartes, Boulogne-Billancourt / LIMSI-CNRS, Orsay.

Filipi, A. and Wales, R. (2004). Perspective-taking and perspective-shifting as socially situated and collaborative actions. *Journal of Pragmatics*, 36, 1851–84.

Fincher-Kiefer, R. (2001). Perceptual components of situation models. *Memory and Cognition*, 29, 336–43.

Fischer, M. H. and Zwaan, R. A. (2008). Embodied language: A review of the role of the motor system in language comprehension. *The Quarterly Journal of Experimental Psychology*, 61(6), 825–50.

Flynn. J. E. (1977). A study of subjective responses to low energy and nonuniform lighting systems. *Lighting Design and Application*, 7(2), 6–15.

Flynn, J. E., Spencer, T. J., Martyniuk, O., and Hendrick, C. (1973). Interim study of procedures for investigating the effect of light on impression and behaviour. *Journal of the Illuminating Engineering Society*, 3(2), 87–94.

Fonseca, F. and Rodriguez, M. A. (2007). From geo-pragmatics to derivation ontologies: New directions for the geospatial semantic web. *Transactions in GIS*, 11(3), 313–16.

Fontaine, S. and Denis, M. (1999). The production of route instructions in underground and urban environments. In C. Freksa and D. M. Mark (eds), *Spatial Information Theory: Cognitive and Computational Foundations of Geographic Information Science* (pp. 83–94). Berlin: Springer.

Frank, A.U., Campari, I. and Formentini, U. (eds) (1992). *Theories and Methods of Spatio-Temporal Reasoning in Geographic Space*. Berlin: Springer.

Franklin, N. and Tversky, B. (1990). Searching imagined environments. *Journal of Experimental Psychology: General*, 119, 63–76.

Franklin, N., Tversky, B., and Coon, V. (1992). Switching points of view in spatial mental models. *Memory & Cognition*, 20, 507–18.

Freksa, C. (1991). Qualitative Spatial Reasoning. In D. Mark and A. Frank (eds), *Cognitive and Linguistic Aspects of Geographic Space* (pp. 361–72). Dordrecht: Kluwer Academic Publishers.

Freksa, C. (1992). Using orientation information for qualitative spatial reasoning. In *Proceedings of the Intl. Conf. GIS, From Space to Territory: Theories and Methods of Spatio-Temporal Reasoning in Geographic Space* (pp. 162–78), Berlin: Springer.

Freksa, C. (2013). Spatial computing—how spatial structures replace computational effort. In M. Raubal, D. Mark, and A. Frank (eds), *Cognitive and Linguistic Aspects of Geographic Space* (pp. 23–42). Berlin: Springer.

Freundschuh, S. and Egenhofer, M. (1997). Human conceptions of spaces: Implications for GIS. *Transactions in GIS* 2(4), 361–75.

Friedman, A., Waller, D., Thrash, T., Greenauer, N., and Hodgson, E. (2011). View combination: A generalization mechanism for visual recognition. *Cognition*, 119, 229–41.

Froese, T., Fischer, M., Grobler, F., Ritzenthaler, J., Yu, K., Sutherland, S. Staub, S., Akinci, B., Akbas, R., Koo, B., Barron, A., and Kunz, J. (1999). Industry foundation classes for project management—a trial implementation. *ITCon (Journal of Information Technology in Construction)*, 4, 17–36.

Fuke, S., Ogino, M., and Asada, M. (2007). Body image constructed from motor and tactile images with visual information. *International Journal of Humanoid Robotics*, 4(2), 347–64.

Fuke, S., Ogino, M., and Asada, M. (2009). VIP neuron model: Head-centered cross-modal representation of the peri-personal space around the face. In *Proceedings of the 7th IEEE International Conference on Developmental Learning* (pp. 145–50).

Gallagher, S. (2005). *How the Body Shapes the Mind*. Oxford: Clarendon Press.

Gallay, M., Denis, M., Parseihian, G., and Auvray, M. (2010). Egocentric and allocentric reference frames in a virtual auditory environment: Differences in navigation skills between blind and sighted individuals. *12th European Workshop on Imagery and Cognition (EWIC)*, June 16–19.

Gallese, V. (2005). Embodied simulation: From neurons to phenomenal experience. *Phenomenology and the Cognitive Sciences*, 4(1), 23–48.

Gapp, K.-P. (1995). Object localization: Selection of optimal reference objects. In A. U. Frank and W. Kuhn (eds), *Spatial Information Theory: A Theoretical Basis for GIS* (pp. 519–35) Berlin, Heidelberg, New York: Springer.

Gero, J. S. (1990). Design prototypes: A knowledge representation schema for design. *AI Magazine*, 11(4), 26–36.

Gero, J. S., Tham, K. W., and Lee, H. S. (1991). Behavior: A link between function and structure in design. In *Intelligent Computer Aided Design*, Proceedings of the IFIP WG 5.2 Working Conference on Intelligent Computer Aided Design (IntCAD91) (pp. 193–225) Columbus, OH, 30 September–3 October 1991.

Gibson, J. J. (1970). *The Senses Considered as Perceptual Systems*. Boston, MA: Houghton Mifflin.

Gibson, J. (1977). The theory of affordances. In R. Shaw and J. Bransford (eds), *Perceiving, Acting and Knowing* (pp. 67–82). New York: Wiley.

Giudice, N. A., Bakdash, J. Z., and Legge, G. E. (2007). Wayfinding with words: Spatial learning and navigation using dynamically updated verbal descriptions. *Psychological Research*, 71, 347–58.

Giudice, N. A., Klatzky, R. L., and Loomis, J. M. (2009). Evidence for amodal representations after bimodal learning: Integration of haptic-visual layouts into a common spatial image. *Spatial Cognition and Computation*, 9, 287–304.

Glenberg, A. M. (1997). What memory is for. *Behavioral and Brain Sciences*, 20, 1–55.

Glenberg, A. M. and Kaschak, M. P. (2002). Grounding language in action. *Psychonomic Bulletin and Review*, 9, 558–65.

Glenberg, A. M. and Robertson, D. A. (1999). Indexical understanding of instructions. *Discourse Processes*, 28, 1–26.

Goerick, C., Wersing, H., Mikhailova, I., and Dunn, M. (2005). Peripersonal space and object recognition for humanoids. In *Proceedings of the IEEE/RSJ International Conference on Humanoid Robots (Humanoids 2005)*, Tsukuba, Japan (pp. 387–92). Washington, DC: IEEE Press.

Golden, B., Bodin, L., Doyle, T., and Stewart, Jr., W. (1980). Approximate traveling salesman algorithms. *Operations Research*, 28, 694–711.

Golding, J. M., Graesser, A. C., and Hauselt, J. (1996). The process of answering direction-giving questions when someone is lost on a university campus: The role of pragmatics. *Applied Cognitive Psychology*, 10, 23–39.

Goldman, Dan B., Curless, B., Salesin, D., Seitz, S. M. (2006). Schematic storyboarding for video visualization and editing. In *Proceedings of SIGGRAPH '06 ACM SIGGRAPH 2006* (pp. 862–71), New York: ACM New York.

Golledge, R. G. (1987). Environmental cognition. In D. Stokols and I. Altman (eds), *Handbook of Environmental Psychology* Vol. I. New York: Wiley.

Golledge, R. (ed.) (1999). *Wayfinding Behavior: Cognitive Mapping and Other Spatial Processes.* Baltimore, MD: Johns Hopkins University Press.

Golledge, R. G., and Spector, A. N. (1978). Comprehending the urban environment: Theory and practice. *Geographical Analysis*, 10, 403–26.

Golledge, R., Klatzky, R., Loomis, J., Spiegle, J., and Tietz, J. (1998). A geographical information system for a GPS based personal guidance system. *International Journal of Geographical Information Science*, 12, 727–49.

Golledge, R. G., Marston, J. R., Loomis, J. M., and Klatzky, R. L. (2004). Stated preferences for components of a personal guidance system for nonvisual navigation. *Journal of Visual Impairment and Blindness*, 98, 135–47.

Gorniak, P. and Roy, D. (2007). Situated language understanding as filtering perceived affordances. *Cognitive Science*, 31, 197–231.

Graesser, A. C., Singer, M., and Trabasso, T. (1994). Constructing inferences during narrative text comprehension. *Psychological Review*, 101(3), 371–95.

Graham, S. M., Joshi, A., and Pizlo, Z. (2000). The traveling salesman problem: A hierarchical model. *Memory and Cognition*, 28, 1101–204.

Gray, J., Breazeal, C., Berlin, M., Brooks, A., and Lieberman, J. (2005). Action parsing and goal inference using self as simulator. In *Proceedings of the Fourteenth IEEE Workshop on Robot and Human Interaction* (pp. 202–9).

Greenauer, N., and Waller, D. (2008). Intrinsic array structure is neither necessary nor sufficient for nonegocentric coding of spatial layouts. *Psychonomic Bulletin & Review*, 15, 1015–21.

Gregory, D. (2010). Visual imagery: Visual format or visual content? *Mind & Language*, 25(4), 394–417.

Grubinger, M., Clough, P., Müller, H., and Deselaers, T. (2006). The IAPR benchmark: A new evaluation resource for visual information systems. In *International Conference on Language Resources and Evaluation*, LREC'06.

Gyselinck, V., De Beni, R., Pazzaglia, F., Meneghetti, C., and Mondoloni, A. (2007). Working memory components and imagery instructions in the elaboration of a spatial mental model. *Psychological Research*, 71, 373–82.

Habel, C., and Tappe, H. (1999). Processes of segmentation and linearization in described events. In V. Stutterheim and R. Klabunde (eds), *Representations and Processes in Language Production* (pp. 117–53). Opladen: Westdeutscher Verlag.

Hägerstrand, T. (1975). Space, time and human conditions. In A. Karlqvist et al. (eds), *Dynamic Allocation of Urban Space*. Lexington: Saxon House Lexington Book.

Hall, E. T. (1966). *The Hidden Dimension*. New York: Anchor Books.

Hanneton, S., Auvray, M., and Durette, B. (2010). The Vibe: A versatile vision-to-audition sensory substitution device. *Applied Bionics and Biomechanics*, 7, 269–76.

Hansen, S., Richter, K.-F., and Klippel, A. (2006). Landmarks in OpenLS—a data structure for cognitive ergonomic route directions. In M. Raubal, H. J. Miller, A. U. Frank, and M. F. Goodchild (eds), *Geographic Information Science* (pp. 128–44). Berlin: Springer.

Haq, S. and Zimring, C. (2003). Just down the road a piece. *Environment and Behavior*, 35(1), 132–60.

Harnad, S. (1990). The symbol grounding problem. *Physica D*, 42, 335–46.

Hauk, O., Johnsrude, I., and Pulvermüller, F. (2004). Somatotopic representation of action words in the motor and premotor cortex. *Neuron*, 41, 301–7.

Haviland, S. E. and Clark, H. H. (1974). What's new? Aquiring new information as a process in comprehension. *Journal of Verbal Learning and Verbal Behavior*, 13, 512–21.

Hawkins, J. and Blakeslee, S. (2004). *On Intelligence*. New York: Holt.

Hayes-Roth, B. and Hayes-Roth, F. (1979). A cognitive model of planning. *Cognitive Science*, 3, 275–310.

Hayes-Roth, B., Hayes-Roth, F., Rosenschein, S., and Cammarata, S. (1979). Modeling planning as an incremental, opportunistic process. In *Proceedings of the Sixth International Joint Conference on Artificial Intelligence*, pp. 375–82.

He, L.-W., Cohen, M. F., and Salesin, D. H. (1996). The virtual cinematographer: A paradigm for automatic real-time camera control and directing. In SIGGRAPH '96: *Proceedings of the 23rd Annual Conference on Computer Graphics and Interactive Techniques* (pp. 217–24, New York: ACM.

Henze, N., Heuten, W., and Boll, S. (2006). Non-intrusive somatosensory navigation support for blind pedestrians. *Proceedings of EuroHaptics*, 459–64.

Hersch, M., Sauser, E., and Billard, A. (2008). Online learning of the body schema. *International Journal of Humanoid Robotics*, 5(2), 161–81.

Herskovits. A. (1998). Schematization. In P. Olivier and K.-P. Gapp (eds), *Representation and Processing of Spatial Expressions* (pp. 149–62). Laurence Earlbaum Associates.

Herzog, G., and Wazinski, P. (1994). VIsual TRAnslator: Linking perceptions and natural language descriptions. *Artificial Intelligence Review*, 8, 175–87.

Heuten, W., Henze, N., Boll, S., and Pielot, M. (2008). Tactile wayfinder: A non-visual support system for wayfinding. *Proceedings of the 5th Nordic Conference on Human-Computer Interaction* (pp. 172–81). 18–22 October, Lund, Sweden.

Hickmann, M. and Robert, S. (eds) (2006). *Space in Languages: Linguistic Systems and Cognitive Categories*. Amsterdam: John Benjamins.

Hirtle, S. C. (1998). The cognitive atlas: Using GIS as a metaphor for memory. In M. Egenhofer and R. Golledge (eds), *Spatial and Temporal Reasoning in Geographic Information Systems* (pp. 263–71). Oxford: Oxford University Press.

Hirtle, S. C. (2008). Landmarks for navigation in humans and robots. In M. E. Jefferies and W.-K. Yeap (eds), *Robotics and Cognitive Approaches to Spatial Mapping* (pp. 203–14). Berlin: Springer.

Hirtle, S. C. and Jonides, J. (1985). Evidence of hierarchies in cognitive maps. *Memory & Cognition*, 13(3), 208–17.

Hirtle, S. C. and Kallman, H. J. (1988). Memory for the locations of pictures: Evidence for hierarchical clustering. *American Journal of Psychology*, 101(2), 150–70.

Hirtle, S. C. and Mascolo, M. F. (1986). Effect of semantic clustering on the memory of spatial locations. *Journal of Experimental Psychology: Learning, Memory, and Cognition*, 12(2), 182–9.

Hirtle, S., Richter, K.-F., Srinivas, S., and Firth, R. (2010). This is the tricky part: When directions become difficult. *Journal of Spatial Information Science*, 1, 53–73.

Hois, J. and Kutz, O. (2008a). Counterparts in language and space: similarity and S-connection. In C. Eschenbach and M. Gruninger (eds), *Formal Ontology in Information Systems* (FOIS'08), (pp. 266–79). Amsterdam: IOS Press.

Hois, J. and Kutz, O. (2008b). Natural Language meets Spatial Calculi. In C. Freksa and N. S. Newcombe (eds), *Spatial Cognition VI: Learning, Reasoning, and Talking about Space* (pp. 266–82). Berlin: Springer.

Holmes, N. and Spence, C. (2004). The body schema and multisensory representation(s) of peripersonal space. *Cognitive Processing*, 5(2), 94–105.

Hölscher, C., Tenbrink, T., and Wiener, J. (2011). Would you follow your own route description? *Cognition*, 121(2), 228–47.

Huang, Z., Eliëns, A., and Visser, C. T. (2003). 'Is it within my reach?'—an agent's perspective. In T. Rist, R. S. Aylett, D. Ballin, J. Rickel, (eds), *Proceedings of the 4th International Workshop Intelligent Virtual Agents* (pp. 150–8). Berlin: Springer (LNAI 2792).

Iachini, T. and Ruggiero, G. (2010). The role of visual experience in mental scanning of actual pathways: Evidence from blind and sighted people. *Perception*, 39, 953–69.

Ishikawa, T. and Montello, D. R. (2006). Spatial knowledge acquisition from direct experience in the environment: Individual differenes in the development of metric knowledge and the integration of separately learned places. *Cognitive Psychology*, 52, 93–129.

Jaffar J. and Maher, M. J. (1994). Constraint logic programming: A survey. *Journal of Logic Programming*, 19/20, 503–81.

Janzen, G. and van Turennout, M. (2004). Selective neural representation of objects relevant for navigation. *Nature Neuroscience*, 7, 673–7.

Janzen, G., Wagensveld, B., and van Turennout, M. (2007). Neural representation of navigational relevance is rapidly induced and long lasting. *Cerebral Cortex*, 17, 975–81.

Johnson, M. (1987). *The Body in the Mind.* Chicago: The University of Chicago Press.

Johnson, M. and Demiris, Y. (2005). Perceptual perspective taking and action recognition. *International Journal of Advanced Robotic Systems*, 2(4), 301–8.

Jones, D., Farrand, P., Stuart, G., and Morris, N. (1995). Functional equivalence of verbal and spatial information in serial short-term memory. *Journal of Experimental Psychology: Learning, Memory, and Cognition*, 21(4), 1008–18.

Karp, R. M. (1972). Reducibility among combinatorial problems. In R. E. Miller and I. Thatcher (eds), *Complexity of Computer Computations* (pp. 85–103). New York: Plenum.

Kaschak, M. P., Zwaan, R. A., Aveyard, M., and Yaxley, R. H. (2006). Perception of auditory motion affects language processing. *Cognitive Science*, 30, 733–44.

Katz, B. F. G., Truillet, P., Thorpe, S., and Jouffrais, C. (2010). NAVIG: Navigation assisted by artificial vision and GNSS. *Proceedings of the Workshop on Multimodal Location Based Techniques for Extreme Navigation*, 17 May, Helsinki.

Kay, L. (1964). An ultrasonic sensing probe as a mobility aid for the blind. *Ultrasonics*, 2, 53–9.

Kelleher, J. D. and Costello, F. J. (2009). Applying computational models of spatial prepositions to visually situated dialog. *Computational Linguistics*, 35(2), 271–306.

Kelleher J. and van Genabith, J. (2004). Visual salience and reference resolution in simulated 3-D environments. *Artificial Intelligence Review*, 21(3–4), 253–67.

Kelleher, J., Sloan, C., and Namee, B. M. (2009). An investigation into the semantics of English topological prepositions. *Cognitive Processing*, 10(2), 233–6.

Kelly, J. W. and Avraamides, M. N. (2011). Cross-sensory transfer of reference frames in spatial memory. *Cognition*, 118, 444–50.

Kelly, J. W., Avraamides, M. N., and Loomis, J. M. (2007). Sensorimotor alignment effects in the learning environment and in novel environments. *Journal of Experimental Psychology: Learning, Memory, & Cognition*, 33, 1092–107.

Kelly, J. W., Avraamides, M. N., and McNamara, T. P. (2010). Reference frames influence spatial memory development within and across sensory modalities. In C. Hölscher, T. F. Shipley, M. O. Belardinelli, J. A. Bateman, and N. S. Newcombe (eds), *Lecture Notes in Artificial Intelligence: Spatial Cognition VII* (pp. 222–33). Berlin: Springer.

Kendon, A. (1990). *Conducting Interaction*. London: Cambridge University Press.

Kessler, K. and Thomson, L. A. (2010). The embodied nature of spatial perspective taking: Embodied transformation versus sensorimotor interference. *Cognition*, 114(1), 72–88.

Key, S. (2009). A computable language of architecture: Description of descriptor language in supporting compound definitions. In *Computation: The New Realm of Architectural Design*, Proceedings of 27th Conference of Education and Research in Computer Aided Architectural Design in Europe (eCAADe), 16–19 September 2009 (pp. 431–38) Istanbul: Istanbul Technical University and Yildiz Technical University.

Khatib, O. (1986). Real-time obstacle avoidance for manipulators and mobile robots. *International Journal of Robotics Research*, 5(1), 90–8.

Klatzky, R. L. (1998). Allocentric and egocentric spatial representations: Definitions, distinctions, and interconnections. In C. Freksa, C. Habel, and K. F. Wender (eds), *Spatial Cognition: An Interdisciplinary Approach to Representation and Processing of Spatial Knowledge* (pp. 1–17). Berlin: Springer.

Klatzky, R. L., Lippa, Y., Loomis, J. M., and Golledge, R. G. (2002). Learning directions of objects specified by vision, spatial audition, or auditory spatial language. *Learning and Memory*, 9, 364–67.

Klatzky, R. L., Lippa, Y., Loomis, J. M., and Golledge, R. G. (2003). Encoding, learning, and spatial updating of multiple object locations specified by 3-D sound, spatial language, and vision. *Experimental Brain Research*, 149, 48–61.

Klatzky, R. L., Marston, J. R., Giudice, N. A., Golledge, R. G., and Loomis, J. M. (2006). Cognitive load of navigating without vision when guided by virtual sound versus spatial language. *Journal of Experimental Psychology: Applied*, 12, 223–32.

Klein, W. (1982). Local deixis in route directions. In R. J. Jarvella and W. Klein (eds), *Speech, Place, and Action* (pp. 161–82). Chichester, UK: Wiley.

Klein, W. (1983). Deixis and spatial orientation in route directions. In H. L. Pick, Jr. and L. P. Acredolog (eds), *Spatial Orientation: Theory, Research, and Application* (pp. 283–311). New York: Plenum.

Klippel, A. (2012). Spatial information theory meets spatial thinking – is topology the Rosetta Stone of spatial cognition? *Annals of the Association of American Geographers*, 6, 1310.

Klippel, A. and Li, R. (2009). The endpoint hypothesis: A topological-cognitive assessment of geographic scale movement patterns. In K. S. Hornsby, C. Claramunt, M. Denis, and G. Ligozat (eds), *Spatial Information Theory*, COSIT'09, (pp. 177–94). Berlin: Springer.

Klippel, A. and Montello, D. R. (2007). Linguistic and nonlinguistic turn direction concepts. In S. Winter, M. Duckham, L. Kulik, and B. Kuipers (eds), *Spatial Information Theory: Proceedings of COSIT '07* (pp. 354–72). Berlin: Springer.

Klippel, A. and Winter, S. (2005). Structural salience of landmarks for route directions. In A. G. Cohn and D. M. Mark (eds), *Spatial Information Theory* (pp. 347–62). Berlin: Springer.

Klippel, A., Tappe, H., Kulik, L., and Lee, P. U. (2005). Wayfinding choremes—a language for modeling conceptual route knowledge. *Journal of Visual Languages and Computing*, 16(4), 311–29.

Kollar, T., Tellex, S., Roy, D., and Roy, N. (2010). Toward understanding natural language directions. In *5th ACM/IEEE International Conference on Human-Robot Interaction*, HRI'10, (pp. 259–66). New York: ACM.

Koller, D. and Friedman, N. (2009). *Probabilistic Graphical Models: Principles and Techniques*. Cambridge, MA: MIT Press.

Kopp, S., Jung, B., Leßmann N., and Wachsmuth, I. (2003). Max—a multimodal assistant in virtual reality construction. *KI-Künstliche Intelligenz*, 4(3), 11–17.

Kordjamshidi, P., van Otterlo, M., and Moens, M. F. (2010a). From language towards formal spatial calculi. In *Workshop on Computational Models of Spatial Language Interpretation (CoSLI'10)*.

Kordjamshidi, P., van Otterlo, M., and Moens, M. F. (2010b). Spatial role labeling: Task definition and annotation scheme. In *7th Conference on International Language Resources and Evaluation (LREC'10)*.

Kordjamshidi, P., Frasconi, P., Otterlo, M. V., Moens, M., and Raedt, L. D. (2011a). Relational learning for spatial relation extraction from natural language. In *The Proceedings of ILP 2011, Lecture Notes in Artificial Intelligence*, Berlin: Springer.

Kordjamshidi, P., Hois, J., van Otterlo, M., and Moens, M. F. (2011b). Machine learning for interpretation of spatial natural language in terms of QSR. *Poster Presentation at the 10th International Conference on Spatial Information Theory (COSIT'11)*.

Kordjamshidi, P., Van Otterlo, M., and Moens, M. (2011c). Spatial role labeling: Towards extraction of spatial relations from natural language. *ACM Transactions on Speech and Language Processing*, 8, 1–36.

Kordjamshidi, P., Bethard, S., and Moens, M. (2012). SemEval-2012 Task 3: Spatial role labeling. In *SEM 2012: The First Joint Conference on Lexical and Computational Semantics, Volume 2: Proceedings of the Sixth International Workshop on Semantic Evaluation* (SemEval 2012) (pp. 365–73), Association for Computational Linguistics.

Kortenkamp, D. (1993). *Cognitive Maps for Mobile Robots: A Representation for Mapping and Navigation*. PhD thesis, The University of Michigan.

Kosslyn, S. M. (1980). *Image and mind*. Cambridge, MA: Harvard University Press.

Kosslyn, S. M., Ball, T. M., and Reiser, B. J. (1978). Visual images preserve metric spatial information: Evidence from studies of image scanning. *Journal of Experimental Psychology: Human Perception & Performance*, 4, 47–60.

Kowadlo, G. and Russell, R. A. (2006). Using naïve physics for odor localization in a cluttered indoor environment. *Autonomous Robots*, 20(3), 215–30.

Kowalski R. and M. Sergot (1986). A logic-based calculus of events. *New Generation Computing*, 4(1), 67–95.

Krueger, J. (2011). Extended cognition and the space of social interaction. *Consciousness and Cognition*, 20(3), 643–57.

Kruijff, G. J. M., Zender, H., Jensfelt, P., and Christensen, H. I. (2007). Situated dialogue and spatial organization: What, where... and why? *International Journal of Advanced Robotic Systems*, 4(1), 125–38.

Kuhn, W. (2007). An image-schematic account of spatial categories. In S. Winter et al. (eds), *COSIT 2007* (pp. 152–68). Berlin: Springer (LNCS 4736).

Kuipers, B. J. (1978). Modeling spatial knowledge. *Cognitive Science*, 2(2), 129–53.

Kuipers, B. (2000). The spatial semantic hierarchy. *Artificial Intelligence*, 119, 191–233.

Kuipers, B., Tecuci, D., and Stankiewicz, B. J. (2003). The skeleton in the cognitive map: A computational and empirical exploration. *Environment and Behavior*, 35(1), 81–106.

Kurata, Y. and Shi, H. (2008). Interpreting motion expressions in route instructions using two projection-based spatial models. In A. R. Dengel, K. Berns, T. M. Breuel, F. Bomarius, and T. R. Roth-Berghofer (eds), *KI 2008: Advances in Artificial Intelligence* (pp. 258–66). Berlin: Springer.

Kurby, C. A., Magliano, J. P., and Rapp, D. N. (2009). Those voices in your head: Activation of auditory images during reading. *Cognition*, 112, 457–61.

Ladd, F. (1970). Black youths view their environment: Neighborhood maps. *Environment and Behavior*, 2, 74–99.

Lafferty, J., McCallum, A., and Pereira, F. (2001). Conditional random fields: Probabilistic models for segmenting and labeling sequence data. In *18th International Conference on Machine Learning*, ICML'01 (pp. 282–9).

Lakoff, G. (1987). *Women, Fire, and Dangerous Things: What Categories Reveal about the Mind.* Chicago: University of Chicago Press.

Lakoff, G. and Johnson, M. (1999). *Philosophy in the Flesh: The Embodied Mind and Its Challenge to Western Thought.* New York: Basic Books.

Landau, B. and Jackendoff, R. (1993). 'What' and 'where' in spatial language and spatial cognition. *Behavioral and Brain Sciences*, 16, 217–65.

Laughlin, W. S. (1968). Hunting: An integrating biobehavior system and its evolutionary importance. In R. B. Lee and I. Devore (eds), *Man The Hunter.* Chicago: Aldine.

Lauria, S., Bugmann, G., Kyriacou, T., Bos, J., and Klein, E. (2002). Converting natural language route instructions into robot executable procedures. In *Proceedings of the 11th IEEE International Workshop on Robot and Human Interactive Communication* (pp. 223–8). Berlin, Germany, September 2002.

Le Yaouanc, J.-M., Saux, E., and Claramunt, C. (2010). A semantic and language-based representation of an environmental scene, *GeoInformatica*, 14(3), 333–52.

Lee, P. U. and Tversky, B. (2005). Interplay between visual and spatial: The effect of landmark descriptions on comprehension of route/survey spatial descriptions. *Spatial Cognition and Computation*, 5, 163–85.

Leiser, D. and Zilbershatz, A. (1989). The traveller: A computational model of spatial network learning. *Environment and Behavior*, 21, 435–63.

Levelt, W. J. M. (1982). Cognitive styles in the use of spatial direction terms. In R. J. Jarvella and W. Klein (eds), *Speech, Place, and Action* (pp. 251–68). London: John Wiley & Sons.

Levelt, W. J. M. (1989). *Speaking: From Intention to Articulation.* Cambridge, MA: MIT Press.

Levinson, S. C. (1996a). Frames of reference and Molyneux's question: Crosslinguistic evidence. In P. Bloom, M. A. Peterson, L. Nadel, and M. F. Garrett (eds), *Language and Space* (pp. 109–69). Cambridge, MA: The MIT Press.

Levinson, S. C. (1996b). Language and space. *Annual Review of Anthropology*, 25, 353–82.

Levinson, S. C. (2003). *Space in Language and Cognition: Explorations in Cognitive Diversity*. Cambridge, UK: Cambridge University Press.

Li, H., Zhao, T., Li, S., and Zhao, J. (2007). The extraction of trajectories from real texts based on linear classification. In NODALIDA 2007 Conference (pp. 128–34). Available at <http://www.cl.ut.ee/Nodalida/Nodalida2007-01.pdf.vanao2>, last accessed 3 May 2013.

Linde, C. and Labov, W. (1975). Spatial networks as a site for the study of language and thought. *Language*, 51, 924–39.

Linderholm, T. and van den Broek, P. (2002). The effect of reading purpose and working memory capacity on the processing of expository text. *Journal of Educational Psychology*, 94(4), 778–84.

Lindgren, I. (2004). *Communication and team performance—a field study of breathing apparatus firefighters' communication during rescue operations*. Master's thesis, Linköping: Linköpings Universitet.

Lindgren, I., Hirsch, R., and Berggren, P. (2007). It takes three points to define a common ground: Breathing apparatus fire-fighters' communication during rescue operations. *Journal of Pragmatics*, 39(9), 1482–502.

Lloyd, D. M. (2009). The space between us: A neurophilosophical framework for the investigation of human interpersonal space. *Neuroscience and Biobehavioral Reviews*, 33(3), 297–304.

Lockwood, K., Forbus, K., and Usher, J. (2005). Spacecase: A model of spatial preposition use. In *Proceedings of the 27th Annual Conference of the Cognitive Science Society* 21–23 July, Stresa, Italy. Cognitive Science Society.

Lockwood, K., Forbus, K. Halstead, D. and Usher, J. (2006). Automatic categorization of spatial prepositions. In *Proceedings of the 28th Annual Conference of the Cognitive Science Society*, 26–29 July, Vancouver, Canada. Cognitive Science Society.

Longo, M. R. and Lourenco, S. F. (2006). On the nature of near space: Effects of tool use and the transition to far space. *Neuropsychologia*, 44, 977–81.

Longo, M. R. and Lourenco, S. F. (2007). Space perception and body morphology: Extent of near space scales with arm length. *Experimental Brain Research*, 177, pp. 285–90.

Loomis, J. M. (1985). Digital map and navigation system for the visually impaired. Unpublished paper. Department of Psychology, University of California, Santa Barbara.

Loomis, J. M., Da Silva, J. A., Fujita, N., and Fukusima, S. S. (1992). Visual space perception and visually directed action. *Journal of Experimental Psychology: Human Perception and Performance*, 18, 906–21.

Loomis, J. M., Klatzky, R. L., Golledge, R. G., Cicinelli, J. G., Pellegrino, J. W., and Fry, P. A. (1993). Nonvisual navigation by blind and sighted: Assessment of path integration ability. *Journal of Experimental Psychology: General*, 122, 73–91.

Loomis, J. M., Golledge, R. G., and Klatzky, R. L. (1998). Navigation system for the blind: Auditory display modes and guidance. *Presence: Teleoperators and Virtual Environments*, 7, 193–203.

Loomis, J. M., Klatzky, R. L., and Golledge, R. G. (2001). Navigating without vision: Basic and applied research. *Optometry and Vision Science*, 78, 282–9.

Loomis, J. M., Lippa, Y., Golledge, R. G., and Klatzky, R. L. (2002). Spatial updating of locations specified by 3-D sound and spatial language. *Journal of Experimental Psychology: Learning, Memory, & Cognition*, 28, 335–45.

Loomis, J. M., Marston, J. R., Golledge, R. G., and Klatzky, R. L. (2005). Personal guidance system for visually impaired people: Comparison of spatial displays for route guidance. *Journal of Visual Impairment and Blindness*, 99, 219–32.

Loomis, J., Klatzky, R. L., Avraamides, M. N., Lippa, Y., and Golledge, R. G. (2007). Functional equivalence of spatial images produced by perception and spatial language. In F. Mast and L. Jäncke (eds), *Spatial Processing in Navigation, Imagery, and Perception* (pp. 29–48). New York: Springer.

Lovelace, K. L., Hegarty, M., and Montello, D. R. (1999). Elements of good route directions in familiar and unfamiliar environments. In C. Freksa and D. M. Mark (eds), *Spatial Information Theory: Cognitive and Computational Foundations of Geographic Information Science* (pp. 65–82). Berlin: Springer.

Lozano-Pérez T. and Wesley, M. A. (1979). An algorithm for planning collision-free paths among polyhedral obstacles. *Communications of the ACM*, 22(10), 560–70.

Lu R. and Zhang, S. (2002). *Automatic Generation of Computer Animation: Using AI for Movie Animation*, vol. 2160 of Lecture Notes in Computer Science. Berlin: Springer.

Lyi, Y., Wang, X., Jin, X., and Wu, L. (2005). On internal cardinal direction relations. In A. G. Cohn and D. M. Mark (eds), *Proceedings of Spatial Information Theory* (pp. 283–99). Berlin: Springer.

Lynch, K. (1960). *The Image of the City*. Cambridge, MA: Technology Press.

McDonald, T. P. and Pelligrino, J. W. (1993). Psychological perspectives on spatial cognition. In T. Gärling and R. G. Golledge (eds), *Behavior and Environment: Psychological and Geographical Approaches* (pp. 47–82). Amsterdam: Elsevier.

MacGregor, J. and Ormerod, T. (1996). Human performance on the traveling salesman problem. *Perception and Psychophysics*, 58, 527–39.

MacGregor, J., Ormerod, T., and Chronicle, E. (2000). A model of human performance on the traveling salesman problem. *Memory and Cognition*, 28, 1183–90.

McNamara, T. P. (1986). Mental representations of spatial relations. *Cognitive Psychology*, 18, 87–121.

McNamara, T. P. (1991). Memory's view of space. In G. H. Bower (ed.), *The Psychology of Learning and Motivation* (pp. 147–86). New York: Academic Press.

McNamara, T. P. (2003). How are the locations of objects in the environment represented in memory? In: C. Freksa, W. Brauer, C. Habel, and K. K. Wender (eds), *Spatial Cognition III: Routes and Navigation, Human Memory and Learning, Spatial Representation and Spatial Reasoning* (pp. 174–91). LNAI 2685. Springer, Berlin.

McNamara, T. P., Ratcliff, R., and McKoon, G. (1984). The mental representation of knowledge acquired from maps. *Journal of Experimental Psychology: Learning, Memory, and Cognition*, 10, 723–32.

McNamara, T. P., Hardy, J. K., and Hirtle, S. C. (1989). Subjective hierarchies in spatial memory. *Journal of Experimental Psychology: Learning, Memory, and Cognition*, 15, 211–27.

McNamara, T. P., Rump, B., and Werner, S. (2003). Egocentric and geocentric frames of reference in memory of large-scale space. *Psychonomic Bulletin and Review*, 10, 589–95.

McNaughton, B. L., Battaglia, F. P., Jensen, O., Moser, E. I., and Moser, M. B. (2006). Path integration and the neural basis of the "cognitive map". *Nature Reviews Neuroscience*, 7(8), 663–78.

McNeill, D. (1992). *Hand and Mind: What Gestures Reveal About Thought*. Chicago: University of Chicago Press.

Maddox, K. B., Rapp, D. N., Brion, S., and Taylor, H. A. (2008). Social influences on spatial memory. *Memory and Cognition*, 36(3), 479–94.

Mainwaring, S. D., Tversky, B., Mohoto Ohgishy, and Schiano, D. J. (2003). Descriptions of simple spatial scenes in English and Japanese. *Spatial Cognition and Computation*, 3(1), 3–43.

Mani, K. and Johnson-Laird, P. N. (1982). The mental representation of spatial descriptions. *Memory and Cognition*, 10, 181–7.

Mani, I. and Pustejovsky, J. (2012). *Interpreting Motion: Grounded Representations for Spatial Language*. Explorations in language and space. Oxford: Oxford University Press.

Mani, I., Hitzeman, J., Richer, J., Harris, D., Quimby, R., and Wellner, B. (2008). SpatialML: annotation scheme, corpora, and tools. In *6th International Language Resources and Evaluation, LREC'08*. Available at <http://www.lrec-conf.org/proceedings/lrec2008/index.html>, last accessed 3 May 2013.

Maravita, A. and Iriki, A. (2004). Tools for the body (schema). *Trends in Cognitive Science*, 8, 79–86.

Mark, D. M., Freksa, C., Hirtle, S. C., Lloyd, R., and Tversky, B. (1999). Cognitive models of geographical space. *International Journal of Geographical Information Science*, 13(8), 747–74.

Màrquez, L., Carreras, X., Litkowski, K. C., and Stevenson, S. (2008). Semantic role labeling: An introduction to the special issue. *Computational Linguistics*, 34(2), 145–59.

Marston, J. R., Loomis, J. M., Klatzky, R. L., Golledge, R. G., and Smith, E. L. (2006). Evaluation of spatial displays for navigation without sight. *ACM Transactions on Applied Perception*, 3, 110–24.

Marston, J. R., Loomis, J. M., Golledge, R. G., and Klatzky, R. L.(2007). Non visual route following with guidance from a simple haptic or auditory display. *Journal of Visual Impairment and Blindness*, 101, 203–11.

Matlock, T. (2004). Fictive motion as cognitive simulation. *Memory and Cognition*, 32, 1389–400.

May, M. (2007). Imaginal repositioning in everyday environments: Effects of testing method and setting. *Psychological Research*, 71, 277–87.

Mellet, E., Briscogne, S., Tzourio-Mazoyer, N., Ghaem, O., Petit, L., Zago, L., et al. (2000). Neural correlates of topographic mental exploration: the impact of route versus survey perspective learning. *Neuroimage*, 12, 588–600.

Mellet, E., Briscogne, S., Crivello, F., Mazoyer, B., Denis, M., and Tzourio-Mazoyer, N. (2002). Neural basis of mental scanning of a topographic representation built from a text. *Cerebral Cortex*, 12, 1322–330.

Melton. A. (1933). Studies of installation at the Pennsylvania Museum of Art. *Museum News*, 10(15), 5–8.

Michon, P.-E. and Denis, M. (2001). When and why are visual landmarks used in giving directions? In D. R. Montello (ed.), *Spatial Information Theory: Foundations of Geographic Information Science* (pp. 292–305). Berlin: Springer.

Michotte, A. (1963). *The Perception of Causality*. New York: Basic Books (Originally published 1946).

Miele, J. A., Landau, S., and Gilden, D. (2006). Talking TMAP: Automated generation of audio-tactile maps using Smith-Kettlewell's TMAP software. *The British Journal of Visual Impairment*, 24, 93–100.

Millar, S. (1994). *Understanding and Representing Space: Theory and Evidence from Studies with Blind and Sighted Children*. Oxford: Clarendon Press.

Miller G. A. and Johnson-Laird, P. N. (1976). *Language and Perception*. Cambridge, MA: Harvard University Press.

Miller, L. E., National Institute of of Standards and Technology (2006). Indoor navigation for first responders: A feasibility study. National Institute of Standards and Technology, Wireless Communication Technologies Group, Gaithersburg, MD.

Minsky, M. (1975). A framework for representing knowledge. In P. H. Winston (ed.), *The Psychology of Computer Vision* (pp. 211–77). New York: McGraw-Hill.

Mix, K. S., Smith, L. B., and Gasser, M. (2009). *The Spatial Foundations of Cognition and Language: Thinking Through Space*. Oxford: Oxford University Press.

Miyake, A. and Shah, P. (1999). *Models of Working Memory: Mechanisms of Active Maintenance and Executive control*. New York: Cambridge University Press.

Miyake, A., Friedman, N. P., Emerson, M. J., Witzki, A. H., Howerter, A. and Wager, T. (2000). The unity and diversity of executive functions and their contributions to complex 'frontal lobe' tasks: A latent variable analysis. *Cognitive Psychology*, 41, 49–100.

Miyake, A., Friedman, N. P., Rettinger, D. A., Shah, P., and Hegarty, M. (2001). Visuospatial working memory, central executive functioning, and psychometric visuospatial abilities: How are they related? *Journal of Experimental Psychology: General*, 130, 621–40.

Montello, D. R. (1993). Scale and multiple psychologies of space. *Spatial Information Theory, a Theoretical Basis for GIS: Lecture Notes in Computer Science*, 716, 312–21.

Montello, D. R. (2009). A conceptual model of the cognitive processing of environmental distance information. In K. S. Hornsby, C. Claramunt, M. Denis, and G. Ligozat (eds), *Spatial Information Theory: Proceedings of COSIT '09* (pp. 1–17). Berlin: Springer.

Moratz, R. and Tenbrink, T. (2008). Affordance-based human-robot interaction. In E. Rome, J. Hertzberg, and G. Dorffner (eds), *Towards Affordance-Based Robot Control* (pp. 63–76). LNAI 4760. Berlin: Springer.

Morrongiello, B. A., Timney, B., Humphrey, K., Anderson, S., and Skory, C. (1995). Spatial knowledge in blind and sighted children. *Journal of Experimental Child Psychology*, 59, 211–33.

Morrow, D. (1994). Spatial models created from text. *Advances in Discourse Processes*, 53, 57–78.

Mou, W. and McNamara, T. P. (2002). Intrinsic frames of reference in spatial memory. *Journal of Experimental Psychology: Learning, Memory, & Cognition*, 28, 162–70.

Mou, W., McNamara, T. P., Valiquette, C. M., and Rump, B. (2004). Allocentric and egocentric updating of spatial memories. *Journal of Experimental Psychology: Learning, Memory, & Cognition*, 30, 142–57.

Nabeshima, C., Kuniyoshi, Y., and Lungarella, M. (2006). Adaptive body schema for robotic tool-use. *Advanced Robotics*, 20(10), 1105–26.

National Federation of the Blind's Access Technology Staff (2006). GPS technology for the blind, a product evaluation. *Braille Monitor*, 49, 101–8.

Naylor-Emlen, S. J. and Taylor, H. A. (2009). Contributing factors to temporal and spatial associations in mental representations of maps. *Spatial Cognition and Computation*, 9(4), 262–86.

Newell, A. (1990). *Unified Theories of Cognition*. Cambridge, MA: Harvard University Press.

Nguyen, N. and Wachsmuth, I. (2009). Modeling peripersonal action space for virtual humans using touch and proprioception. In Z. Ruttkay, M. Kipp, A. Nijholt, and H. H. Vilhjálmsson (eds), *Proceedings of the 9th Conference on Intelligent Virtual Agents* (pp. 63–75). Berlin: Springer (LNAI 5773).

Nguyen, N. and Wachsmuth, I. (2011). From body space to interaction space—modeling spatial cooperation for virtual humans. In Tumer, Yolum, Sonenberg, and Stone (eds), *Proceedings of the 10th International Conference on Autonomous Agents and Multiagent Systems* (AAMAS 2011), (pp. 1047–54) Taipei, Taiwan.

Noordzij, M. L. and Postma, A. (2005). Categorical and metric distance information in mental representations derived from route and survey descriptions. *Psychological Research*, 69, 221–32.

Noordzij, M. L., Zuidhoek, S., and Postma, A. (2006). The influence of visual experience on the ability to form spatial mental models based on route and survey descriptions. *Cognition*, 100, 321–42.

Norman, J. (2002). Two visual systems and two theories of perception: An attempt to reconcile the constructivist and ecological approaches. *Behavioral and Brain Sciences* 25, 73–144.

Nothegger, C., Winter, S., and Raubal, M. (2004). Selection of salient features for route directions. *Spatial Cognition and Computation*, 4, 113–36.

O'Keefe, J. and Nadel, L. (1978). *The Hippocampus as a Cognitive Map*. London: Oxford University Press.

O'Neill. M. J. (1991a). Effects of signage and floor plan configuration on wayfinding accuracy. *Environment and Behavior*, 23(5), 553–74.

O'Neill. M. J. (1991b). Evaluation of a conceptual model of architectural legibility. *Environment and Behavior*, 23(3), 259–84.

Page, M. P. A. and Norris, D. (1998). The primacy model: A new model of immediate serial recall. *Psychological Review*, 105(4), 761–81.

Paivio, A. (1990). *Mental Representations: Dual Coding Approach*. Oxford: Oxford University Press.

Pantelides, S. (2010). *Spatial updating: are described objects automatically updated?* Unpublished master's thesis, University of Cyprus, Nicosia, Cyprus.

Parkes, D. (1988). 'Nomad': An audio-tactile tool for the acquisition, use and management of spatially distributed information by visually impaired people. In A. F. Tatham and A. G. Dodds (eds), *Proceedings of the Second International Symposium on Maps and Graphics for Visually Impaired People, London, 1988*, pp. 24–9.

Parmentier, F. B. R., Andrés, P., Elford, G., and Jones, D. M. (2006). Organization of visuo-spatial serial memory: Interaction of temporal order with spatial and temporal grouping. *Psychological Research*, 70, 200–17.

Pasqualotto, A. and Proulx, M. J. (2012). The role of visual experience for the neural basis of spatial cognition. *Neuroscience and Biobehavioral Reviews*, 36(4), 1179–87.

Pazzaglia, F., De Beni, R., and Meneghetti, C. (2007). The effects of verbal and spatial interference in the encoding of spatial and nonspatial texts. *Psychological Research*, 71(4), 484–94.

Pedica, C. and Vilhjálmsson, H. H. (2009). Spontaneous avatar behavior for human territoriality. In Z. Ruttkay, M. Kipp, A. Nijholt, and H. H. Vilhjálmsson (eds), *Proceedings of the 9th Conference on Intelligent Virtual Agents* (pp. 344–57). Berlin: Springer (LNAI 5773).

Peponis, J., C. Zimring, C., and Choi, Y. K. (1990). Finding the building in wayfinding. *Environment and Behavior*, 22(5), 555–90.

Perky, C. W. (1910). An experimental study of imagination. *American Journal of Psychology*, 21(3), 422–52.

Perrig, W., and Kintsch, W. (1985). Propositional and situational representations of text. *Journal of Memory and Language*, 24, 503–18.

Peruch, P., Chabanne, V., Nesa, M., Thinus-Blanc, C., and Denis, M. (2006). Comparing distances in mental images constructed from visual experience or verbal descriptions: The impact of survey versus route perspective. *The Quarterly Journal of Experimental Psychology*, 59(11), 1950–67.

Piaget, J, and Inhelder, B. (1967). *The Child's Conception of Space*. New York: W.W. Norton.

Pizlo, Z., Stefanov, E., Saalwaechter, J., Li, Z., Haxhimusa, Y., and Kropatsch, W. G. (2006). Traveling salesman problem: A foveating pyramid model. *Memory and Cognition*, 28, 1101–204.

Plumert, J. M., Carswell, C., DeVet, K., and Ihrig, D. (1995). The content and organization of communication about object locations. *Journal of Memory and Language*, 34, 477–98.

Plumert, J. M., Spalding, T. L., and Nichols-Whitehead, P. (2001). Preferences for ascending and descending hierarchical organization in spatial communication. *Memory and Cognition*, 29, 274–84.

Prabhakaran, V., Narayanan, K., Zhao, Z., and Gabrieli, J. D. E. (2000). Integration of diverse information in working memory within the frontal lobe. *Nature Neuroscience*, 3, 85–90.

Previc, F. H. (1998). The neuropsychology of 3-D space. *Psychological Bulletin*, 124(2), 123–64.

Pulvermüller, F. (2005). Brain mechanisms linking language and action. *Nature Reviews Neuroscience*, 6(7), 576–82.

Pustejovsky, J. (ed.) (2010). *Airlie ISO-Space Workshop*, September 2010. <http://iso-space.org/airlie-workshop-2010>, last accessed 6 May 2013.

Pustejovsky, J., Moszkowicz, J., and Verhagen, M. (2011). ISO-Space: The annotation of spatial information in language. In *ACL-ISO International Workshop on Semantic Annotation* (ISA'06).

Pylyshyn, Z. W. (2006). *Seeing and Visualizing: It's Not What You Think*. Cambridge, MA: MIT Press.

Rabinowitz, J. C., Craik, F. I., and Acherman, B. P. (1982). A processing resource account of age differences in recall. *Canadian Journal of Psychology*, 36, 325–44.

Ranck, J. B. J. (1985). Head direction cells in the deep cell of dosal presubiculum in freely moving rats. In G. Buzsáki and C. H. Vanderwolf (eds), *Electrical Activity of the Archicortex*. Akademiai Kiado, Budapest.

Randell, D. A., Cui, Z., and Cohn, A. G. (1992). A spatial logic based on regions and connection. In *3rd International Conference on the Principles of Knowledge Representation and Reasoning*, KR'92, pp. 165–76.

Rapp, D. N., Klug, J. L., and Taylor, H. A. (2006). Character movement and the representation of space during narrative comprehension. *Memory and Cognition*, 34(6), 1206–20.

Raubal, M. and Egenhofer, M. (1998). Comparing the complexity of wayfinding tasks in built environments. *Environment & Planning B*, 25(6), 895–913.

Raubal, M. and Winter, S. (2002). Enriching wayfinding instructions with local landmarks. In M. J. Egenhofer and D. M. Mark (eds), *Geographic Information Science* (pp. 243–59). Berlin: Springer.

Regier, T. (1996). *The Human Semantic Potential: Spatial Language and Constrained Connectionism*. Cambridge, MA: MIT Press.

Regier, T. and Zheng, M. (2007). Attention to endpoints: A cross-linguistic constraint on spatial meaning. *Cognitive Science*, 31(4), 705–19.

Rehrl, K., Leitinger, S., Gartner, G., and Ortag, F. (2009). An analysis of direction and motion concepts in verbal descriptions of route choices. In K. Stewart Hornsby, C. Claramunt, M. Denis, and G. Ligozat (eds), *Spatial Information Theory. 9th International Conference, COSIT 2009* (pp. 471–88). Aber Wrac'h, France. Berlin: Springer.

Renz, J. and Nebel, B. (2007). Qualitative spatial reasoning using constraint calculi. In M. Aiello, I. Pratt-Hartmann, and J. van Benthem (eds), *Handbook of Spatial Logics* (pp. 161–215). Berlin: Springer.

Richter, K.-F., Tomko, M., and Winter, S. (2008). A dialog-driven process of generating route directions. *Computers, Environment and Urban Systems*, 32, 233–45.

Rickel, J. and Johnson, W. L. (2000). Task-oriented collaboration with embodied agents in virtual worlds. In J. Cassell, J. Sullivan, S. Prevost, and E. Churchill (eds), *Embodied Conversational Agents* (pp. 95–122). Cambridge, MA: MIT Press.

Rieser, J. J.and Pick, H. L. J. (2007). Using locomotion to update spatial orientation: What changes with learning and development? In J. M. Plumert and J. P. Spencer (eds), *The Emerging Spatial Mind* (pp. 77–103). New York: Oxford University Press.

Rieser, J. J., Guth, D. A., and Hill, E. W. (1986). Sensitivity to perspective structure while walking without vision. *Perception*, 15, 173–88.

Rieser, J. J., Garing, A. E., and Young, M. F. (1994). Imagery, action, and young children's spatial orientation: It's not being there that counts, it's what one has in mind. *Child Development*, 65, 1262–78.

Rinck, M., Hähnel, A., Bower, G. H., and Glowalla, U. (1997). The metrics of spatial situation models. *Journal of Experimental Psychology: Learning, Memory, and Cognition*, 23(3), 622–37.

Roentgen, U. R., Gelderblom, G., J., Soede, M., and de Witte, L. P. (2008). Inventory of electronic mobility aids for persons with visual impairments: A literature review. *Journal of Visual Impairment and Blindness*, 102, 702–24.

Rohrer, T. (2005). Image schemata in the brain. In *From Perception to Meaning: Image Schemas in Cognitive Linguistics* (pp. 165–96). In B. Hampe and J. Grady (eds), Berlin: Mouton de Gruyter.

Rouhizadeh, M., Bauer, D., Coyne, B., Rambow, O., and Sproat, R. (2011). Collecting spatial information for locations in a text-to-scene conversion system. In J. Hois, R. Ross, J. Kelleher, and J. Bateman (eds), *Computational Models of Spatial Language Interpretation and Generation (CoSLI'11)* (pp. 16–23). CEUR-WS Proceedings.

Roy. D. K. (2002). Learning visually-grounded words and syntax for a scene description task. *Computer Speech and Language*, 16(3).

Roy, D. and Reiter, E. (2005). Connecting language to the world. *Artificial Intelligence*, 167(1–2), 1–12.

Ruby, P., and Decety, J. (2001). Effect of subjective perspective taking during simulation of action: A PET investigation of agency. *Nature Neuroscience*, 4, 546–50.

Ruddle, R. A., Payne, S. J., and Jones, D. M. (1997). Navigating buildings in 'desk-top' virtual environments: Experimental investigations using extended navigational experience. *Journal of Experimental Psychology: Applied*, 3, 143–59.

Russell, S. and Norvig, P. (2003). *Artificial Intelligence: A Modern Approach*. Upper Saddle River, NJ: Prentice Hall.

Sadeghian, P. and Kantardzic, M. (2008). The new generation of automatic landmark detection systems: Challenges and guidelines. *Spatial Cognition and Computation*, 8, 252–87.

Sadoski, M. and Paivio, A. (2001). *Imagery and Text: A Dual Coding Theory of Reading and Writing*. Mahwah, NJ: Erlbaum.

Schank, R. C. and Abelson, R. P. (1977). *Scripts, Plans, Goals and Understanding: An Inquiry into Human Knowledge Structures*. Hilsdale, NJ: Erlbaum.

Schneider, L. F. and Taylor, H. A. (1999). How do you get there from here? Mental representations of route descriptions. *Applied Cognitive Psychology*, 13, 415–41.

Schultz, C. and Bhatt, M. (2010). A multi-modal data access framework for spatial assistance systems. In *Proceedings of the Second ACM SIGSPATIAL International Workshop on Indoor Spatial Awareness*, in conjunction with ACM SIGSPATIAL GIS, (pp. 39–46). New York: ACM.

Seifert, I. (2009). *Spatial Planning Assistance: A Cooperative Approach*. Doctoral Dissertation, Akademische Verlagsgesellschaft AKA.

Shanon, B. (1984). Room descriptions. *Discourse Processes*, 7, 225–55.

Shelton, A. L. and McNamara, T. P. (2001). Systems of spatial reference in human memory. *Cognitive Psychology*, 43, 274–310.

Shelton, A. L. and McNamara, T. P. (2004). Orientation and perspective dependence in route and survey learning. *Journal of Experimental Psychology: Learning, Memory, and Cognition*, 30(1), 158–70.

Shemyakin, F. N. (1962). General problems of orientation in space and space representations. In B. G. Anayev (ed.), *Psychological Science in the USSR* Vol. 1 (pp. 186–251). Washington, DC: U.S. Joint Publications Research Service.

Shen, Q., Zhang, X., and Jiang, W. (2009). Annotation of spatial relations in natural language. In *International Conference on Environmental Science and Information Application Technology*.

Shepherd, R. N. and Metzler, J. (1971). Mental rotation of three-dimensional objects. *Science*, 171, 701–3.

Shi, H. and Tenbrink, T. (2009). Telling Rolland where to go: HRI dialogues on route navigation. In K. R. Coventry, T. Tenbrink, and J. Bateman (eds.), *Spatial Language and Dialogue*, Vol. 3 (pp. 177–89). Oxford: Oxford University Press.

Sholl, J. S. (2001). The role of a self-reference system in spatial navigation. In D. Montello (ed.), *Spatial Information Theory: Foundations of geographic information science COSIT 2001 Proceedings)* (Vol. 2205, pp. 217–32). Berlin: Springer, Lecture Notes in Computer Science.

Sholl, M. J. (1987). Cognitive maps as orienting schemata. *Journal of Experimental Psychology: Learning, Memory, and Cognition*, 13, 615–28.

Siegel, A. W., and White, S. H. (1975). The development of spatial representations of large-scale environments. In H. W. Reese (ed.), *Advances in Child Development and Behavior* (Vol. 10, pp. 9–55). New York: Academic Press.

Sisbot, E. A., Marin, L. F., Alami, R., and Simeon, T. (2006). A mobile robot that performs human acceptable motion. In *Proceedings International Conference on Intelligent Robots and Systems* (pp. 1811–16).

Skaggs, W. E., Knierim, J. J., Kudrimoti, H. S., and McNaughton, B. L. (1995). A model of the neural basis of the rat's sense of direction. In G. Tesauro, D. Touretzky, and T. Leen (eds). *Advances in Neural Information Processing Systems* (Vol. 7, pp. 173–80). Cambridge, MA: MIT Press.

Soh, B. K. and Smith-Jackson, T. L. (2004). Influence of map design, individual differences, and environmental cues on wayfinding performance. *Current Psychology of Cognition*, 16(4), 409–58.

Sorrows, M. E. and Hirtle, S. C. (1999). The nature of landmarks for real and electronic spaces. In C. Freksa and D. M. Mark (eds), *Spatial Information Theory: Cognitive and Computational Foundations of Geographic Information Science* (pp. 37–50). Berlin: Springer.

Squire, L. R. (1992). Memory and the hippocampus: A synthesis from findings with rats, monkeys, and humans. *Psychological Review*, 99, 195–231.

Staats, A. W. (1991). Unified positivism and unification psychology: Fad or new field? *American Psychologist*, 46, 899–912.

Stevens, A. and Coupe, P. (1978). Distortions in judged spatial relations. *Cognitive Psychology*, 10, 422–37.

Stuckenschmidt, H., Parent, C., and Spaccapietra, S. (eds), (2009). *Modular Ontologies—Concepts, Theories and Techniques for Knowledge Modularization*. Berlin: Springer.

Sweller, J. (1988). Cognitive load during problem solving: Effects on learning. *Cognitive Science*, 12, 257–85.

Talmy L. (2000). *Toward a Cognitive Semantics*. Cambridge, MA: MIT Press.

Talmy, L. (1983). How language structures space. In H. L. Pick, Jr. and L. P. Acredolo (eds), *Spatial Orientation: Theory, Research and Application* (pp. 225–82). New York: Plenum.

Talmy, L. (2006). The fundamental system of spatial schemas in language. In B. Hampe (ed.), *From Perception to Meaning: Image Schemas in Cognitive Linguistics* (pp. 37–47). Berlin: Mouton de Gruyter.

Taylor, H. A. and Tversky, B. (1992a). Descriptions and depictions of environments. *Memory and Cognition*, 20, 483–96.

Taylor, H. A. and Tversky, B. (1992b). Spatial mental models derived from survey and route descriptions. *Journal of Memory and Language*, 31(2), 261–92.

Taylor, H. A. and Tversky, B. (1996). Perspective in spatial descriptions. *Journal of Memory and Language*, 35, 371–91.

Taylor, H. A., Naylor, S. J., and Chechile, N. A. (1999). Goal-directed influences on the representation of spatial perspective. *Memory and Cognition*, 27(2), 309–19.

Tellex, S. and Roy, D. (2006). Spatial routines for a simulated speech-controlled vehicle. In *Proceedings of the Human-Robot Interaction Conference*. Salt Lake City, UT, March 2006.

Tenbrink. T. (2005). Identifying objects on the basis of spatial contrast: An empirical study. In C. Freksa, M. Knauff, B. Krieg-Brückner, B. Nebel, and T. Barkowsky (eds), *Spatial Cognition IV: Reasoning, Action, Interaction*. International Conference on Spatial Cognition 2004 (pp. 124–46). Berlin, Heidelberg: Springer.

Tenbrink, T. (2011). Reference frames of space and time in language. *Journal of Pragmatics*, 43, 704–22.

Tenbrink, T. and Kuhn, W. (2011). A model of spatial reference frames in language. In M. Egenhofer, N. Giudice, R. Moratz, and M. Worboys (eds), *Conference on Spatial Information Theory (COSIT'11)* (pp. 371–90). Berlin: Springer.

Tenbrink, T. and Seifert, I. (2011). Conceptual layers and strategies in tour planning. *Cognitive Processing*, 12, 109–25.

Tenbrink, T. and Wiener, J. (2009). The verbalization of multiple strategies in a variant of the traveling salesperson problem. *Cognitive Processing*, 10(2), 143–61.

Tenbrink, T. and Winter, S. (2009). Variable granularity in route directions. *Spatial Cognition and Computation*, 9(1), 64–93.

Tenbrink, T., Coventry, K. R., and Andonova, E. (2011). Spatial strategies in the description of complex configurations. *Discourse Processes*, 48(4), 237–66.

Tettamanti, M., Buccino, G., Saccuman, M. C., Gallese, V., Danna, M., Scifo, P., et al. (2005). Listening to action-related sentences activates frontoparietal motor circuits. *Journal of Cognitive Neuroscience*, 17(2), 273–81.

Tezuka, T. and Tanaka, K. (2005). Landmark extraction: A web mining approach. In A. G. Cohn and D. M. Mark (eds), *Spatial Information Theory* (pp. 379–96). Berlin: Springer.

Thelen, E. (1995). Time-scale dynamics and the development of an embodied cognition. In R. F. Port and T. Van Gelder, T. (eds), *Mind as Motion: Explorations in the Dynamics of Cognition*. Cambridge, MA: Bradford.

Thomas, A. K., Bonura, B. M., and Taylor, H. A. (2012). The influence of semantic relationships on older adult map memory. *Psychology and Aging*, 27(3), 657–65.

Thomas, N. J. T. (1999). Are theories of imagery theories of imagination? An active perception approach to conscious mental content. *Cognitive Science*, 23, 207–45.

Thorndyke, P. W. and Hayes-Roth, B. (1982). Differences in spatial knowledge acquired from maps and navigation. *Cognitive Psychology*, 14, 560–89.

Tinti, C., Adenzato, M., Tamietto, M., and Cornoldi, C. (2006). Visual experience is not necessary for efficient survey spatial cognition: Evidence from blindness. *Quarterly Journal of Experimental Psychology*, 59, 1306–28.

Tlauka, M. and Wilson, P. N. (2006). Mental representations acquired through language: A study of orientation specificity. *Spatial Cognition and Computation*, 6(2), 161–78.

Tlauka, M., Keage, H., and Clark, C. R. (2005). Viewing a map versus reading a description of a map: Modality-specific encoding of spatial information. *Cognitive Science*, 29, 39–50.

Tlauka, M., Clark, C. R., Liu, P., and Conway, M. (2009). Encoding modality and spatial memory retrieval. *Brain and Cognition*, 70, 116–22.

Tom, A. and Denis, M. (2003). Referring to landmark or street information in route directions: What difference does it make? In W. Kuhn, M. F. Worboys, and S. Timpf (eds), *Spatial Information Theory: Foundations of Geographic Information Science* (pp. 384–97). Berlin: Springer.

Tom, A. and Denis, M. (2004). Language and spatial cognition: Comparing the roles of landmarks and street names in route instructions. *Applied Cognitive Psychology*, 18, 1213–30.

Tostoes, A., Carapinha, A. and Corte-Real. Gulbenkian, P. (2006). *Architecture and Landscape*. London: Calouste Gulbenkian Foundation.

Tversky, B. (1991). Spatial mental models. In G. H. Bower (ed.), *The Psychology of Learning and Motivation: Advances in Research and Theory* (Vol. 27, pp. 109–45). New York: Academic Press.

Tversky, B. (1993). Cognitive maps, cognitive collages, and spatial mental models. In A. U. Frank and I. Campari (eds), *Spatial Information Theory: A Theoretical Basis for GIS—Proceedings of Conference on Spatial Information Theory (COSIT)* (pp. 14–24). Berlin: Springer.

Tversky, B. (2003). Structures on mental spaces: How people think about space. *Environment and Behavior*, 35, 66–80.

Tversky, B. and Lee, P. U. (1998). How space structures language. In C. Freksa, C. Habel, and K. F. Wender (eds.), *Spatial Cognition: An Interdisciplinary Approach to Representing and Processing Spatial Knowledge* (pp. 157–75). Berlin: Springer.

Tversky, B. and Lee, P. U. (1999). Pictorial and verbal tools for conveying routes. In C. Freksa and D. M. Mark (eds), *Spatial Information Theory: Cognitive and Computational Foundations of Geographic Information Science* (pp. 51–64). Berlin: Springer.

Tversky, B., Franklin, N., Taylor, H. A., and Bryant, D. J. (1994). Spatial mental models from descriptions. *Journal of the American Society for Information Science*, 45(9), 656–68.

Tyler, A. and Evans, V. (2003). *The Semantics of English Prepositions: Spatial Scenes, Embodied Meaning and Cognition*, Cambridge: Cambridge University Press.

Ullmer-Ehrich, V. (1982). The structure of living space descriptions. In R. J. Jarvella and W. Klein (eds), *Speech, Place, and Action* (pp. 219–49). New York: Wiley.

Ungar, S., Blades, M., Spencer, C., and Morsley, K. (1994). Can visually impaired children use tactile maps to estimate directions? *Journal of Visual Impairment and Blindness*, 88, 221–33.

Ungerleider, I. G. and Mishkin, M. (1982). Two cortical visual systems. In D. J. Ingle, M. A. Goodale, and R. J. W. Mansfield (eds), *Analysis of Visual Behavior*. Cambridge, MA: MIT Press.

US GSA. (2007). US Courts Design Guide, 2007. Judicial Conference of the United States. US General Services Administration (GSA). <http://www.gsa.gov/graphics/pbs/Courts_Design_Guide_07.pdf>, last accessed 23 April 2010.

Vallar, G. and Maravita, A. (2009). Personal and extrapersonal spatial perception. In G. G. Berntson and J. T. Cacioppo (eds), *Handbook of Neuroscience for the Behavioral Sciences* (pp. 322–36). Hoboken, NJ: John Wiley & Sons, Inc.

van Rijsbergen, C.J. (1979). *Information Retrieval.* Oxford: Butterworths.

Vanetti, E. J. and Allen, G. L. (1988). Communicating environmental knowledge: The impact of verbal and spatial abilities on the production and comprehension of route directions. *Environment and Behavior*, 20, 667–82.

Varges, S. (2004). Overgenerating referring expressions involving relations. In *Proceedings of the Third International Conference on Natural Language Generation* (INLG-04) Brockenhurst, UK (pp. 171–81). Berlin: Springer.

Walker, B. N. and Lindsay, J. (2006). Navigation performance with a virtual auditory display: Effects of beacon sound, capture radius, and practice. *Human Factors*, 48, 265–78.

Waller, D. and Hodgson, E. (2006). Transient and enduring spatial representations under disorientation and self-rotation. *Journal of Experimental Psychology: Learning, Memory, & Cognition*, 32, 867–82.

Waller, D., Friedman, A., Hodgson, E., and Greenauer, N. (2009). Learning scenes from multiple views: Novel views can be recognized more efficiently than learned views. *Memory & Cognition*, 37, 90–9.

Wang, C. C., Thorpe, C., Thrun, S., Hebert, M., and Durrant-Whyte (2007). Simultaneous localization, mapping and moving object tracking. *The International Journal of Robotics Research*, 26(9), 889–916.

Wang, Q., Taylor, H. A., and Brunyé, T. T. (2012). When going the right way is hard to do: Distinct phases of action compatibility in spatial knowledge development. *Acta Psychologica*, 139(3), 449–57.

Wang, R. F. (2004). Between reality and imagination: When is spatial updating automatic? *Perception & Psychophysics*, 66, 68–76.

Wang, R. F. and Brockmole, J. R. (2003a). Human navigation in nested environments. *Journal of Experimental Psychology: Learning, Memory, & Cognition*, 29, 398–404.

Wang, R. F. and Brockmole, J. R. (2003b). Simultaneous spatial updating in nested environments. *Psychonomic Bulleting & Review*, 10, 981–6.

Wang, R. F. and Spelke, E. S. (2000). Updating egocentric representations in human navigation. *Cognition*, 77, 215–50.

Watson, J. B. (1925). *Behaviorism*. New York: People's Institute Publishing Company.

Werner, S. and Schindler, L. E. (2004). The role of spatial reference frames in architecture. *Environment and Behavior*, 36(4), 461–82.

Werner, S., Krieg-Brückner, B., Mallot, H. A., Schweizer, K., and Freksa, C. (1997). Spatial cognition: The role of landmark, route, and survey knowledge in human and robot navigation. In M. Jarke, K. Pasedach, and K. Pohl (eds), *Informatik '97* (pp. 41–50). Berlin: Springer.

Werner, S., Krieg-Brückner, B. and Herrmann, T. (2000). Modelling navigational knowledge by route graphs. In *Spatial Cognition II, Integrating Abstract Theories, Empirical Studies, Formal Methods, and Practical Applications* (pp. 295–316). Berlin: Springer.

White, P. A. (2006). The role of activity in visual impressions of causality. *Acta Psychologica*, 123, 166–85.

Wiener, J. M. and Mallot, H. A. (2003). Fine-to-coarse route planning and navigation in regionalized environments. *Spatial Cognition and Computation*, 3(4), 331–58.

Wiener, J. M. and Tenbrink, T. (2008). Traveling salesman problem: The human case. *Künstliche Intelligenz: KI und Kognition*, 1(08), 18–22.

Wiener, J. M., Ehbauer, N., and Mallot, H. A. (2009). Planning paths to multiple targets: Memory involvement and planning heuristics in spatial problem solving. *Psychological Research*, 73(5), 644–58.

Wilcoxon, F. (1945). Individual comparisons by ranking methods. *Biometrics*, 1, 80–3.

Wineman J. D. and Peponis, J. (2010). Constructing spatial meaning. *Environment and Behavior*, 42(1), 86–109.

Winter, S. (2003). Route adaptive selection of salient features. In W. Kuhn, M. F. Worboys, and S. Timpf (eds), *Spatial Information Theory: Foundations of Geographic Information Science* (pp. 349–61). Berlin: Springer.

Winter, S. and Wu, Y. (2008). Towards a conceptual model of talking to a route planner. In M. Bertolotto, C. Ray, and X. Li (eds), *W2GIS 2008* (pp. 107–23). Berlin: Springer.

Wolbers, T., Hegarty, M., Buchel, C., and Loomis, J. M. (2008). Spatial updating: How the brain keeps track of changing object locations during observer motion. *Nature Neuroscience*, 11(10), 1223–30.

Yamamoto, N. and Shelton, A. L. (2005). Visual and proprioceptive representations in spatial memory. *Memory & Cognition*, 33, 140–50.

Yamaoka, F., Kanda, T., Ishiguro, H., and Hagita, N. (2008). How close? Model of proximity control for information-presenting robots. In *Proceedings of the 3rd ACM/IEEE International Conference on Human–Robot Interaction* (pp. 140–4). New York: ACM.

Yaxley, R. H. and Zwaan, R. A. (2007). Simulating visibility during language comprehension. *Cognition*, 105, 229–36.

Yeap, W. K. (1988). Towards a computational theory of cognitive maps. *Artificial Intelligence*, 34, 297–360.

Zacks, J. M., Speer, N. K., Swallow, K. M., Braver, T. S., and Reynolds, J. R. (2007). Event perception: A mind/brain perspective. *Psychological Bulletin*, 133(2), 273–93.

Zacks, J. M., Kumar, S., Abrams, R., and Mehta, R. (2009). Using movement and intentions to understand human activity. *Cognition*, 112, 201–16.

Zhang, C., Zhang, X., Jiang, W., Shen, Q., and Zhang, S. (2009). Rule-based extraction of spatial relations in natural language text. In *International Conference on Computational Intelligence and Software Engineering*. New York: IEEE.

Zhao, L., Liu, Y., and Badler, N. I. (2005). Applying empirical data on upper torso movement to real-time collision-free reach tasks. In *Proceedings of the 2005 SAE Digital Human Modeling for Design and Engineering Conference and Exhibition* (pp. 2885–90). Berlin: Springer.

Zlatev, J. (2007). Spatial semantics. In D. Geeraerts and H. Cuyckens (eds.), *The Oxford Handbook of Cognitive Linguistics* (pp. 318–50). Oxford: Oxford University Press.

Zwaan, R. A. (1999). Embodied cognition, perceptual symbols, and situation models. *Discourse Processes*, 28, 81–8.

Zwaan, R. A. (2004). The immersed experiencer: Toward an embodied theory of language comprehension. In B. H. Ross (ed.), *The Psychology of Learning and Motivation* (Vol. 44, pp. 35–62). New York: Academic Press.

Zwaan, R. A. and Radvansky, G. A. (1998). Situation models in language comprehension and memory. *Psychological Bulletin*, 123, 162–85.

Zwaan, R. A. and Taylor, L. J. (2006). Seeing, acting, understanding: Motor resonance in language comprehension. *Journal of Experimental Psychology: General*, 135, 1–11.

Zwaan, R. A. and Van Oostendorp, H. (1993). Do readers construct spatial representations in naturalistic story comprehension? *Discourse Processes*, 16(1&2), 125–43.

Zwaan, R. A., Stanfield, R. A., and Yaxley, R. H. (2002). Language comprehenders mentally represent the shapes of objects. *Psychological Science*, 13, 168–71.

Zwarts, J. (2005). Prepositional aspect and the algebra of paths. *Linguistics and Philosophy*, 28(6), 739–79.

Index of authors

Index of subjects